ESSENTIAL CALCULUS
with Applications

RICHARD A. SILVERMAN

DOVER PUBLICATIONS, INC.

New York

In Memory of
A. G. S.

Copyright © 1977, 1989 by Richard A. Silverman.
All rights reserved under Pan American and International Copyright Conventions.

Published in Canada by General Publishing Company, Ltd., 30 Lesmill Road, Don Mills, Toronto, Ontario.
Published in the United Kingdom by Constable and Company, Ltd.

This Dover edition, first published in 1989, is a corrected, slightly enlarged republication of the work first published by the W. B. Saunders Company, Philadelphia, 1977. The section "Supplementary Hints and Answers," originally issued in a separate instructor's manual, has been added to the Dover edition by the author, who has also corrected a number of errors in the original text.

Manufactured in the United States of America
Dover Publications, Inc., 31 East 2nd Street, Mineola, N.Y. 11501

Library of Congress Cataloging-in-Publication Data

Silverman, Richard A.
  Essential calculus with applications / by Richard A. Silverman.
    p.   cm.
  "Corrected, slightly enlarged republication of the work first published by the W. B. Saunders Company, Philadelphia, 1977. The section 'Supplementary Hints and Answers,' originally issued in a separate instructor's manual, has been added to the Dover edition by the author, who has also corrected a number of errors in the original text."
  Includes index.
  ISBN 0-486-66097-4
  1. Calculus. I. Title.
QA303.S55432   1989
515—dc20
                                        89-33584
                                          CIP

TO THE INSTRUCTOR

The attributes and philosophy of this book are best described by giving a running synopsis of each of the six chapters. This summary is accompanied by open expressions of my pedagogical preferences. Like most authors, I tend to regard these not as idiosyncrasies, but as the only reasonable way to do things! If you disagree in spots, I hope you will attribute this lack of modesty to an excess of enthusiasm, an occupational hazard of those with the effrontery to write books.

The contents of Chapter 1 are often called "precalculus," and are in fact just what that term implies, namely, material that ought to be at one's mathematical fingertips before attempting the study of calculus proper. Opinions differ as to what such a background chapter should contain. Some authors cannot wait to get on with the main show, even at the risk of talking about derivatives to students who are still struggling with straight lines, while others seem unwilling to venture into the heartland of calculus without a year's supply of mathematical rations. I have tried to strike a happy medium by travelling light, but well-equipped. Thus there is a brief section on sets, a larger one on numbers, a little bit on mathematical induction, and quite a lot on inequalities and absolute values, two topics that always seem to give students trouble despite their precalculus character. There is a whole section on intervals, both finite and infinite. The last three sections of the chapter administer a modest dose of analytic geometry, with the emphasis on straight lines and their equations. It should not take long to bring all the students up to the mathematical level of Chapter 1, regardless of their starting points, and those few who are there already can spend their spare time solving extra problems while the others catch up!

The class is now ready to attack Chapter 2, and with it the study of differential calculus. The chapter begins with a rather leisurely and entirely concrete discussion of the function concept. It is my belief that many books adopt too abstract an approach to this important subject. Thus I do not hesitate to use terms like "variable" and "argument," which some may regard as old-fashioned, relegating the mapping and ordered pair definitions of function to the problems. At the same time, I find this a natural juncture to say a few words about functions of several variables. After all, why should one have to wait until the very end of the book to write a simple equation like $F(x, y) = 0$? And what's wrong with a few examples of nonnumerical functions, which crop up all the time in the social sciences? While still in the first three sections of Chapter 2, the student encounters one-to-one functions and inverse functions, and then composite functions and sequences after specializing to numerical functions of a single variable. Graphs of equations and functions are treated in terms of solution sets, with due regard for parity of functions and its consequences for the symmetry of their graphs.

Having mastered the concept of function, in all its various manifestations, the student now arrives at Sec. 2.4, where derivatives and limits are introduced *simultaneously.* I am of the opinion that the novice can hardly develop any respect for the machinery of limits, without first being told that limits are needed to define *derivatives.* Here the development of the individual's understanding must recapitulate the actual historical evolution of the subject. For the same reason, I feel that no time should be wasted in getting down to such brass tacks as difference quotients, rates of change, and increments. Moreover, after defining the tangent to a curve, I find it desirable to immediately say something about differentials. This is a small price to pay for the ability to motivate the ubiquitous "*d* notation," and differentials have many other uses too (for example, in Secs. 4.6 and 6.2).

It is now time for the student to learn more about limits. This is done in Sec. 2.6, where a number of topics are presented in quick order, namely, algebraic operations on limits, one-sided limits, the key concept of continuity, algebraic operations on continuous functions, and the fact that differentiability implies continuity. Armed with this information, one can now become a minor expert on differentiation, by mastering the material in Secs. 2.7 and 2.8. After establishing the basic differentiation formula $(x^r)' = rx^{r-1}$ for r a positive or negative integer, I authorize the student to make free use of the same formula for r an arbitrary real number. Why waste time justifying special cases when the "master formula" itself will be proved once and for all in Sec. 4.4? (However, in a concession to tradition, the validity of the formula for r a rational number is established in the problems, in the usual two ways.) Following a brief discussion of higher derivatives, the student arrives next at the rule for differentiating an inverse function and the all-important chain rule. Unlike most authors, I use a proof of the chain rule which completely avoids the spurious difficulty stemming from the possibility of a vanishing denominator, and which has the additional merit of generalizing at once to the case of functions of several variables (see Sec. 6.3). The method of implicit differentiation is treated as a corollary of the chain rule, and I do not neglect to discuss what can go wrong with the method if it is applied blindly. Chapter 2, admittedly a long one, closes with a comprehensive but concise treatment of limits of other kinds, namely, limits involving infinity, asymptotes, the limit of an infinite sequence, and the sum of an infinite series. Once having grasped the concept of the limit of a function at a point, the student should have little further difficulty in assimilating these variants of the limit concept, and this seems to me the logical place to introduce them.

In Chapter 3 differentiation is used as a tool, and the book takes a more practical turn. I feel that the concept of velocity merits a section of its own, as do related rates and the concept of marginality in economic theory. It is then time to say more about the properties of continuous functions and of differentiable functions, and I do so in that order since the student is by now well aware that continuity is a weaker requirement than differentiability. The highly plausible fact that a continuous image of a closed interval is itself a closed interval leads to a quick proof of the existence of global extrema for a continuous function defined in a closed interval, with the intermediate value theorem as an immediate consequence. The connection between the sign of the derivative of a function at a point and its behavior in a neighborhood of the point is then used to prove Rolle's theorem and the mean value theorem, in turn. With the mean value theorem now available, I immediately exploit the opportunity to introduce the antiderivative and the indefinite integral, which will soon be needed to do integral calculus.

The chapter goes on to treat local extrema, including the case where the function under investigation may fail to be differentiable at certain points. Both the first and second derivative tests for a strict local extremum are proved in a straight-forward way, with the help of the mean value theorem. The next section, on concavity and inflection points, is somewhat of an innovation, in that it develops a complete parallelism between the theory of monotonic functions and critical points, on the one hand, and the theory of concave functions and inflection points, on the other. The chapter ends with a discussion of concrete optimization problems, and the three solved examples in Sec. 3.8 are deliberately chosen to be nontrivial, so that the student can have a taste of the "real thing."

It is now Chapter 4, and high time for integral calculus. Here I prefer to use the standard definition of the Riemann integral, allowing the points ξ_i figuring in the approximating sum σ to be *arbitrary* points of their respective subintervals. Students seem to find this definition perfectly plausible, in view of the interpretation of σ as an approximation to the area under the graph of the given function. Once the definite integral is defined, it is immediately emphasized that all continuous functions are integrable, and this fact is henceforth used freely. After establishing a few elementary properties of definite integrals, I prove the mean value theorem for integrals and interpret it geometrically. It is then a simple matter to prove the fundamental theorem of calculus. Next the function $\ln x$ is defined as an integral, in the usual way, and its properties and those of its inverse function e^x are systematically explored. The related functions $\log_a x$, a^x and x^r are treated on the spot, and the validity of the formula $(x^r)' = rx^{r-1}$ for arbitrary real r is finally proved, as promised back in Chapter 2. The two main techniques of integration, namely, integration by substitution and integration by parts, are discussed in detail. The chapter ends with a treatment of improper integrals, both those in which the interval of integration is infinite and those in which the integrand becomes infinite.

There are various ways in which integration can be used as a tool, but foremost among these is certainly the use of integration to solve differential equations. It is for this reason that I have made Chapter 5 into a brief introduction to differential equations and their applications. All the theory needed for our purposes is developed in Sec. 5.1, both for first-order and second-order equations. The next section is then devoted to problems of growth and decay, a subject governed by simple first-order differential equations. The standard examples of population growth, both unrestricted and restricted, are gone into in some detail, as is the topic of radioactive decay. The last section of this short chapter is devoted to problems of motion, where second-order differential equations now hold sway. Inclusion of this material may be regarded as controversial in a book like this, but I for one do not see anything unreasonable in asking even a business or economics student to devote a few hours to the contemplation of Newton's mechanics, a thought system which gave birth first to modern industrial society and then to the space age. In any event, those who for one reason or another still wish to skip Sec. 5.3 hardly need my permission to do so.

The last of the six chapters of this book is devoted to the differential calculus of functions of several variables. Here my intent is to highlight the similarities with the one-dimensional case, while not neglecting significant differences. For example, this is why I feel compelled to say a few words about the distinction between differentiable functions of several variables and those that merely have partial derivatives. However, I do not dwell on such matters. It turns out that much of the theory of Chapters 2 and 3 can be generalized almost effortlessly to the n-dimensional case, without doing violence to the elementary character of the book. In particular, as already noted, the proof of the chain rule in Sec. 6.3 is virtually the same as the one in Sec. 2.8. Chapter 6 closes with a concise treatment of extrema in n dimensions, including the test for strict local extrema and the use of Lagrange multipliers to solve optimization problems subject to constraints. I stop here, because unlike some authors I see no point in reproducing the standard examples involving indifference curves, budget lines, marginal rates of substitution, and the like, to be found in every book on microeconomic theory. I conceive of this book as one dealing primarily with the common mathematical ground on which many subjects rest, and the applications chosen here are ones which shed most light on the kind of mathematics we are trying to do, not those which are most intriguing from other points of view.

The idea of writing this book in the first place was proposed to me by John S. Snyder, Jr. of the W. B. Saunders Co. Without his abiding concern, I find it hard to imagine that the book would ever have arrived at its present form. In accomplishing a total overhaul of an earlier draft, I was guided by helpful suggestions from a whole battery of reviewers, notably, Craig Comstock of the Naval Postgraduate School, John A. Pfaltzgraff of the University of North Carolina, J. H. Curtiss of the University of Miami, Carl M. Bruns of Florissant Valley Community College, David Brown of the University of Pittsburgh, and Maurice Beren of the Lowell Technological Institute. The last of these reviewers played a particularly significant role in my revision of Chapter 1. I would also like to thank my friend Neal Zierler for checking all the answers to the problems in the first draft of the book, and my copy-editor Lloyd Black for his patience in dealing with the kind of author who keeps reading proof, looking for trouble, until it is finally taken away from him once and for all. It has been a pleasure to work with all these fine people.

TO THE STUDENT

Calculus cannot be learned without solving lots of problems. Your instructor will undoubtedly assign you many problems as homework, probably from among those that do not appear in the Selected Hints and Answers section at the end of the book. But, at the same time, every hint or answer in that section challenges you to solve the corresponding problem, whether it has been assigned or not. This is the only way that you can be sure of your command of the subject. Problems marked with stars are either a bit harder than the others, or else they deal with side issues. However, there is no reason to shun these problems. They're neither that hard nor that far off the main track.

The system of cross references used in this book is almost self-explanatory. For example, Theorem 1.48 refers to the one and only theorem in Sec. 1.48, Example 2.43b refers to the one and only example in Sec. 2.43b, and so on. Any problem cited without a further address will be found at the end of the section where it is mentioned. The book has a particularly complete index to help you find your way around. Use it freely.

Mathematics books are not novels, and you will often have to read the same passage over and over again before you grasp its meaning. Don't let this discourage you. With a little patience and fortitude, you too will be doing calculus before long. Good luck!

Chapter 1

MATHEMATICAL
BACKGROUND

1.1 INTRODUCTORY REMARKS

1.11. You are about to begin the study of calculus, a branch of mathematics which dates back to the seventeenth century, when it was invented by Newton and Leibniz independently and more or less simultaneously. At first, you will be exposed to ideas that you may find strange and abstract, and that may not seem to have very much to do with the "real world." After a while, though, more and more applications of these ideas will put in an appearance, until you finally come to appreciate just how powerful a tool calculus is for solving a host of practical problems in fields as diverse as physics, biology and economics, just to mention a few.

Why this delay? Why can't we just jump in feet first, and start solving practical problems right away? Why must the initial steps be so methodical and careful?

The reason is not hard to find, and it is a good one. You are in effect learning a new language, and you must know the meaning of key words and terms before trying to write your first story in this language, that is, before solving your first nonroutine problem. Many of the concepts of calculus are unfamiliar, and were introduced, somewhat reluctantly, only after it gradually dawned on mathematicians that they were in fact indispensable. This is certainly true of the central concept of calculus, namely, the notion of a "limit," which has been fully understood only for a hundred years or so, after having eluded mathematicians for millennia. Living as we do in the modern computer age, we can hardly expect to learn calculus in archaic languages, like that of "infinitesimals," once so popular. We must also build up a certain amount of computational facility, especially as involves *inequalities*, before we are equipped to tackle the more exciting problems of calculus. And we must become accustomed to think both algebraically and geometrically at the same time, with the help of rectangular coordinate systems. All this "tooling up" takes time, but nowhere near as much as in other fields, like music, with its endless scales and exercises. After all, in calculus we need only train our minds, not our hands!

It is also necessary to maintain a certain generality in the beginning, especially in connection with the notion of a "function." The power of calculus is intimately related to its great generality. This is why so many different kinds of problems can be solved by the methods of calculus. For example, calculus deals with "rates of change" in general, and not just special kinds of rates of change, like "velocity," "marginal cost" and "rate of cooling," to mention only three. From the calculus point of view, there are often deep similarities between things that appear superficially unrelated.

In working through this book, you must always have your pen and scratch pad at your side, prepared to make a little calculation or draw a rough figure at a moment's notice. Never go on to a new idea without understanding the old ideas on which it is based. For example, don't try to do problems involving "continuity" without having mastered the idea of a "limit." This is really a workshop course, and your only objective is to learn how to solve calculus problems. Think of an art class, where there is no premium on anything except making good drawings. That will put you in the right frame of mind from the start.

1.12. Two key problems.   Broadly speaking, calculus is the mathematics of change. Among the many problems it deals with, two play a particularly prominent role, in ways that will become clearer to you the more calculus you learn. One problem is

> (1) Given a relationship between two changing quantities, what is the rate of change of one quantity with respect to the other?

And the other, so-called "converse" problem is

> (2) Given the rate of change of one quantity with respect to another, what is the relationship between the two quantities?

Thus, from the very outset, we must develop a language in which "relationships," whatever they are, can be expressed precisely, and in which "rates of change" can be defined and calculated. This leads us straight to the basic notions of "function" and "derivative." In the same way, the second problem leads us to the equally basic notions of "integral" and "differential equation." It is the last concept, of an equation involving "rates of change," that unleashes the full power of calculus. You might think of it as "Newton's breakthrough," which enabled him to derive the laws of planetary motion from a simple differential equation involving the force of gravitation. Why *does* an apple fall?

We will get to most of these matters with all deliberate speed. But we must first spend a few sections reviewing that part of elementary mathematics which is an indispensable background to calculus. Admittedly, this is not the glamorous part of our subject, but first things first! We must all stand on some common ground. Let us begin, then, from a starting point where nothing is assumed other than some elementary algebra and geometry, and a little patience.

1.2 SETS

A little set language goes a long way in simplifying the study of calculus. However, like many good things, sets should be used sparingly and only when the occasion really calls for them.

1.21.   A collection of objects of any kind is called a *set*, and the objects themselves are called *elements* of the set. In mathematics the elements are usually numbers or symbols. Sets are often denoted by capital letters and their elements by small letters. If x is an element of a set A, we may write $x \in A$, where the symbol $\in$ is read "is an element of." Other ways of reading $x \in A$ are "x is a member of A," "x belongs to A," and "A contains x." For example, the set of all Portuguese-speaking countries in Latin America contains a single element, namely Brazil.

1.22. If every element of a set A is also an element of a set B, we write $A \subset B$, which reads "A is a *subset* of B." If A is a subset of B, but B is not a subset of A, we say that A is a *proper subset* of B. In simple language, this means that B not only contains all the elements of A, but also one or more extra elements. For example, the set of all U.S. Senators is a proper subset of the set of all members of the U.S. Congress.

1.23. a. One way of describing a set is to write its elements between curly brackets. Thus the set $\{a, b, c\}$ is made up of the elements a, b and c. Changing the order of the elements does *not* change the set. For example, the set $\{b, c, a\}$ is the same as $\{a, b, c\}$. Repeating an element does not change a set. For example, the set $\{a, a, b, c, c\}$ is the same as $\{a, b, c\}$.

b. We can also describe a set by giving properties that uniquely determine its elements, often using the colon: as an abbreviation for the words "such that." For example, the set $\{x: x = x^2\}$ is the set of all numbers x which equal their own squares. You can easily convince yourself that this set contains only two elements, namely 0 and 1.

1.24. Union of two sets. The set of all elements belonging to at least one of two given sets A and B is called the *union* of A and B. In other words, the union of A and B is made up of all the elements which are in the set A or in the set B, or possibly in both. We write the union of A and B as $A \cup B$, which is often read "A cup B," because of the shape of the symbol $\cup$. For example, if A is the set $\{a, b, c\}$ and B is the set $\{c, d, e\}$, then $A \cup B$ is the set $\{a, b, c, d, e\}$.

1.25. Intersection of two sets. The set of all elements belonging to both of two given sets A and B is called the *intersection* of A and B. In other words, the intersection of A and B is made up of only those elements of the sets A and B which are in both sets; elements which belong to only one of the sets A and B do not belong to the intersection of A and B. We write the intersection of A and B as $A \cap B$, which is often read "A cap B," because of the shape of the symbol $\cap$. For example, if A is the set $\{a, b, c, d\}$ and B is the set $\{b, d, e, f, g\}$, then $A \cap B$ is the set $\{b, d\}$.

1.26. Empty sets. A set which has no elements at all is said to be an *empty set* and is denoted by the symbol $\varnothing$. For example, the set of unicorns in the Bronx Zoo is empty.

By definition, an empty set is considered to be a subset of every set. This is just a mathematical convenience.

1.27. Equality of sets. We say that two sets A and B are *equal* and we write $A = B$ if A and B have the same elements. If A is empty, we write $A = \varnothing$. For example, $\{x: x = x^2\} = \{0, 1\}$, as already noted, while $\{x: x \neq x\} = \varnothing$ since no number x fails to equal itself!

PROBLEMS

1.   Find all the proper subsets of the set $\{a, b, c\}$.

2.   Write each of the following sets in another way, by listing elements:
     (a) $\{x: x = -x\}$;   (b) $\{x: x + 3 = 8\}$;   (c) $\{x: x^2 = 9\}$;
     (d) $\{x: x^2 - 5x + 6 = 0\}$;   (e) $\{x: x \text{ is a letter in the word "calculus"}\}$.

3.   Let $A = \{1, 2, \{3\}, \{4, 5\}\}$. Which of the following are true?
     (a) $1 \in A$;   (b) $3 \in A$;   (c) $\{2\} \in A$.
     How many elements does A have?

4.   Which of the following are true?
     (a) If $A = B$, then $A \subset B$ and $B \subset A$;   (b) If $A \subset B$ and $B \subset A$, then $A = B$;
     (c) $\{x: x \in A\} = A$;   (d) $\{\text{all men over 80 years old}\} = \varnothing$.

5.   Find the union of the sets A and B if
     (a) $A = \{a, b, c\}$, $B = \{a, b, c, d\}$;   (b) $A = \{1, 2, 3, 4\}$, $B = \{-1, 0, 2, 3\}$.

6.   Find the intersection of the sets A and B if
     (a) $A = \{1, 2, 3, 4\}$, $B = \{3, 4, 5, 6\}$;   (b) $A = \{a, b, c, d\}$, $B = \{f, g, h\}$.

7.   Given any set A, verify that $A \cup A = A \cap A = A$.

8.   Given any two sets A and B, verify that both A and B are subsets of $A \cup B$, while $A \cap B$ is a subset of both A and B.

9.   Given any two sets A and B, verify that $A \cap B$ is always a subset of $A \cup B$. Can $A \cap B$ ever equal $A \cup B$?

10. Given any two sets A and B, by the *difference* $A - B$ we mean the set of all elements which belong to A but not to B. Let $A = \{1, 2, 3\}$. Find $A - B$ if
     (a) $B = \{1, 2\}$;   (b) $B = \{4, 5\}$;   (c) $B = \varnothing$;   (d) $B = \{1, 2, 3\}$.

11. Which of the following sets are empty?
     (a) $\{x: x \text{ is a letter before } c \text{ in the alphabet}\}$;
     (b) $\{x: x \text{ is a letter after } z \text{ in the alphabet}\}$;
     (c) $\{x: x + 7 = 7\}$;
     (d) $\{x: x^2 = 9 \text{ and } 2x = 4\}$.

*12. Which of the following sets are empty?
     (a) The set of all right triangles whose side lengths are whole numbers;
     (b) The set of all right triangles with side lengths in the ratio 5:12:13;
     (c) The set of all regular polygons with an interior angle of 45 degrees;
     (d) The set of all regular polygons with an interior angle of 90 degrees;
     (e) The set of all regular polygons with an interior angle of 100 degrees.
     Explain your reasoning.
          Comment. A polygon is said to be *regular* if all its sides have the same length and all its interior angles are equal.

*13. Let $A = \{a, b, c, d\}$, and let B be the set of all subsets of A. How many elements does B have?

1.3 NUMBERS

     In this section we discuss numbers of various kinds, beginning with integers and rational numbers and moving on to irrational numbers and real numbers. The set of all real numbers is called the *real number system*. It is the number system needed to carry out the calculations called for in calculus.

     1.31. The number line. Suppose we construct a straight line L through a point

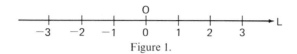
Figure 1.

O and extend it indefinitely in both directions. Selecting an arbitrary unit of measurement, we mark off on the line to the right of O first 1 unit, then 2 units, 3 units, and so on. Next we do the same thing to the left of O. The marks to the right of O correspond to the *positive integers* 1, 2, 3, and so on, and the marks to the left of O correspond to the *negative integers* -1, -2, -3, and so on. The line L, "calibrated" by these marks, is called the *number line*, and the point O is called the *origin* (of L). The direction from negative to positive numbers along L is called the *positive direction*, and is indicated by the arrowhead in Figure 1.

1.32. Integers

   a.   The set of positive integers is said to be *closed* under the operations of addition and multiplication. In simple language, this means that if we add or multiply two positive integers, we always get another positive integer. For example, $2 + 3 = 5$ and $2 \cdot 3 = 6$, where 5 and 6 are positive integers. On the other hand, the set of positive integers is *not* closed under subtraction. For example, $2 - 3 = -1$, where -1 is a negative integer, rather than a positive integer.

   The number 0 corresponding to the point O in Figure 1 is called *zero*. It can be regarded as an integer which is neither positive nor negative. Following mathematical tradition, we use the letter Z to denote the set of all integers, positive, negative and zero. The set Z, unlike the set of positive integers, is closed under subtraction. For example, $4 - 2 = 2$, $3 - 3 = 0$ and $2 - 5 = -3$, where the numbers 2, 0 and -3 are all integers, whether positive, negative or zero.

   b.   An integer n is said to be an *even number* if $n = 2k$, where k is another integer, that is, if n is divisible by 2. On the other hand, an integer n is said to be an *odd number* if $n = 2k + 1$, where k is another integer, that is, if n is not divisible by 2, or equivalently leaves the remainder 1 when divided by 2. It is clear that every integer is either an even number or an odd number.

   1.33. Rational numbers.   The set Z is still too small from the standpoint of someone who wants to be able to *divide* any number in Z by any other number in Z and still be sure of getting a number in Z. In other words, the set Z is not closed under division. For example, $2 \div 3 = \frac{2}{3}$ and $-4 \div 3 = -\frac{4}{3}$, where $\frac{2}{3}$ and $-\frac{4}{3}$ are fractions, not integers. Of course, the quotient of two integers is *sometimes* an integer, and this fact is a major preoccupation of the branch of mathematics known as *number theory*. For example, $8 \div 4 = 2$ and $10 \div -5 = -2$. However, to make division possible in general, we need a bigger set of numbers than Z. Thus we introduce *rational numbers*, namely fractions of the form m/n, where m and n are both integers and n *is not zero*. Note that every integer m, including zero, is a rational number, since $m/1 = m$.

   Let Q (for "quotient") denote the set of all rational numbers. Then the set Q is closed under the four basic arithmetical operations of addition, subtraction, multiplication and division, *provided that we never divide by zero*. It cannot be emphasized too strongly that *division by zero is a forbidden operation* in this course. These matters are considered further in Problems 3 and 13.

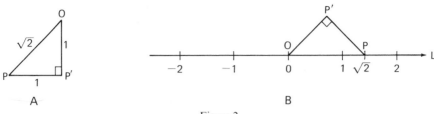

Figure 2.

1.34. Irrational numbers

a. With respect to the number line, the rational numbers fill up the points corresponding to the integers and many but *not all* of the points in between. In other words, there are points of the number line which do *not* correspond to rational numbers. To see this, suppose we construct a right triangle $PP'O$ with sides PP' and $P'O$ of length 1, as in Figure 2A. Then, by elementary geometry, the side OP is of length $\sqrt{1^2 + 1^2} = \sqrt{2}$ (use the familiar Pythagorean theorem). Suppose we place the side OP on the number line, as in Figure 2B, with the point O coinciding with the origin of the line. Then the point P corresponds to the number $\sqrt{2}$. But, as mathematicians concluded long ago, the number $\sqrt{2}$ cannot be rational, and therefore P is a point of the number line which does not correspond to a rational number.

b. By an *irrational number* we simply mean a number, like $\sqrt{2}$, which is not rational. To demonstrate that $\sqrt{2}$ is irrational, we argue as follows. First we digress for a moment to show that the result of squaring an odd number (Sec. 1.32b) is always an odd number. In fact, every odd number is of the form $2k + 1$, where k is an integer, and, conversely, every number of this form is odd. But, squaring the expression $2k + 1$, we get

$$(2k + 1)^2 = 4k^2 + 4k + 1 = 2(2k^2 + 2k) + 1,$$

which is odd, since $2k^2 + 2k$ is itself an integer (why?).

Now, returning to the main argument, suppose $\sqrt{2}$ is a rational number. Then $\sqrt{2}$ must be of the form m/n, where m and n are positive integers and we can assume that the fraction m/n has been reduced to lowest terms, so that m and n are no longer divisible by a common factor other than 1. (For example, the fraction $\frac{12}{18}$ is not in lowest terms, but the equivalent fraction $\frac{2}{3}$ is.) We can then write

$$\sqrt{2} = \frac{m}{n}. \tag{1}$$

Squaring both sides of (1), we have

$$2 = \frac{m^2}{n^2},$$

or equivalently

$$m^2 = 2n^2. \tag{2}$$

Thus m^2 is an even number, being divisible by 2, and therefore the number m itself must be even, since if m were odd, m^2 would also be odd, as shown in the preceding paragraph. Since m is even, we can write m in the form

$$m = 2k, \tag{3}$$

where k is a positive integer. Squaring both sides of (3), we have

$$m^2 = 4k^2. \tag{4}$$

Substituting (4) into (2), we get

$$4k^2 = 2n^2,$$

or equivalently

$$n^2 = \frac{4k^2}{2} = 2k^2.$$

But then n^2 is an even number, and hence so is n, for the reason just given in connection with m^2 and m.

Thus we have managed to show that m and n are both even numbers. Therefore m and n are both divisible by 2. But this contradicts the original assumption that the fraction m/n has been reduced to lowest terms. Since we run into a contradiction if we assume that $\sqrt{2}$ is a rational number, we must conclude that $\sqrt{2}$ is an irrational number. This fact was known to the ancient Greeks, who proved it in just the same way.

c. There are many other irrational numbers. For example, the square roots $\sqrt{3}$, $\sqrt{5}$ and $\sqrt{7}$ are all irrational, and so is π, the ratio of the circumference of a circle to its diameter. For convenience, we use the letter I to designate the set of all irrational numbers.

1.35. The real number system. Let R be the set made up of Q, the set of all rational numbers, and I, the set of all irrational numbers. In other words, let R be the union of Q and I, in the language of sets. Thus

$$R = Q \cup I$$

in symbolic notation (Sec. 1.24). The set R is called the *real number system*, and its elements are called the *real numbers*. From now on, when we use the word "number" without further qualification, we will always mean a *real* number.

1.36. Properties of the real numbers

Next we list several useful facts about real numbers. The student who finds these things interesting is encouraged to pursue them further by visiting the library and looking up a more detailed treatment of the subject.

a. There is one and only one point on the number line corresponding to any given real number, and, conversely, there is one and only one real number corresponding to any given point on the number line. For this reason, the number line is often called the *real line*. In mathematical language, we say that there is a *one-to-one correspondence* between the real numbers and the points of the real line, or between the real number system and the real line itself.

b. Let N be any positive integer, *no matter how large*. Then between any two distinct real numbers there are N other real numbers. Since N is as large as we please, this can be expressed mathematically by saying that between any two distinct real numbers there are *arbitrarily many* real numbers, or better still, *infinitely many* real numbers. In view of the one-to-one correspondence between the real numbers and the points of the real line, this fact is geometrically obvious, since between any two distinct points of the real line we can clearly pick as many other points as we please.

    c. If a rational number is expressed in decimal form, the decimal either terminates in some digit from 1 to 9 or else the decimal does not terminate, but continues indefinitely with groups of repeated digits after a certain decimal place. For example, the rational numbers $\frac{1}{2}, \frac{1}{5}, \frac{1}{8}, \frac{1}{10}$ and $\frac{1}{16}$ are represented by the terminating decimals 0.5, 0.2, 0.125, 0.1 and 0.0625, respectively, while the rational numbers $\frac{1}{3}$, $\frac{1}{6}$ and $\frac{1}{7}$ are represented by the repeating decimals $0.3333\ldots = 0.\overline{3}, 0.1666\ldots = 0.1\overline{6}$ and $0.142857142857\ldots = 0.\overline{142857}$. Here the dots $\ldots$ mean "and so on forever," and the digits written beneath the horizontal line repeat over and over again. Actually, a terminating decimal can be regarded as a special kind of repeating decimal, namely, one with an endless run of zeros after a certain decimal place. Thus $0.125 = 0.125\overline{0}$, $0.0625 = 0.0625\overline{0}$, and so on.

    d. Conversely, if a number in decimal form is a repeating decimal (which includes the case of a terminating decimal), then the number is a rational number, and it can be put in the form of a fraction m/n.

    e. If an irrational number is expressed in decimal form, the decimal does not terminate, but continues indefinitely with *no* groups of repeated digits. For example, $\sqrt{2} = 1.414213562373\ldots$, where the dots $\ldots$ again mean "and so on forever," but this time with no groups of repeated digits. Conversely, if a number in decimal form is this kind of nonrepeating decimal, then the number is an irrational number.

    f. It follows from the foregoing that there is a one-to-one correspondence between the real number system and the set of all decimals, repeating and nonrepeating.

1.37. Mathematical induction

    a. In mathematics we often encounter assertions or formulas involving an arbitrary positive integer n. As an example, consider the formula

$$1 + 3 + 5 \cdots + (2n - 1) = n^2, \tag{5}$$

which asserts that the sum of the first n odd integers equals the square of n. Here the dots $\cdots$ indicate the missing terms, if any, and it is understood that the left side of (5) reduces to simply 1 if $n = 1$, $1 + 3$ if $n = 2$, and $1 + 3 + 5$ if $n = 3$. To prove a formula like (5), we can use the following important technique, known as the principle of *mathematical induction*. Suppose that the formula (or assertion) is known to be true for $n = 1$, and suppose that as a result of assuming that it is true for $n = k$, where k is an arbitrary positive integer, we can prove that it is also true for $n = k + 1$. *Then the formula is true for all k.*

    The reason why mathematical induction works is perfectly clear: First we choose $k = 1$ and use the truth of the formula $n = k = 1$ to deduce its truth for $n = k + 1 = 2$. This shows that the formula is true for $n = 2$. Playing the same game again, we now choose $k = 2$ and use the truth of the formula for $n = k = 2$ to deduce its truth for $n = k + 1 = 3$. Doing this over and over again, we can prove the truth of the formula for every positive integer n, *no matter how large.*

    b. Thus, to prove formula (5), we first note that (5) is certainly true for $n = 1$, since it then reduces to the trivial assertion that

$$1 = 1^2.$$

Suppose (5) holds for $n = k$, so that we have

$$1 + 3 + 5 + \cdots + (2k - 1) = k^2. \tag{6}$$

Then, adding $2k + 1$ to both sides of (6), where $2k + 1$ is the next odd number after $2k - 1$, we get

$$1 + 3 + 5 + \cdots + (2k - 1) + (2k + 1) = k^2 + 2k + 1. \tag{7}$$

The expression on the right clearly equals $(k + 1)^2$, so that (7) takes the form

$$1 + 3 + 5 + \cdots + (2k - 1) + (2k + 1) = (k + 1)^2.$$

But this is just the form taken by (5) if $n = k + 1$, since then

$$2n - 1 = 2(k + 1) - 1 = 2k + 1.$$

In this way, we have shown that if (5) is true for $n = k$, it is also necessarily true for $n = k + 1$. Therefore, by the principle of mathematical induction, (5) is true *for all n* starting from $n = 1$.

  c. The truth of the assertion for $n = 1$ is only needed to "get the induction started." This condition can be relaxed. For example, to give a rather wild example, suppose the assertion is known to be true for $n = 8$, and suppose its truth for $n = k$ implies its truth for $n = k + 1$. Then the assertion is true for all $n = 8, 9, \ldots,$ that is, for all n starting from 8. This is actually the situation in Problem 20.

PROBLEMS

1.   Is the set of negative integers closed under the operation of addition? Give numerical examples to show that the set of negative integers is not closed under the operations of subtraction, multiplication and division.

2.   Give numerical examples to show that
    (a)   The sum of two rational numbers is a rational number;
    (b)   The difference of two rational numbers is a rational number;
    (c)   The product of two rational numbers is a rational number;
    (d)   The quotient of two rational numbers is a rational number.

3.   Show algebraically that the set of rational numbers is closed under the operation of addition. Do the same for the operations of subtraction, multiplication and division.

4.   Which of the following exist?
    (a)   A largest positive integer;
    (b)   A smallest positive integer;
    (c)   A largest positive integer less than 100;
    (d)   A smallest positive integer greater than 100.

5.   Is the number $1 - \sqrt{2}$ rational or irrational? Explain your answer.

6.   Give an example to show that the sum of two irrational numbers can be a rational number. How about the difference of two irrational numbers?

7.   Give an example to show that the product of two irrational numbers can be a rational number. How about the quotient of two irrational numbers?

8.   What conclusions can you draw from Problems 6 and 7 about whether or not the set of irrational numbers is closed under the operations of addition, subtraction, multiplication and division?

9.   Prove that $0 \cdot c = 0$ for every real number c.

10.   Give examples other than those in the text of rational numbers which terminate when expressed in decimal form.

11. Give examples other than those in the text of rational numbers which continue indefinitely with groups of repeated digits when expressed in decimal form.

12. Use mathematical induction to prove that

$$1 + 2 + 3 + \cdots + n = \frac{n(n+1)}{2}$$

for all $n = 1, 2, \ldots$

13. Let a be any real number, possibly zero. Why is the expression $a/0$ meaningless? In other words, why is division by zero impossible?

   Comment.   On the other hand, if $a \neq 0$, then $0/a$ is a perfectly respectable number, equal to 0.

14. It can be shown (Sec. 1.4, Prob. 12) that between any two rational numbers there is another rational number. Illustrate this statement by inserting another rational number between $\frac{31}{100}$ and $\frac{3111}{10000}$.

*15. Verify that $\frac{1}{3} + \frac{1}{6} = \frac{1}{2}$ by adding the corresponding decimals. What conclusions can you draw from this about any decimal with an endless run of nines after a certain decimal place?

*16. Which rational number (in lowest terms) is expressed by the following repeating decimal?
   (a) $0.91\overline{9}$;   (b) $0.\overline{31}$;   (c) $1.2\overline{31}$;   (d) $-0.\overline{417}$.

*17. Explain why a rational number, when expressed in decimal form, either terminates or continues indefinitely with groups of repeated digits.

*18. It can be shown (Sec. 1.5, Prob. 13) that between any two irrational numbers there is another irrational number. Illustrate this statement by inserting another irrational number between $\sqrt{2} = 1.414213562\ldots$ and $1.414215784\ldots$

*19. Use mathematical induction to prove that

$$1^2 + 2^2 + 3^2 + \cdots + n^2 = \frac{2n^3 + 3n^2 + n}{6}$$

for all $n = 1, 2, \ldots$

*20. Verify that every integer greater than seven can be written as a sum made up of threes and fives exclusively. For example, $8 = 3 + 5, 9 = 3 + 3 + 3,$ $10 = 5 + 5, 11 = 3 + 3 + 5,$ and so on.

1.4 INEQUALITIES

1.41. Let a and b be any two numbers. Then there are only three, mutually exclusive possibilities:

   (1) Either a *equals* b, written $a = b$;
   (2) Or a *is greater than* b, written $a > b$;
   (3) Or a *is less than* b, written $a < b$.

On the *real line*, $a > b$ simply means that the point corresponding to the number a lies to the right of the point corresponding to the number b, or, in simpler language, that "the point a" lies to the right of "the point b" (Sec. 1.56). Similarly, $a < b$ means that the point a lies to the left of the point b. Note that $a > b$ and $b < a$ mean exactly the same thing, and so do $a < b$ and $b > a$.

Another way of saying that a is greater than b is to say that if b is subtracted from a, we get a *positive* number, that is, a number greater than zero, while if a is subtracted from b, we get a *negative* number, that is, a number less than zero. In other words, $a > b$, $a - b > 0$ and $b - a < 0$ all mean exactly the same thing, and similarly, so do $a < b$, $b - a > 0$ and $a - b < 0$. We regard it as a known fact that if a and b are both positive numbers, then so are the sum $a + b$ and the product ab.

Formulas involving the symbols $>$ and $<$ (or the symbols $\geqslant$ and $\leqslant$ to be introduced in Sec. 1.47) are called *inequalities*. There are several theorems about inequalities which are both intuitively reasonable and very easy to prove. We now prove some of these which are particularly useful.

1.42. THEOREM. *Adding the same number to each side of an inequality does not change the sense of the inequality. That is,*

$$\text{If } a > b, \quad \text{then } a + c > b + c, \tag{1}$$

where c is any number at all, while

$$\text{If } a < b, \quad \text{then } a + c < b + c. \tag{1'}$$

Proof. To prove (1), we need only show that $(a + c) - (b + c) > 0$, which means exactly the same thing as $a + c > b + c$. But

$$(a + c) - (b + c) = (a - b) + (c - c) = (a - b) + 0 = a - b > 0,$$

since $a > b$. The proof of (1') is just as easy, and is left as an exercise. □

The symbol □ is a modern way of indicating the end of a proof. The old-fashioned way is "Q.E.D.," which you may recall from elementary geometry.

1.43. THEOREM. *Multiplying both sides of an inequality by the same positive number does not change the sense of the inequality. That is,*

$$\text{If } a > b \text{ and } c > 0, \quad \text{then } ac > bc, \tag{2}$$

while

$$\text{If } a < b \text{ and } c > 0, \quad \text{then } ac < bc. \tag{2'}$$

Proof. To prove (2), we need only show that $ac - bc > 0$, which means exactly the same thing as $ac > bc$. But $a - b$ is positive, since $a > b$, and c is positive, by hypothesis. Therefore the product $(a - b)c = ac - bc$ is also positive, since the product of two positive numbers is a positive number. Thus $ac - bc > 0$, as required. The proof of (2') is just as easy, and is left as an exercise. □

1.44. THEOREM. *Multiplying both sides of an inequality by the same **negative** number **changes** the sense of the inequality. That is,*

$$\text{If } a > b \text{ and } c < 0, \quad \text{then } ac < bc, \tag{3}$$

while

$$\text{If } a < b \text{ and } c < 0, \quad \text{then } ac > bc. \tag{3'}$$

Proof. To prove (3), we need only show that $ac - bc < 0$, which means exactly the same thing as $ac < bc$. But $a - b$ is positive, since $a > b$, and c is negative, by hypothesis. Therefore the product $(a - b)c = ac - bc$ is negative, since the product of a positive number and a negative number is a negative number. Thus $ac - bc < 0$, as required. The proof of (3') is just as easy, and is left as an exercise. $\square$

1.45. THEOREM. *If $a > b$ and $b > c$, then $a > c$.*
Proof. By hypothesis, $a - b > 0$ and $b - c > 0$. But then

$$(a - b) + (b - c) = a - c > 0,$$

since the sum of two positive numbers is positive. Alternatively, the theorem follows at once by examining the relative positions of a, b and c, regarded as points on the real line (give the details). $\square$

1.46. THEOREM. *Let a and b be positive numbers such that $a > b$. Then*

$$\frac{1}{a} < \frac{1}{b}. \tag{4}$$

Proof. To prove (4), we need only show that

$$\frac{1}{b} - \frac{1}{a} > 0.$$

Writing

$$\frac{1}{b} - \frac{1}{a} = \frac{a}{ab} - \frac{b}{ab} = \frac{a - b}{ab}, \tag{5}$$

we note that $a - b > 0$, since $a > b$, while $ab > 0$, since $a > 0$ and $b > 0$. It follows that the expression on the right in (5) is positive, being the quotient of two positive numbers. $\square$

1.47. In dealing with inequalities, it is a great convenience to introduce the symbol $\geqslant$, which means "is either greater than or equal to," and the symbol $\leqslant$, which means "is either less than or equal to." Thus $a \geqslant b$ means "a is either greater than or equal to b," while $a \leqslant b$ means "a is either less than or equal to b." It is possible for both inequalities $a \geqslant b$ and $a \leqslant b$ to be valid simultaneously, but only if a is actually equal to b, since the three possibilities listed in Sec. 1.41 are mutually exclusive. In other words, $a \geqslant b$ and $a \leqslant b$ together imply $a = b$.

Clearly $a \geqslant b$ means exactly the same thing as $a - b \geqslant 0$, while $a \leqslant b$ means exactly the same thing as $a - b \leqslant 0$.

1.48. Here is another theorem on inequalities, this time involving the symbol $\leqslant$:
THEOREM. *If $a \leqslant b$ and $c \leqslant d$, then*

$$a + c \leqslant b + d. \tag{6}$$

Proof. As just noted, $a \leqslant b$ means exactly the same thing as $a - b \leqslant 0$, while $c \leqslant d$ means exactly the same thing as $c - d \leqslant 0$. But the sum of two numbers which are negative or zero is itself a number which is negative or zero. In other words, the sum of two nonpositive numbers is a nonpositive number. It follows that

$$(a - b) + (c - d) = (a + c) - (b + d) \leqslant 0,$$

which means exactly the same thing as (6). $\square$

1.49. Inequalities are often combined. For example, $a \geqslant b > c$ means that both inequalities $a \geqslant b$ and $b > c$ hold simultaneously. Similarly, $d < e \leqslant f$ means that both $d < e$ and $e \leqslant f$ hold simultaneously. Thus we have

$$2 \geqslant \sqrt{4} \geqslant 1, \qquad 2 \geqslant \sqrt{4} > 1, \qquad 1 < \sqrt{2} < 2, \qquad 1 < \sqrt{2} \leqslant 2.$$

Give other examples involving the same numbers. Bear in mind that by $\sqrt{x}$, where x is a positive number, we always mean the *positive* square root of x. Thus, for example, $\sqrt{4}$ equals 2, never -2.

PROBLEMS

1. Show that
   (a) If $a > b$, then $-a < -b$;
   (b) If $a > b$ and $c > d$, then $a + c > b + d$.
2. Given two unequal rational numbers $p = m/n$ and $p' = m'/n'$, written with positive denominators (as is always possible), show that $p > p'$ is equivalent to $mn' > m'n$, while $p < p'$ is equivalent to $mn' < m'n$.
3. Which is larger?
   (a) $\frac{33}{10}$ or $\frac{10}{3}$;   (b) $-\frac{33}{10}$ or $-\frac{10}{3}$;   (c) $\frac{167}{50}$ or $\frac{10}{3}$.
4. Verify that if $a > b > 0$ and $c > d > 0$, then $ac > bd > 0$.
5. Verify that if $a > 0$, $b > 0$ and $b^2 > a^2$, then $b > a$. Use this to confirm that $\sqrt{3} > \sqrt{2}$.
6. Show that $a^2 > a$ if $a > 1$, while $a^2 < a$ if $0 < a < 1$. When does $a^2 = a$?
7. Verify that

$$\sqrt{2} \geqslant 1^2 \geqslant 1 \geqslant \frac{1 - 1}{2} \geqslant 0 \geqslant -3.$$

Write this in another way, using the symbol $\leqslant$ instead.
8. Show that
   (a) If $a \geqslant b$, then $-a \leqslant -b$;
   (b) If $a \geqslant b$ and $b \geqslant c$, then $a \geqslant c$;
   (c) If $a \geqslant b$ and $b > c$ or if $a > b$ and $b \geqslant c$, then $a > c$.
9. Verify that if $a \geqslant b$ and $c > 0$, then $ac \geqslant bc$, while if $a \geqslant b$ and $c < 0$, then $ac \leqslant bc$.
10. Given a number x, the largest integer less than or equal to x is called the *integral part* of x and is denoted by $[x]$, not to be confused with $\{x\}$, the set whose only element is x. Find
    (a) $[\frac{1}{2}]$;   (b) $[1]$;   (c) $[\frac{4}{3}]$;   (d) $[\sqrt{2}]$;   (e) $[-\frac{1}{2}]$;
    (f) $[-\sqrt{2}]$.
11. Let n be an integer. Find
    (a) $[n]$;   (b) $[n + \frac{1}{2}]$;   (c) $[n - \frac{1}{2}]$.
*12. Let p and q be two rational numbers such that $p < q$. Show that the number $\frac{1}{2}(p + q)$ is also rational and lies between p and q. Use this to show that there is no largest rational number less than 1, and no smallest rational number greater than 0. Can we change the word "rational" to "real" here?
*13. Verify that
    (a) $a^2 + b^2 \geqslant 2ab$;   (b) $(a + b)^2 \geqslant 4ab$;

    (c) If $a > 0$, then $a + \frac{1}{a} \geqslant 2$.

*14.   The *arithmetic mean* of two positive numbers x and y is defined as

$$a = \frac{x + y}{2}$$

and the *geometric mean* as

$$g = \sqrt{xy}.$$

Verify that $g < a$ unless $x = y$, in which case $g = a$.

*15.   Use the preceding problem to show that of all rectangles with a given perimeter (combined side length), the square has the greatest area.

1.5   THE ABSOLUTE VALUE

1.51.   By the *absolute value* of a number x we mean the number which equals x if x is nonnegative and $-x$ if x is negative. If x is expressed in decimal form, then the absolute value of x is just the decimal without its minus sign if it has one. The absolute value of x is denoted by $|x|$, with two vertical lines (never with brackets). In other words,

$$|x| = \begin{cases} x & \text{if } x \geqslant 0, \\ -x & \text{if } x < 0. \end{cases} \tag{1}$$

Thus, for example, $|2.2| = 2.2, |-3.1| = -(-3.1) = 3.1, |0| = 0$. Note that $|-x| = |x|$ for any number x. For example, $|-3.1| = |3.1| = 3.1$.

1.52. Theorem.   *The inequalities*

$$-|x| \leqslant x \leqslant |x| \tag{2}$$

and

$$|x + y| \leqslant |x| + |y| \tag{3}$$

hold for arbitrary numbers x and y.

   Proof.   To prove (2), we merely note that, by (1), $x = |x|$ if $x \geqslant 0$, while $x = -|x|$ if $x < 0$. Therefore (2) holds in either case.

   To prove (3), we write

$$-|y| \leqslant y \leqslant |y|. \tag{4}$$

as well as (2). It then follows from Theorem 1.48, with $a = x, b = |x|, c = y, d = |y|$, that

$$x + y \leqslant |x| + |y|, \tag{5}$$

and from the same theorem, this time with $a = -|x|, b = x, c = -|y|, d = y$, that

$$-|x| - |y| \leqslant x + y. \tag{5'}$$

But (5) and (5') together imply (3). In fact, if $x + y \geqslant 0$, then $x + y = |x + y|$, so that (5) is equivalent to (3), while if $x + y < 0$, then $x + y = -|x + y|$, so that (5') becomes $-|x| - |y| \leqslant -|x + y|$, which is again equivalent to (3).   □

   1.53.   According to (3), the absolute value of the sum of any two numbers is either less than or equal to the sum of the absolute values of the numbers. More concisely, "the absolute value of a sum cannot exceed the sum of the absolute values."

You should convince yourself by testing the various possibilities that (3) reduces to the equality $|x + y| = |x| + |y|$ when x and y have the same sign or at least one of the numbers x and y is zero, and that (3) can be replaced by the "strict" inequality $|x + y| < |x| + |y|$ when x and y have opposite signs. Formula (3) is often called the "triangle inequality," for a reason we do not go into here.

1.54. The coordinate of a point on the real line. As we have seen in Sec. 1.36a, there is one and only one point on the real line corresponding to any given real number, and, conversely, there is one and only one real number corresponding to any given point on the real line. Thus, to specify a point P on the line, we need only give the real number corresponding to P. This number is called the *coordinate* of P.

Let d be the distance between the origin O, namely the point with coordinate zero, and the point P. Then P has the coordinate d if P lies to the right of O (see Figure 3A) and the coordinate $-d$ if P lies to the left of O (see Figure 3B). If the point P coincides with the origin O, its distance from O is zero, and therefore so is its coordinate.

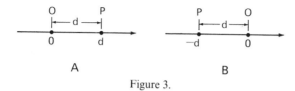

Figure 3.

Conversely, suppose P has the coordinate x, where x is any real number. Then P is just the point at distance $|x|$ from O, lying to the right of O if x is positive (see Figure 4A) and to the left of O if x is negative (see Figure 4B). If $x = 0$, then P clearly coincides with O.

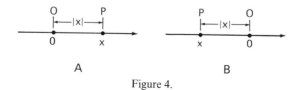

Figure 4.

1.55. The distance between two points on the real line

THEOREM. *Let P_1 and P_2 be two points on the real line with coordinates x_1 and x_2, respectively, and let d be the distance between P_1 and P_2. Then*

$$d = |x_1 - x_2|. \tag{6}$$

Proof. Let P_1 and P_2 both lie to the right of the origin O. Then formula (6) follows from Figure 5A if P_1 lies to the right of P_2, and from Figure 5B if P_1 lies

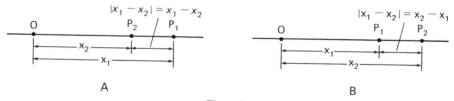

Figure 5.

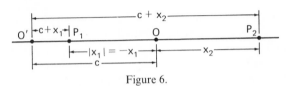

Figure 6.

to the left of P_2. On the other hand, if P_1 or P_2 (or both) lies to the left of O, as in Figure 6, then we can replace O by a new origin O', a distance c to the left of O, such that P_1 and P_2 both lie to the right of O'. With respect to the new origin O', the points P_1 and P_2 have coordinates $c + x_1$ and $c + x_2$, as is apparent from the figure, where the point P_1 lies to the left of O and the distance from O' to P_1 equals $c - |x_1| = c + x_1$. Therefore, by the first part of the proof, we now have

$$d = |(c + x_1) - (c + x_2)| = |(c - c) + (x_1 - x_2)| = |x_1 - x_2|,$$

so that formula (6) is still valid. □

1.56. In talking about real numbers, we will make free use of geometrical language whenever it seems appropriate. In particular, we will usually say "the point x" instead of "the point with coordinate x." The distance between the points P_1 and P_2 will often be denoted by $|P_1P_2|$, as suggested by the absolute value in (6). The same "double vertical line notation" will also be used for the distance between points in the plane and in space.

PROBLEMS

1. Verify that $|xy| = |x|\,|y|$.
   Comment. Thus "the absolute value of a product equals the product of the absolute values."
2. Show that
$$x + |x| = \begin{cases} 2x & \text{if } x \geqslant 0, \\ 0 & \text{if } x < 0. \end{cases}$$
3. Verify that $|x|^2 = x^2$ and $|x| = \sqrt{x^2}$ for every number x, regardless of the sign of x.
4. Show that
$$|x + y + z| \leqslant |x| + |y| + |z|$$
   for arbitrary numbers x, y, z.
5. More generally, show that
$$|x_1 + x_2 + \cdots + x_n| \leqslant |x_1| + |x_2| + \cdots + |x_n|$$
   for arbitrary numbers $x_1, x_2, \ldots, x_n$.
6. Which points are at distance 2 from the point -1?
7. Find the two points which are four times closer to the point -1 than to the point 4.
8. When does the point x^2 lie to the right of the point x? When does it lie to the left? When do the two points coincide?
9. If P_1 and P_2 are the points with coordinates x_1 and x_2, verify that the point with coordinate $\frac{1}{2}(x_1 + x_2)$ is the midpoint of the segment P_1P_2.

*10. Show that

$$|x - y| \geqslant ||x| - |y||$$

for arbitrary x and y. When does equality occur?

*11. Solve the equation

(a) $|x - 1| = 2$;   (b) $|x - 1| = |3 - x|$;   (c) $|x + 1| = |3 + x|$;
(d) $|2x| = |x - 2|$.

In other words, find the set of all numbers x satisfying the equation.

*12. What happens to the point $(1 - x)a + xb$ as x varies from 0 to 1?

*13. Verify that between any two real numbers x_1 and x_2

(a) There is a rational number and an irrational number;
(b) There are infinitely many rational numbers and infinitely many irrational numbers.

1.6 INTERVALS AND NEIGHBORHOODS

1.61. Intervals

a. Let a and b be any two real numbers such that $a < b$, and consider the set I of all real numbers x such that x is greater than a but less than b. Then I is called an *open interval* and is denoted by the symbol (a, b). Note that I does *not* include the points $x = a$ and $x = b$, called the *end points* of the interval. We can also denote (a, b) by writing $a < x < b$, it being understood that $a < x < b$ is shorthand for the set $I = \{x : a < x < b\}$.

b. Suppose we enlarge the open interval (a, b) by including the end points $x = a$ and $x = b$. Then the resulting set is called a *closed interval* and is denoted by the symbol $[a, b]$, with square brackets instead of round brackets (parentheses). Since $[a, b]$ is clearly the set of all x such that x is greater than or equal to a but less than or equal to b, we can also denote $[a, b]$ by writing $a \leqslant x \leqslant b$, this being shorthand for the set $\{x : a \leqslant x \leqslant b\}$.

c. Sometimes it is convenient to speak of intervals which include one end point but not the other. Thus we have the interval $[a, b)$, which includes the left end point a but excludes the right end point b, or the interval $(a, b]$, which includes the right end point b but excludes the left end point a. Note the crucial difference between the meaning of a round bracket (or) and a square bracket [or]. We can also write $[a, b)$ as $a \leqslant x < b$ and $(a, b]$ as $a < x \leqslant b$. These intervals, which are neither open nor closed, might be regarded as "half open," but they might just as well be regarded as "half-closed." The intervals (a, b), $[a, b]$, $[a, b)$ and $(a, b]$ are all assigned the same *length*, namely $b - a$.

d. The geometrical meaning of the various kinds of intervals is shown in Figure 7, where included end points are indicated by solid dots and excluded end points by hollow dots.

1.62. Examples

a. Find the open interval $a < x < b$ or (a, b) equivalent to

$$-5 < x + 2 < 3 \tag{1}$$

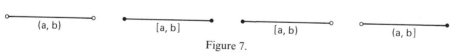

(a, b)   [a, b]   [a, b)   (a, b]

Figure 7.

SOLUTION.   Each of the two inequalities in (1), namely

$$-5 < x + 2 \qquad (2)$$

and

$$x + 2 < 3, \qquad (3)$$

remains valid if we add the same number to both sides. Adding -2 to both sides of (2), or equivalently subtracting 2 from both sides, we get

$$-7 < x. \qquad (2')$$

Similarly, subtracting 2 from both sides of (3), we get

$$x < 1. \qquad (3')$$

Combining (2') and (3'), we find that the open interval equivalent to (1) is

$$-7 < x < 1.$$

This is just the interval $(-7, 1)$ in bracket form.

   b.   Find the closed interval $a \leqslant x \leqslant b$ or $[a, b]$ equivalent to

$$1 \leqslant x - 5 \leqslant 4. \qquad (4)$$

SOLUTION.   Again, each of the inequalities in (4), namely

$$1 \leqslant x - 5, \qquad x - 5 \leqslant 4, \qquad (5)$$

remains valid if we add the same number to both sides. The only sensible choice of the number to be added is 5, of course, and this converts the inequalities (5) to

$$6 \leqslant x, \qquad x \leqslant 9. \qquad (5')$$

Combining these inequalities, we find that the closed interval equivalent to (4) is

$$6 \leqslant x \leqslant 9,$$

or $[6, 9]$ in bracket form.

1.63. Neighborhoods

   a.   By a *neighborhood* of a point c we mean any open interval with c as its midpoint. Thus, for example, the intervals $(-1, 1)$, $(-2, 2)$ and $(-3, 3)$ are all neighborhoods of the origin of the real line, that is, of the point $x = 0$. If we exclude the midpoint c from any neighborhood of c, the resulting set is called a *deleted neighborhood* of c. Note that a deleted neighborhood is the union of *two* open intervals, rather than a single open interval.

   b. · Let δ (the Greek letter delta) denote any positive number. Then by the δ-*neighborhood* of a point c we mean the neighborhood of c of length 2δ. In other words, the δ-neighborhood of c is the open interval $c - \delta < x < c + \delta$, or equivalently $(c - \delta, c + \delta)$, shown in Figure 8A. The corresponding *deleted δ-neighborhood*

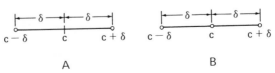

A   B

Figure 8.

of c is the union of intervals $(c - \delta, c) \cup (c, c + \delta)$, shown in Figure 8B, where the hollow dot indicates the missing point c. Note that $(c - \delta, c + \delta)$ can also be described as the set of all x whose distance from c is less than δ, and hence, by Theorem 1.55, as the set of all x such that $|x - c| < \delta$. Similarly, $(c - \delta, c) \cup (c, c + \delta)$ can be described as the set of all x such that $0 < |x - c| < \delta$.

There is nothing sacred about the use of the letter δ in this context, apart from mathematical tradition, and we could use any other letter as well. A common choice is ε (the Greek letter epsilon).

c. Example. Find the 1-neighborhood of the point 2.

SOLUTION. Here $c = 2$, $\delta = 1$, and the neighborhood is just the open interval $2 - 1 < x < 2 + 1$, namely $1 < x < 3$ or $(1, 3)$. The corresponding deleted 1-neighborhood is the set $(1, 2) \cup (2, 3)$.

1.64. Infinite intervals

a. In discussing intervals, it is convenient to introduce two new symbols. These are ∞, called (*plus*) *infinity*, and $-\infty$, called *minus infinity*. The symbols ∞ and $-\infty$ must not be thought of as numbers, even though they appear in inequalities. Using ∞ and $-\infty$, we now define the following kinds of intervals, where c is an arbitrary number:

(1) The set of all numbers x such that $x < c$, denoted by $-\infty < x < c$;
(2) The set of all numbers x such that $x \leqslant c$, denoted by $-\infty < x \leqslant c$;
(3) The set of all numbers x such that $x > c$, denoted by $c < x < \infty$;
(4) The set of all numbers x such that $x \geqslant c$, denoted by $c \leqslant x < \infty$;
(5) The set of *all* numbers x, namely the whole real number system, denoted by $-\infty < x < \infty$

In bracket notation, we denote these five kinds of intervals by $(-\infty, c)$, $(-\infty, c]$, (c, ∞), $[c, \infty)$ and $(-\infty, \infty)$, respectively, with a round bracket for an excluded end point and a square bracket for an included end point, just as before. These intervals, involving ∞ and $-\infty$, are said to be *infinite*, as opposed to the *finite* intervals (a, b), $[a, b]$, $[a, b)$ and $(a, b]$.

b. Since ∞ and $-\infty$ are not numbers, we cannot allow either $x = \infty$ or $x = -\infty$. Therefore it is meaningless to write intervals like $-\infty \leqslant x < c$, $c \leqslant x \leqslant \infty$, $-\infty \leqslant x \leqslant \infty$, etc., and no infinite interval written in bracket form can have a square bracket next to the symbol ∞ or $-\infty$.

c. Example. Find the set of all x such that

$$|x - 1| - |x - 2| = 1. \tag{6}$$

SOLUTION. According to Theorem 1.55, equation (6) means that the distance between the point x and the point 1 minus the distance between the point x and the point 2 equals 1. This happens when $x \geqslant 2$ and only then (what goes wrong if $x < 2$?). Therefore the set of all x satisfying (6) is the infinite interval $2 \leqslant x < \infty$, or $[2, \infty)$ in bracket form.

PROBLEMS

1. What is the open interval, in bracket form, equivalent to $-3 < x - 3 < -1$? The closed interval equivalent to $-2 \leqslant x + 1 \leqslant 4$?

2. What is the half-open interval, in bracket form, equivalent to $1 \leqslant x - 1 < 7$? Equivalent to $\sqrt{2} < x + 2 \leqslant \sqrt{3}$?

3. Find the set of all x such that $|x - 1| + |x - 2| = 1$.

4. Find the $\sqrt{2}$-neighborhood of the point 3. Write the corresponding deleted neighborhood as a union of open intervals.

5. What is the open interval, in bracket form, equivalent to $-6 < 3x < 3$? The closed interval equivalent to $-6 \leqslant -3x \leqslant 3$?

6. Find a simpler way of writing
(a) $(1, 3) \cup \{3\}$; (b) $[1, 3) \cup [3, \infty)$; (c) $(-\infty, 1) \cup (U, \infty)$.

7. Find a simpler way of writing
(a) $[-2, 3] \cap [-1, 1]$; (b) $[-1, 1] \cap [1, 2]$; (c) $(-\infty, 1] \cap (-1, \infty)$.

*8. In what interval is the expression

$$\frac{1}{x + |x|}$$

defined? How about $\sqrt{1 - x}$?

1.7 RECTANGULAR COORDINATES

As shown in Sec. 1.54, any given point on a line can be uniquely specified by giving a single real number, called the *coordinate* of the point. We now show how to uniquely specify any given point in a *plane*. This can be done by giving *two* real numbers, again called the *coordinates* of the point.

1.71. Suppose that at a convenient point in a plane (this page, say) we construct a pair of perpendicular lines, known as *coordinate lines* or *coordinate axes*, intersecting in a point O, called the *origin* (of coordinates). For convenience, we choose one of the lines parallel to the short dimension of the page, calling it the *x-axis* and labelling it Ox, and the other line parallel to the long dimension of the page, calling it the *y-axis* and labelling it Oy. Each line is regarded as extending indefinitely in both directions, and each is equipped with a *positive direction*, pointing to the right in the case of the *x*-axis and upward in the case of the *y*-axis, as indicated by the arrowheads in Figure 9A.

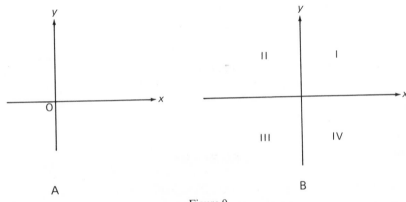

Figure 9.

The plane containing the pair of perpendicular lines Ox and Oy is called the *xy-plane*. The two coordinate lines divide the *xy*-plane into four regions, called *quadrants*. These are indicated by the Roman numerals in Figure 9B, where I refers to the *first quadrant*, II to the *second quadrant*, and so on. Note that the quadrants are arranged in the *counterclockwise direction*, that is, in the direction opposite to that in which the hands of a clock move.

1.72. We are now able to associate a pair of numbers with any given point P in the *xy*-plane, by making the following construction which you have probably already encountered before: Through the point P we draw two straight lines, one perpendicular to the *x*-axis, the other perpendicular to the *y*-axis. Suppose that, as in Figure 10, the first line intersects the *x*-axis in the point with coordinate a, where the *x*-axis is regarded as a number line with the indicated positive direction. Suppose also that the second line intersects the *y*-axis in the point with coordinate b, where the *y*-axis is regarded as another number line with the indicated positive direction. Then the numbers a and b are called the *rectangular coordinates*, or simply the *coordinates*, of the point P. More exactly, a is called the *abscissa* or *x-coordinate* of P, while b is called the *ordinate* or *y-coordinate* of P.

Conversely, given any pair of numbers a and b, to "plot" (that is, to find) the point P in the *xy*-plane with abscissa a and ordinate b, we simply reverse the above construction: We draw two straight lines, one perpendicular to the *x*-axis through the point of the *x*-axis with coordinate a, the other perpendicular to the *y*-axis through the point of the *y*-axis with coordinate b. Then, as is immediately apparent, the point of intersection of these two lines is just the point P with abscissa a and ordinate b.

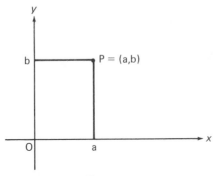

Figure 10.

1.73. The point P with abscissa a and ordinate b may also be denoted by (a, b). The symbol (a, b) is called an *ordered pair*, and is a special kind of two-element set of real numbers, namely one in which *the order of the elements matters*. Thus, although the ordinary sets $\{a, b\}$ and $\{b, a\}$ are identical (Sec. 1.23a), the ordered pairs (a, b) and (b, a) are different unless $a = b$.

Do not confuse the ordered pair (a, b) with the same symbol (a, b) used to designate an open interval with end points a and b. The context will always show which meaning is to be attached to the symbol (a, b). Although it would be nice to have different symbols for different things, this is a case where mathematical tradition must be respected.

Note that the origin O has abscissa zero and ordinate zero. In other words, O is the point $(0, 0)$.

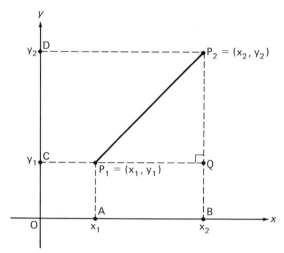

Figure 11.

1.74. The distance between two points in the plane

THEOREM.   *Let $P_1 = (x_1, y_1)$ and $P_2 = (x_2, y_2)$ be two points in the xy-plane, and let $|P_1P_2|$ be the distance between them. Then*

$$|P_1P_2| = \sqrt{|x_1 - x_2|^2 + |y_1 - y_2|^2}. \qquad (1)$$

Proof. Dropping perpendiculars from P_1 and P_2 to the x- and y-axes, we find that P_1P_2 is the hypotenuse of the right triangle P_1QP_2 shown in Figure 11. Moreover, it is obvious that $|P_1Q| = |AB|$ and $|QP_2| = |CD|$, where A and B have coordinates x_1 and x_2 regarded as points of the x-axis, while C and D have coordinates y_1 and y_2 regarded as points of the y-axis. Therefore, by the Pythagorean theorem,

$$|P_1P_2|^2 = |P_1Q|^2 + |QP_2|^2 = |AB|^2 + |CD|^2,$$

and hence

$$|P_1P_2| = \sqrt{|AB|^2 + |CD|^2}. \qquad (2)$$

But, according to Theorem 1.55,

$$|AB| = |x_1 - x_2|, \qquad |CD| = |y_1 - y_2|. \qquad (3)$$

Combining (2) and (3), we get (1) at once. □

Since $|x|^2 = x^2$ for every number x, regardless of the sign of x, we can just as well write (1) in the equivalent form

$$|P_1P_2| = \sqrt{(x_1 - x_2)^2 + (y_1 - y_2)^2} = \sqrt{(x_2 - x_1)^2 + (y_2 - y_1)^2}. \qquad (4)$$

1.75. Examples

a.   Find the distance between the points $P_1 = (-1, 2)$ and $P_2 = (1, -2)$.
SOLUTION.   According to (4),

$$|P_1P_2| = \sqrt{(-1 - 1)^2 + (2 - (-2))^2} = \sqrt{(-2)^2 + 4^2}$$
$$= \sqrt{4 + 16} = \sqrt{20} = 2\sqrt{5}.$$

b.   Find the distance between two points P_1 and P_2 lying on the same line parallel to the x-axis. On the same line parallel to the y-axis.

SOLUTION. If P_1 and P_2 lie on the same line parallel to the x-axis, then P_1 and P_2 have the same ordinate, say b, so that $P_1 = (x_1, b)$ and $P_2 = (x_2, b)$. Therefore, in this case, formula (4) reduces to

$$|P_1P_2| = \sqrt{(x_1 - x_2)^2 + (b - b)^2} = \sqrt{(x_1 - x_2)^2} = |x_1 - x_2|. \qquad (5)$$

This is just what we would expect from Theorem 1.55, which deals with the case where the common ordinate b equals zero.

In the same way, it is easy to show that if P_1 and P_2 lie on the same line parallel to the y-axis, so that P_1 and P_2 have the same *abscissa*, then formula (4) reduces to

$$|P_1P_2| = |y_1 - y_2|. \qquad (6)$$

Give the details.

c. Is the triangle ABC with vertices $A = (0, 0)$, $B = (3, 3)$, $C = (-1, 7)$ a right triangle?

SOLUTION. Yes. To see this, we first calculate the squares of the side lengths of ABC, with the help of formula (4). As a result, we obtain

$$|AB|^2 = 3^2 + 3^2 = 18,$$
$$|BC|^2 = (-4)^2 + 4^2 = 32,$$
$$|AC|^2 = (-1)^2 + 7^2 = 50,$$

so that

$$|AC|^2 = |AB|^2 + |BC|^2$$

for the given triangle ABC. It follows from the converse of the Pythagorean theorem (explain) that ABC is a right triangle with the side AC as its hypotenuse.

PROBLEMS

1. Plot the points $A = (2, 0)$, $B = (2, 2)$, $C = (0, 3)$, $D = (-2, 2)$, $E = (-2, 0)$, $F = (0, -1)$ on ordinary graph paper. Then join A to C, B to D, C to E, D to F, E to A, and finally F to B. What is the resulting figure?
2. Suppose the figure in the preceding problem is shifted one unit to the right and two units upward. Then A, B, C, D, E, F go into new points A', B', C', D', E', F'. What are these new points?
3. If the point (x, y) lies in the first quadrant, then $x > 0$, $y > 0$. Write similar conditions characterizing the other three quadrants.
4. Find the distance between the pair of points
   (a) $(1, 3), (5, 7)$;   (b) $(-2, -3), (1, 1)$;   (c) $(1, 3), (1, 4)$;   (d) $(2, 4), (5, 4)$.
5. Give an example of four points, each in a different quadrant, whose distances from the origin are all equal.
6. Given two points $P_1 = (x_1, y_1)$ and $P_2 = (x_2, y_2)$, verify that the point with abscissa $\frac{1}{2}(x_1 + x_2)$ and ordinate $\frac{1}{2}(y_1 + y_2)$ is the midpoint of the segment P_1P_2.
7. Two vertices A and B of an isosceles triangle ABC lie at the points $(0, 1)$ and $(10, 1)$. Find the abscissa of the point C if $|AC| = |BC|$.
8. Locate the points $A = (4, 1)$, $B = (3, 5)$, $C = (-1, 4)$ and $D = (0, 0)$. Show that $ABCD$ is a square. What is the side length of the square?

9. Find the midpoints of the sides of the square $ABCD$ in the preceding problem.

*10. Find all points which are equidistant from the x-axis, the y-axis and the point $(3, 6)$.

*11. How many points of the form (m, n), where m and n are integers, lie inside the circle of radius $\frac{5}{2}$ with its center at the origin?

*12. Given three noncollinear points $A = (0, 0)$, $B = (x, y)$ and $D = (x', y')$, what choice of the point C makes the quadrilateral $ABCD$ a parallelogram?

1.8 STRAIGHT LINES

1.81. The slope of a line

a. Let L be any nonvertical straight line in the xy-plane, and let $P_1 = (x_1, y_1)$ and $P_2 = (x_2, y_2)$ be any two distinct points of L. Then by the *slope* of L we mean the ratio

$$m = \frac{y_2 - y_1}{x_2 - x_1} \tag{1}$$

To interpret the slope geometrically, we draw the line through P_1 parallel to the x-axis and the line through P_2 parallel to the y-axis, intersecting in the point $A = (x_2, y_1)$, as shown in Figure 12. Then the slope m is just the ratio

$$m = \frac{|P_2 A|}{|P_1 A|} \tag{2}$$

of the length of the side $P_2 A$ to the length of the side $P_1 A$ in the right triangle $P_1 A P_2$.

b. It is important to note that the slope of a line L does not depend on the particular choice of the points used to define the slope. To see this, let P_3 and P_4 be any two points on L other than P_1 and P_2, and suppose the line through P_3 parallel to the x-axis intersects the line through P_4 parallel to the y-axis in the point B, as shown in Figure 13, where L, P_1, P_2 and A are exactly the same as in Figure 12. Then the right triangles $P_1 A P_2$ and $P_3 B P_4$ are similar (why?), and therefore

$$\frac{|P_4 B|}{|P_3 B|} = \frac{|P_2 A|}{|P_1 A|},$$

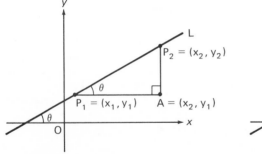

Figure 12.

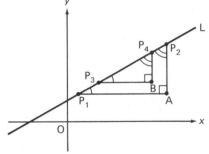

Figure 13.

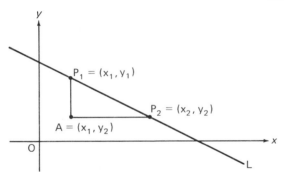

Figure 14.

so that the formula

$$m = \frac{|P_4B|}{|P_3B|}$$

leads to exactly the same value of m as formula (2).

    c. The slope of the line L in Figure 12 is clearly positive, since in this case both $x_2 - x_1$ and $y_2 - y_1$ are positive, and hence so is the ratio (1). However, a line may well have a *negative* slope. For example, the ratio (1) is negative for the line L shown in Figure 14. To see this, we need only note that $y_2 - y_1$ is now negative, while $x_2 - x_1$ is still positive.

    Thus, in brief, if a line slopes *up* to the right, its slope is *positive*, while if a line slopes *down* to the right, its slope is *negative.* *

1.82. The inclination of a line

    a. By the *inclination* of a straight line L in the xy-plane we mean the smallest angle between the x-axis and L, as measured from the x-axis to L in the counterclockwise direction. The inclination, which we denote by θ (the Greek letter theta), will be measured in *degrees*, denoted by the symbol $^\circ$. Moreover, any line parallel to the x-axis, including the x-axis itself, will be regarded as having the inclination 0° (zero degrees). Since vertical angles are equal (see Figure 15), it doesn't matter

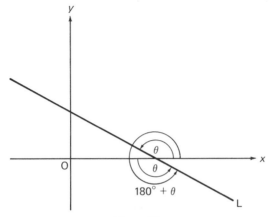

Figure 15.

*Note that in interpreting the slope geometrically (and in proving Theorem 1.74), we have tacitly assumed that the line L slopes *up* to the right and that the point P_1 lies to the *left* of the point P_2. As an exercise, consider the modifications required in the other cases.

whether the measurement of θ is started to the right or to the left of the point in which L intersects the x-axis. It is also clear from the figure that if L makes the angle $180° + \theta$ with the x-axis, then L also makes the smaller angle θ with the x-axis. Therefore the inclination θ of any line whatsoever lies in the half-open interval $0° \leqslant \theta < 180°$.

    b.   Now, according to formula (2), the slope m of the line L is just the ratio of the length of the side P_2A to the length of the side P_1A in the right triangle P_1AP_2. Students who have had some elementary trigonometry will recall that this ratio ("the opposite side over the adjacent side") is also called the *tangent* of the interior angle at P_1 in the triangle P_1AP_2. But this angle is precisely the inclination θ of the line L, since the side P_1A is parallel to the x-axis. Thus, in the notation of trigonometry, we have the formula

$$m = \tan \theta, \tag{3}$$

which is read as "m equals the tangent of θ," giving the relation between the inclination of a line and its slope. The same argument as in Sec. 1.81b, based on the use of similar triangles, shows that the tangent of an angle depends only on the size of the given angle and not on the size of the right triangle containing the angle.

    Figure 16 shows various lines, together with their inclinations and slopes, as related by formula (3). Note that although a vertical line has inclination 90°, its slope is undefined, since setting $x_1 = x_2$ in formula (1) would make the denominator zero. It is for this reason, of course, that L was assumed to be nonvertical in Sec. 1.81a.

    c.   In connection with formula (3), it should be noted that if the angle θ lies between 90° and 180°, then $\tan \theta$ is negative, as we would expect since the line L then slopes down to the right and has negative slope. To calculate tangents between 90° and 180°, we use the formula

$$\tan (180° - \theta) = -\tan \theta, \tag{4}$$

established in every course on trigonometry. For example,

$$\tan 135° = \tan (180° - 45°) = -\tan 45° = -1,$$

$$\tan 150° = \tan (180° - 30°) = -\tan 30° = -\frac{1}{\sqrt{3}},$$

and so on.

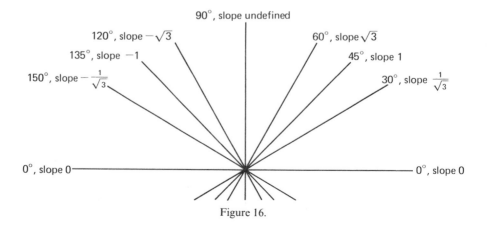

Figure 16.

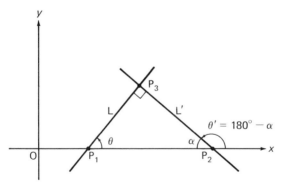

Figure 17.

1.83. Examples

a. Find the slope m of the line going through the points $(1, 3)$ and $(4, 6)$.
SOLUTION.   According to (1),

$$m = \frac{y_2 - y_1}{x_2 - x_1} = \frac{6 - 3}{4 - 1} = \frac{3}{3} = 1.$$

b.   Find the inclination θ of the line going through the same points.
SOLUTION.   According to (3) and the definition of inclination, θ is the smallest angle whose tangent is 1, namely $45°$.

c.   Find the slope m of the line whose inclination is $15°$.
SOLUTION.   Here (3) gives

$$m = \tan 15° = 0.26795,$$

where we consult a table of tangents or use a pocket scientific calculator.

d.   How are the slopes of a pair of perpendicular lines related?
SOLUTION.   Let the lines be L and L', as in Figure 17, where L has slope m and inclination θ, while L' has slope m' and inclination $\theta' = 180° - \alpha$. Here α (the Greek letter alpha) is the other acute interior angle of the triangle $P_1 P_2 P_3$. Then

$$m = \tan \theta = \frac{|P_2 P_3|}{|P_1 P_3|}. \qquad (5)$$

On the other hand, by (4),

$$m' = \tan \theta' = \tan (180° - \alpha) = -\tan \alpha,$$

so that

$$m' = -\frac{|P_1 P_3|}{|P_2 P_3|}. \qquad (5')$$

Comparing (5) and (5′), we find that

$$m' = -\frac{1}{m},$$

or, equivalently,

$$m = -\frac{1}{m'}.$$

In other words, *the slope of either of two perpendicular lines is the negative of the reciprocal of the slope of the other line.*

   e.   Find the slope of the line L' perpendicular to the line L going through the points $(2, 3)$ and $(4, 6)$.

   SOLUTION.   Let m be the slope of L and m' the slope of L'. Then

$$m = \frac{6 - 3}{4 - 2} = \frac{3}{2},$$

so that

$$m' = -\frac{1}{m} = -\frac{2}{3}.$$

PROBLEMS

1. Find the slope of the line going through the pair of points
   (a) $(-2, 4), (-3, -7)$;   (b) $(-2, 6), (1, 5)$;   (c) $(2, 3), (2, 5)$;
   (d) $(1, \sqrt{2}), (2, \sqrt{3})$.
2. Let L be a line with slope m and L' a line with slope m'. When are L and L' parallel?
3. Find the inclination of the line going through the pair of points
   (a) $(2, 4), (4, 6)$;   (b) $(2, 3), (2, 5)$;   (c) $(2, -4), (4, -6)$;
   (d) $(2, \sqrt{2}), (-2, \sqrt{2})$.
4. Find the slope of the line with inclination
   (a) $20°$;   (b) $100°$;   (c) $165°$.
5. Find the slope of every line parallel to the line going through the points $(1, -4)$ and $(-2, 5)$.
6. Find the slope of every line perpendicular to the line going through the points $(1, 3)$ and $(-3, -1)$.
7. Show that the line L going through the points $(1, 3)$ and $(2, 5)$ is perpendicular to the line L' going through the points $(4, 6)$ and $(2, 7)$.
*8. How is the line going through the points $(0, -\sqrt{3})$ and $(1, \sqrt{2})$ related to the line going through the points $(-1, \sqrt{3})$ and $(0, \sqrt{2})$?

1.9   MORE ABOUT STRAIGHT LINES

1.91. The equation of a straight line

   a.   By the term "equation of a straight line" we mean a mathematical statement, in equation form, which expresses the relationship between the x-coordinate and the y-coordinate of every point on the line. To find such a statement, we reason as follows: Let L be a nonvertical straight line with slope m, going through a fixed point $P_1 = (x_1, y_1)$, and let $P = (x, y)$ be an arbitrary point on L, as in Figure 18. Then, expressing the slope of L in terms of the coordinates of P_1 and P, we get

$$\frac{y - y_1}{x - x_1} = m.$$

It follows that

$$y - y_1 = m(x - x_1),$$

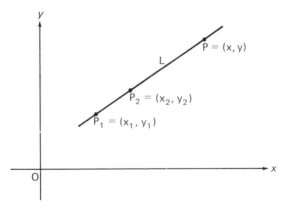

Figure 18.

or equivalently

$$y = mx + (y_1 - mx_1). \tag{1}$$

b. Despite its appearance, the right side of (1) does not actually depend on the particular choice of the fixed point $P_1 = (x_1, y_1)$ on the line L. To see this, suppose we replace P_1 by another fixed point $P_2 = (x_2, y_2)$ on L, as in the figure. Then we get

$$y = mx + (y_2 - mx_2), \tag{1'}$$

instead of (1). But we can easily show that *the two expressions* $y_1 - mx_1$ *and* $y_2 - mx_2$ *are equal.* In fact, suppose we calculate the slope of L with the help of the points P_1 and P_2, relying on the fact, proved in Sec. 1.81b, that the slope of a line L does not depend on the particular choice of the points used to define the slope. The result is

$$\frac{y_2 - y_1}{x_2 - x_1} = m,$$

so that

$$y_2 - y_1 = m(x_2 - x_1),$$

or

$$y_2 - mx_2 = y_1 - mx_1,$$

as claimed.

Since the expression $y_1 - mx_1$ in (1) does not depend on the particular choice of the point $P_1 = (x_1, y_1)$ on L, we might just as well denote $y_1 - mx_1$ by a single symbol, say b. Equation (1) then becomes simply

$$y = mx + b, \tag{2}$$

which is the desired *equation of a straight line with slope m.* The geometric meaning of the term b will be given in a moment (Sec. 1.92b).

1.92. The intercepts of a straight line

a. Let L be a straight line *other than the coordinate axes themselves,* and suppose L intersects the x-axis in a point $(a, 0)$. Then a is called an *x-intercept* of L. Similarly, if L intersects the y-axis in a point $(0, b)$, we call b a *y-intercept* of L. By an *intercept*

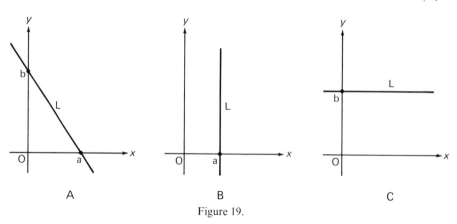

A

B

C

Figure 19.

we mean either an x-intercept or a y-intercept. Consulting Figure 19, we see that

(1) If L is neither vertical nor horizontal, then L has exactly two intercepts, namely an x-intercept and a y-intercept (as in Figure 19A);

(2) If L is vertical or horizontal, then L has just one intercept, an x-intercept but no y-intercept if L is vertical (as in Figure 19B) and a y-intercept but no x-intercept if L is horizontal (as in Figure 19C).

b. Now let L be the straight line with equation (2), where we assume that L is neither horizontal nor vertical, so that L has both an x-intercept and a y-intercept. To find these intercepts, we need only note that the substitution $x = 0$ in (2) gives

$$y = m \cdot 0 + b = b,$$

while the substitution $y = 0$ gives

$$0 = mx + b$$

or

$$mx = -b,$$

so that

$$x = -\frac{b}{m}$$

(why is m nonzero in this case?). Therefore L intersects the x-axis in the point $(-b/m, 0)$ and the y-axis in the point $(0, b)$. But this just means that the line L has $-b/m$ as its x-intercept and b as its y-intercept.

c. If the line L is vertical and hence parallel to the y-axis, every point of L has the same abscissa regardless of its ordinate. Suppose this abscissa is a. Then every point $P = (x, y)$ on L satisfies the simple equation

$$x = a, \tag{3}$$

in which the ordinate y does not appear at all. Note that (3) is *not* a special case of (2). This is hardly surprising, since in deriving (2) the case where L is vertical, and hence has no slope (recall Sec. 1.82b), was excluded from the outset.

d. If the line L is horizontal and hence parallel to the x-axis, every point of L has the same ordinate regardless of its abscissa. Suppose this ordinate is b. Then every point $P = (x, y)$ on L satisfies the simple equation

$$y = b, \tag{4}$$

in which the abscissa x does not appear at all. This time (4) *is* a special case of (2). In fact, to get (4) from (2) we need only make the substitution $m = 0$. This is hardly surprising, since every horizontal line has zero slope.

e. All three equations (2), (3) and (4) can be combined into the single "master equation"

$$Ax + By + C = 0, \tag{5}$$

where A, B and C are *constants*, that is, fixed numbers, and at least one of the numbers A and B is nonzero. In fact, if $A \neq 0$ and $B = 0$, equation (5) becomes

$$Ax + C = 0,$$

or

$$x = -\frac{C}{A},$$

which is of the form (3) with $a = -C/A$, while if $A = 0$ and $B \neq 0$, (5) becomes

$$By + C = 0,$$

or

$$y = -\frac{C}{B},$$

which is of the form (4) with $b = -C/B$. In any case, if $B \neq 0$ we can divide both sides of (5) by B, obtaining

$$\frac{A}{B}x + y + \frac{C}{B} = 0,$$

or

$$y = -\frac{A}{B}x - \frac{C}{B},$$

which is of the form (2) with slope $m = -A/B$ and y-intercept $b = -C/B$.

f. Note that none of the equations for a straight line involves powers of x and y higher than the first power. In mathematics such an equation is said to be *linear* (in x and y). This term stems from the fact that every linear equation in x and y is the equation of some straight line in the xy-plane. Another way of writing the equation of a straight line, involving the intercepts, is given in Problem 12.

By "the line $y = mx + b$" we mean, of course, the line *with equation* $y = mx + b$. Similarly, "the line $Ax + By + C = 0$" means the line *with equation* $Ax + By + C = 0$, and so on.

1.93. Examples

a. Find the line with slope 2 going through the point $(3, 1)$.

SOLUTION   Using (1) with $m = 2$, $x_1 = 3$ and $y_1 = 1$, we get

$$y = 2x + (1 - 2 \cdot 3) = 2x - 5.$$

b.   Find the line going through the points $(1, 3)$ and $(4, 6)$.

SOLUTION.   The line has slope

$$m = \frac{6 - 3}{4 - 1} = \frac{3}{3} = 1.$$

Therefore, choosing $(1, 3)$ as the point (x_1, y_1) in (1), we get

$$y = x + (3 - 1 \cdot 1) = x + 2.$$

Naturally, the same result is obtained if we choose $(4, 6)$ as the fixed point on the line (check this).

c.   Find the slope and intercepts of the line

$$y = 3x + 2. \tag{6}$$

SOLUTION.   Since (6) is of the form (2), with $m = 3$ and $b = 2$, the line has slope 3, y-intercept 2 and x-intercept $-b/m = -\frac{2}{3}$.

d.   Find the line L which goes through the point $(1, 3)$ and is parallel to the line (6).

SOLUTION.   Being parallel to (6), L has the same slope as (6), namely 3. Using (1), with $m = 3, x_1 = 1$ and $y_1 = 3$, we get

$$y = 3x + (3 - 3 \cdot 1) = 3x. \tag{7}$$

Note that L goes through the origin of the xy-plane (why?).

e.   Find the line L' which goes through the point $(1, 3)$ and is *perpendicular* to the line (6).

SOLUTION.   Being perpendicular to (6), L' must have slope $-\frac{1}{3}$ (recall Sec. 1.83d). Thus, instead of (7), we now have

$$y = -\frac{1}{3}x + \left(3 + \frac{1}{3} \cdot 1\right) = -\frac{1}{3}x + \frac{10}{3}. \tag{8}$$

We can also write (8) as

$$x + 3y - 10 = 0,$$

which is of the form (5), with $A = 1, B = 3, C = -10$.

f.   Find the point of intersection of the lines (6) and (8).

SOLUTION.   The abscissa x_1 of the point of intersection of the lines (6) and (8) is characterized by the fact that both lines have the same ordinate y_1 when $x = x_1$. It follows that x_1 is the solution of the equation

$$3x + 2 = -\frac{1}{3}x + \frac{10}{3}, \tag{9}$$

obtained by setting the right side of (6) equal to the right side of (8). Solving (9) for x, we get

$$\frac{10}{3}x = \frac{4}{3},$$

or

$$x = x_1 = \frac{4}{10} = \frac{2}{5}. \tag{10}$$

To get y_1, the ordinate of the point of intersection, we substitute (10) into (6), or into (8), obtaining

$$y_1 = 3x_1 + 2 = 3 \cdot \frac{2}{5} + 2 = \frac{16}{5}.$$

Therefore the lines (6) and (8) intersect in the point

$$(x_1, y_1) = \left(\frac{2}{5}, \frac{16}{5}\right).$$

PROBLEMS

1. Find the line with slope 2 going through the point
    (a) $(1, 0)$;   (b) $(0, 1)$;   (c) $(1, 1)$;   (d) $(0, 0)$;   (e) $(1, -1)$.
2. Find the line going through the pair of points
    (a) $(2, -5), (3, 2)$;   (b) $(-\frac{1}{2}, 0), (0, \frac{1}{4})$;   (c) $(-3, 1), (7, 11)$;
    (d) $(5, 3), (-1, 6)$.
3. Find the line with slope m and y-intercept b if
    (a) $m = -1, b = 1$;   (b) $m = 3, b = 0$;   (c) $m = 0, b = -2$;
    (d) $m = -\frac{1}{2}, b = \frac{3}{2}$.
4. Find the slope m, x-intercept a and y-intercept b of the line
    (a) $y = 3x - 6$;   (b) $y = 2x + 4$;   (c) $y = -x + 3$;   (d) $y = 2$.
5. Do the same for the line
    (a) $5x - y + 4 = 0$;   (b) $3x + 2y = 0$;   (c) $2y - 6 = 0$;
    (d) $x + y + 1 = 0$.
6. Find the line which goes through the point $(2, -4)$ and is parallel to the line $y = 2x + 3$.
7. Find the line which goes through the point $(1, 2)$ and is perpendicular to the line going through the points $(2, 4)$ and $(3, 5)$. What is the point of intersection of these two lines?
8. What is the area of the triangle lying between the coordinate axes and the line $2x + 5y - 20 = 0$?
9. Does the point $(2, 3)$ lie above the line $y = 2x + 1$ or below it?
10. What is the relationship between the lines $2x + 3y - 1 = 0$ and $4x + 6y + 3 = 0$? Between the lines $2x + 5y - 4 = 0$ and $15x - 6y + 5 = 0$?
11. Find the line joining the origin to the point of intersection of the lines $x + 2y - 3 = 0$ and $x - 3y + 7 = 0$.
*12. Show that the equation of the line with x-intercept a and y-intercept b is

$$\frac{x}{a} + \frac{y}{b} = 1,$$

    assuming that a and b are both nonzero.
*13. Find the line with x-intercept a and y-intercept b if
    (a) $a = 1, b = 2$;   (b) $a = -3, b = -1$;   (c) $a = \frac{1}{4}, b = \frac{1}{8}$.
*14. There are two lines, each with equal intercepts, going through the point $(2, 3)$. Find them.
*15. What is the (perpendicular) distance between the parallel lines $3x - 4y - 10 = 0$ and $6x - 8y + 5 = 0$?

*16. Let d be the (perpendicular) distance between the point $P_1 = (x_1, y_1)$ and the line $Ax + By + C = 0$. Show that

$$d = \frac{|Ax_1 + By_1 + C|}{\sqrt{A^2 + B^2}}. \tag{11}$$

*17. Find the distance between
   (a) The point $(3, 1)$ and the line $3x + 4y - 3 = 0$;
   (b) The point $(1, 1)$ and the line $5x - 12y + 72 = 0$;
   (c) The point $(1, -2)$ and the line $x - 2y - 5 = 0$.

Chapter 2

DIFFERENTIAL CALCULUS

2.1 FUNCTIONS

2.11. Constants and variables. The quantities encountered in mathematics fall into two broad categories, namely "fixed quantities," called *constants*, and "changing quantities," called *variables*. A constant "takes only one value" in the course of a given problem, while a variable "takes two or more values" in the course of one and the same problem.

For example, let L be the straight line with slope m and y-intercept b. Then, according to Sec. 1.9, L has the equation

$$y = mx + b. \tag{1}$$

Here m and b are constants characterizing the given line L, while x and y are variables, namely the abscissa and ordinate of a point which is free to change its position along L.

As another example, suppose a stone is dropped from a high tower. Let s be the distance fallen by the stone and t the elapsed time after dropping the stone. Then, according to elementary physics, s and t are variables which are related, at least for a while, by the formula

$$s = \frac{1}{2} gt^2,$$

where g is a constant known as the "acceleration due to gravity." To a good approximation, this formula becomes

$$s = 16t^2, \tag{2}$$

if s is measured in feet and t in seconds.

2.12. Related variables and the function concept

a. A great many problems arising in mathematics and its applications involve *related variables*. This means that there are at least two relevant variables, and the value of one of them depends on the value of the other, or on the values of the others if there are more than two. For example, the position of a spy satellite depends on the elapsed time since launching, the cost of producing a commodity depends on the quantity produced, the area of a rectangle depends on both its length and its width, and so on.

Actually, the situation we have in mind is where knowledge of the values of all but one of the variables *uniquely* determines the value of the remaining variable, in a crucial sense to be spelled out in a moment (Sec. 2.12b). The variables whose values are chosen in advance are called *independent variables*, and the remaining variable, whose value is determined by the values of the other variables, is called the *dependent variable*. The "rule" or "procedure" leading from the values of the independent variables to the value of the dependent variable, regardless of how this is accomplished, is called a *function*. We then say that the dependent variable "is a function of" the independent variables. The independent variables are often called the *arguments* of the function.

Thus, for example, if x and y are the abscissa and ordinate of a variable point on the line with slope m and y-intercept b, then y is a function of x, as described by formula (1). Similarly, the distance traversed by a falling stone is a function of time, as described by formula (2).

b. Consider a function of one independent variable, say x, and let y be the dependent variable. Then the function assigns a value of y to each value of x. Expressed somewhat differently, the function establishes a correspondence between the values of x and those of y. This correspondence must be such that each value of x *uniquely* determines the corresponding value of y. This simply means that *to each value of x there corresponds one and only one value of y.* On the other hand, the same value of y may well correspond to more than one value of x. The situation is the same for several independent variables $x_1, x_2, \ldots, x_n$ if for "value of x" we read "set of values of $x_1, x_2, \ldots, x_n$."

Thus if

$$y = x^2,$$

then y is a function of x, since to each value of x there corresponds one and only one value of y. On the other hand, each positive value of y corresponds to *two* values of x. For example, the value $y = 4$ corresponds to the two values $x = 2$ and $x = -2$. This clearly prevents x from being a function of y, since every positive value of y fails to uniquely determine the corresponding value of x.

As another example, let A be the area of the rectangle of length l and width w. Then A is a function of l and w, since the relation between these variables is described by the simple formula

$$A = lw, \tag{3}$$

leading to one and only one value of A for any given pair of values of l and w. Solving (3) for l, we get

$$l = \frac{A}{w},$$

which shows that l is a function of A and w. Similarly,

$$w = \frac{A}{l},$$

which shows that w is a function of A and l.

c. Do not jump to the conclusion that all related variables are numbers, and that all functions involve the use of formulas or numerical calculations. For example, the name of a car's owner is a function of the inscription on the car's license plate, and the rule or procedure describing this function is just this: Look up the plate in the motor vehicle records of the state in question, and find the owner's name.

On the other hand, it is true that examples like this are a bit "offbeat." Almost all the functions considered in this book involve variables which take only numerical values.

2.13. Function notation

a. The fact that one variable, say y, is a function of another variable, say x, can be indicated by writing

$$y = f(x),$$

pronounced "y equals f of x." Here f is a letter denoting the function, that is, the rule or procedure (usually, but not always, involving some formula) leading from the values of x to the values of y. Suppose we give the independent variable x the value c. Then the corresponding value of y is denoted by $f(c)$, and is called the *value of the function f at c*. This is a bit fussy, and it is simpler to use the same letter to denote both the independent variable and its values. We can then call $f(x)$ the *value of f at x*.

Although, strictly speaking, $f(x)$ is the value of the function f at x, we will often talk about "the function $y = f(x)$" or simply "the function $f(x)$." Suppose $y = f(x) = x^2$, for example. Then we might talk about "the function $f(x) = x^2$," "the function $y = x^2$," or simply "the function x^2."

b. There is nothing sacred about the use of the letter f to denote a function, apart from its being the first letter of the word "function," and other letters will do just as well. Common choices are Latin letters like g, h, F, etc., or Greek letters like φ (phi), ψ (psi), Φ (capital phi), etc. Sometimes the letter is chosen to suggest a geometrical or physical quantity under discussion. Thus A is often used for area, V for volume, t for time, and so on.

c. Functions of several variables are indicated in the same way. Thus $f(s, t)$ means a function of two independent variables s and t, $\Phi(u, v, w)$ means a function of three independent variables u, v and w, and so on.

2.14. The domain and range of a function

a. There is still one thing missing in our definition of a function, for we have yet to specify the set of values taken by the independent variable (or variables). This set is called the *domain* of the function. For example, returning to the problem of the falling stone, we observe that formula (2) does not describe the motion of the stone for all values of t, but only until the stone hits the ground. If the stone is dropped from a height of 64 feet, say, it hits the ground after falling for 2 seconds ($64 = 16 \cdot 2^2$) and is subsequently motionless. In other words, formula (2) is valid only during the time interval $0 \leqslant t \leqslant 2$, a fact we can make explicit by writing

$$s = 16t^2 \qquad (0 \leqslant t \leqslant 2) \tag{4}$$

instead of (2). The subsequent behavior of the stone is described by the formula

$$s = 64,$$

or, more exactly, by

$$s = 64 \qquad (t > 2). \tag{5}$$

We can also write (5) as

$$s = 64 \qquad (2 < t < \infty), \tag{5'}$$

in terms of the infinite interval $2 < t < \infty$. Incidentally, this shows the desirability of considering *constant functions*, that is, functions which take only one value.

   b.   Formulas (4) and (5) can be combined into the single formula

$$s = \begin{cases} 16t^2 & \text{if } 0 \leqslant t \leqslant 2, \\ 64 & \text{if } t > 2. \end{cases} \tag{6}$$

Moreover, noting that the stone is motionless before it is dropped, as well as after it hits the ground, we have the even more comprehensive formula

$$s = \begin{cases} 0 & \text{if } t < 0, \\ 16t^2 & \text{if } 0 \leqslant t \leqslant 2, \\ 64 & \text{if } t > 2. \end{cases} \tag{7}$$

Formulas (4), (6) and (7) all describe different functions, in the sense that in each case the domain of the function, that is, the set of allowed values of the independent variable t, is different.

   c.   Another way of saying that a function f has the domain D is to say that f *is defined in D*. Thus the function (6) is defined in the interval $0 \leqslant t < \infty$, that is, for all nonnegative t, while the function (7) is defined in the interval $-\infty < t < \infty$, that is, for all t, positive, negative and zero.

   d.   By the *range* of a function we mean the set of all values taken by the function, or, equivalently, the set of all values taken by the *dependent* variable. For example, all three functions (4), (6) and (7) have the same range, namely the interval $0 \leqslant s \leqslant 64$. On the other hand, the range of the constant function (5) is the set whose only element is the number 64.

2.15. Examples

   a.   Let

$$f(x) = \sqrt{1 - x^2}. \tag{8}$$

Find $f(0)$, $f(1)$ and $f(2)$.

   SOLUTION.   To find $f(0)$, we merely substitute $x = 0$ into (8), obtaining

$$f(0) = \sqrt{1 - 0^2} = \sqrt{1} = 1,$$

and similarly

$$f(1) = \sqrt{1 - 1^2} = \sqrt{0} = 0.$$

On the other hand, the quantity

$$f(2) = \sqrt{1 - 2^2} = \sqrt{-3}$$

"does not exist," since there is no real number whose square is negative. There is a sense in which meaning can be ascribed to "imaginary numbers" like $\sqrt{-3}$, but such an extension of the concept of number lies beyond the scope of this book, in which all numbers are assumed to be *real* (Sec. 1.35).

   Whenever a function $f(x)$ is specified by an explicit formula like (8), we will understand the domain of $f(x)$ to be the *largest* set of numbers x for which the formula makes sense. In the present case, this set is just the interval $-1 \leqslant x \leqslant 1$, since $1 - x^2$ is negative for any other value of x and we do not take square roots of negative numbers. Note that any smaller set can serve as the domain of a function whose values are given by the same formula (8), but in such cases we will always explicitly indicate the domain, as in the formula

$$f(x) = \sqrt{1 - x^2} \qquad (0 < x < 1),$$

where the domain is now the *smaller* interval $0 < x < 1$.

b. Is the area of a rectangle a function of its perimeter?

SOLUTION. No, since knowledge of the perimeter of a rectangle does not uniquely determine its area. For example, the rectangle of length 15 and width 3 has perimeter $15 + 3 + 15 + 3 = 36$ and area $15 \cdot 3 = 45$, while the square of side 9 has the same perimeter $9 + 9 + 9 + 9 = 36$ and a different area $9^2 = 81$.

c. Turning to a function of two variables, let

$$g(x, y) = \frac{x + y}{x - y}.$$

Find $g(1, 2)$, $g(2, 1)$ and $g(1, 1)$. What is the domain of $g(x, y)$?

SOLUTION. Easy substitutions give

$$g(1, 2) = \frac{1 + 2}{1 - 2} = \frac{3}{-1} = -3$$

and

$$g(2, 1) = \frac{2 + 1}{2 - 1} = \frac{3}{1} = 3.$$

On the other hand, $g(1,1)$ fails to exist, since

$$g(1, 1) = \frac{1 + 1}{1 - 1} = \frac{2}{0},$$

and division by zero is impossible. The domain of $g(x, y)$ is the set of all pairs of numbers x and y such that $x \neq y$, since $x = y$ leads to division by zero. Regarding each such pair of numbers as the rectangular coordinates of a point (x, y) in the xy-plane, we see that the domain of $g(x, y)$ is the set of all points in the xy-plane except those on the line $y = x$.

2.16. One-to-one functions and inverse functions

a. Let y be a function of x, or, in symbols, $y = f(x)$. Then, as in Sec. 2.12b, x *uniquely determines* y, that is, *to each value of x there corresponds one and only one value of y.* On the other hand, there is nothing so far to prevent more than one value of x from corresponding to one and the same value of y. However, suppose we now impose an *extra requirement* on the function $y = f(x)$, namely that not only should x uniquely determine y but also that y *should uniquely determine x.* Then not only does there correspond one and only one value of y to each value of x, but also *to each value of y there corresponds one and only one value of x.* A function $y = f(x)$ of this special type is called a *one-to-one function.*

b. Let $y = f(x)$ be a one-to-one function. Then there is a simple rule leading from the *dependent* variable y back to the *independent variable* x. In fact, let y_1 be any given value of y. Then x_1, the corresponding value of x, is just the unique value of x to which the function $y = f(x)$ assigns the value y_1. Thus the rule leading from y to x is just as much a function as the original rule leading from x to y. This new function, leading from y to x, is denoted by $x = f^{-1}(y)$ and is called the *inverse function,* or simply the *inverse,* of the original function $y = f(x)$. Never make the mistake of confusing the inverse function $f^{-1}(y)$ with the reciprocal $1/f(y)$.

Let $f(x)$ be a one-to-one function defined in some interval I. Then we simply say that $f(x)$ is *one-to-one in* I.

   c. Example.   The function

$$y = f(x) = x^2, \tag{9}$$

with domain $-\infty < x < \infty$ and range $0 \leqslant y < \infty$, is not one-to-one, since to each positive value of y there correspond two values of x, namely $\sqrt{y}$ and $-\sqrt{y}$ (as always, $\sqrt{y}$ means the *positive* square root of y). But suppose we restrict the domain of (9) to nonnegative values of x. Then to each value of y there corresponds precisely one value of x, namely $\sqrt{y}$. In other words, the function

$$y = f(x) = x^2 \qquad (0 \leqslant x < \infty)$$

is one-to-one, with inverse

$$x = f^{-1}(y) = \sqrt{y} \qquad (0 \leqslant y < \infty).$$

More generally, it is easy to see that the function $y = x^2$ is one-to-one in any interval in which x is of fixed sign, but not in any interval in which x changes sign.

PROBLEMS

1. If $f(x) = x^2 + 3x + 6$, find $f(0)$, $f(1)$, $f(2)$ and $f(\sqrt{2})$.
2. If $\varphi(t) = |t| + 3t^2$, find $\varphi(-2)$, $\varphi(-1)$, $\varphi(0)$ and $\varphi(\sqrt{3})$.
3. Let

$$g(x) = \frac{2x + 1}{3x^2 - 1}.$$

   Find $g(-1)$, $g(0)$, $g(1)$, $g(1/\sqrt{2})$ and $g(1/\sqrt{3})$.
4. Find the domain and range of the function

   (a) $y = \sqrt{x^2 - 9}$;   (b) $y = \sqrt{9 - x^2}$;   (c) $y = \dfrac{1}{x - 3}$;

   (d) $y = \dfrac{1}{x + 5}$.
5. Let

$$f(x,y) = \frac{x - 2y}{2x - y}.$$

   Find $f(3, 1)$, $f(0, 1)$, $f(1, 0)$, $f(a, a)$ and $f(a, -a)$.
6. Find the domain and range of the function

$$z = \frac{1}{x} + \frac{1}{y}.$$

7. Is the number of hairs on your head a function of time?
8. Is a man's birthday a function of his first name? Of his Social Security number, assuming that he has one?
9. Let P be the closing price of a given security traded on the stock exchange, and let d be the date. Then P is a function of d. Given d, how do you find P? Is the function ever undefined?
10. Is the weight of a first-class letter a function of its postage?
11. Is the area of a square a function of its perimeter?

12. Is the number of a page of this book a function of the number of commas on the page?
13. "The area of a right triangle is a function of two variables." True or false?
14. "The area of a parallelogram is a function of two variables." True or false?
15. Express the volume V of a brick as a function of its length l, width w and height h.
16. Let

$$f(x, y, z) = \frac{1}{\sqrt{x}} + \frac{1}{\sqrt{y}} + \frac{1}{\sqrt{z}}.$$

Find $f(1, 1, 1)$, $f(4, 1, 9)$, $f(1, 9, 1)$ and $f(4, 9, 16)$.

17. The function given by the table

x	0	20		60	80	100
y	32	68	104	140		212

is familiar from everyday life. What is it? Fill in the missing entries in the table. Find a formula relating y to x and one relating x to y.

18. Is the inverse of a one-to-one function always a one-to-one function?
19. Let X be the domain and Y the range of a one-to-one function $y = f(x)$. What are the domain and range of the inverse function $x = f^{-1}(y)$?
20. Which of the following functions are one-to-one?

(a) $y = x$;   (b) $y = \dfrac{1}{x}$;   (c) $y = \dfrac{1}{x - 1}$;   (d) $y = \sqrt{x}$;

(e) $y = |x|$.

Find the inverse of each one-to-one function.

21. "The position of a clock's hands is a one-to-one function of the time of day." True or false?
*22. Find the domain and range of the function $y = [x]$, where $[x]$ is the integral part of x (Sec. 1.4, Prob. 10).
*23. Verify that the following formal definitions of function, domain, value and range agree in all essentials with those given in the text:
   Given any two nonempty sets X and Y, let f be a set of ordered pairs (x, y) with $x \in X$ and $y \in Y$ such that for every $x \in X$ there is one and only one ordered pair $(x, y) \in f$ with x as its first element. Then f is said to be a *function* defined in X, and X is called the *domain* of f. If (x, y) is an ordered pair in f, then y, the second element of the pair, is called the *value* of f at x, written $f(x)$. The set of all values of a function f, that is, the set $\{f(x): x \in X\}$, is called the *range* of f.
*24. Is the set Y in the preceding problem always the range of f?
*25. How many different functions are there with domain $X = \{1, 2, \ldots, n\}$ and range $Y = \{a, b\}$?
*26. Let the function f be defined as a set of ordered pairs (x, y), as in Problem 23. When is f one-to-one? If f is one-to-one, how is the inverse f^{-1} obtained?
*27. We can exhibit the behavior of a function f with domain X and range Y by drawing a diagram like Figure 1, where X and Y are represented by disks, the values of the independent and dependent variables x and y are represented by points inside the disks, and each value of x is connected by an arrow to the

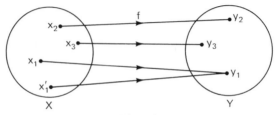

Figure 1.

corresponding value of y. The arrows show quite explicitly how f "maps" or "carries" the values of x into those of y.

The function represented in Figure 1 is not one-to-one, since two arrows terminate on the same point y_1. Modify Figure 1 in such a way as to make it represent a one-to-one function.

*28. What is the simplest way of converting the "mapping diagram" of a one-to-one function f into the analogous diagram for the inverse function f^{-1}?

*29. We say that there is a *one-to-one correspondence* between two sets A and B if there is a one-to-one function with domain A and range B. Two sets are said to *have the same number of elements* if there is a one-to-one correspondence between them. Show that the set of all even numbers has the same number of elements as the set of all odd numbers.

*30. A set A is said to *have n elements* if there is a one-to-one correspondence between A and the set $\{1, 2, \ldots, n\}$ made up of the first n positive integers. If a set A contains n elements, where n is some positive integer, we say that A is *finite*; otherwise A is said to be *infinite*. (An empty set is regarded as finite.) Which of the following sets are finite and which infinite?
(a) The set of all cells in a human body;
(b) The set of all integers less than 1,000;
(c) The set of all integers greater than 1,000,000;
(d) The set of all right triangles whose side lengths are integers.

2.2 MORE ABOUT FUNCTIONS

A variable whose values are all real numbers is called a *real variable*, and a function whose values are all real numbers is called a *numerical function*. Calculus is primarily concerned with numerical functions of one or more real variables, and these are the only functions to be considered in the rest of this book. Thus, from now on, when we use the words "function" and "variable" without further qualification, we will always mean a *numerical* function and a *real* variable, just as the word "number" always means a *real* number.

2.21. Two functions $f(x)$ and $g(x)$ are said to be *identically equal*, and we write $f(x) \equiv g(x)$, if the functions have the same domain X and if $f(x) = g(x)$ for all x in X. Note the distinction between the ordinary equals sign $=$ and the sign $\equiv$ with three bars.

For example, the two functions

$$f(x) = \frac{1}{2}(x + 1) - \frac{1}{2}(x - 1)$$

and

$$g(x) = \frac{1}{2}(1 + x) + \frac{1}{2}(1 - x)$$

are both identically equal to the constant function 1. An equation like

$$\frac{1}{2}(1 + x) + \frac{1}{2}(1 - x) \equiv 1,$$

involving the sign $\equiv$, is called an *identity*.

2.22. Composite functions

a. Functions are often combined by letting the arguments of one function equal the values of another. In this way, we get *composite functions* like $f(g(x))$ and $g(f(x))$. For example, suppose

$$f(x) = 1 + x, \qquad g(x) = x^2. \tag{1}$$

Then straightforward substitution shows that

$$f(g(x)) = 1 + g(x) = 1 + x^2 \tag{2}$$

and

$$g(f(x)) = [f(x)]^2 = (1 + x)^2.$$

In the same way,

$$f(f(x)) = 1 + f(x) = 2 + x$$

and

$$g(g(x)) = [g(x)]^2 = x^4.$$

Things are not always this simple. For example, suppose

$$f(x) = \sqrt{x}, \qquad g(x) = -|x|.$$

Then

$$f(g(x)) = \sqrt{-|x|}$$

fails to exist for every value of x except $x = 0$. This shows that a composite function is defined only for values of the independent variable such that the values of the "inner function" belong to the domain of the "outer function."

Never make the mistake of confusing the composite function $f(g(x))$ with the product function $f(x)g(x)$. For example, the product of the functions (1) is

$$f(x)g(x) = (1 + x)x^2 = x^2 + x^3,$$

which is not the same as the composite function (2).

b. Let $y = f(x)$ be a one-to-one function, with inverse $x = f^{-1}(y)$. Substituting $y = f(x)$ into $x = f^{-1}(y)$, and then substituting $x = f^{-1}(y)$ into $y = f(x)$, we get the important pair of identities

$$f^{-1}(f(x)) \equiv x, \qquad f(f^{-1}(y)) \equiv y, \tag{3}$$

involving the composite functions $f^{-1}(f(x))$ and $f(f^{-1}(y))$. These formulas tell us that each of the functions f and f^{-1} "nullifies" the action of the other.

2.23. Sequences

a. A function f whose domain is the set $\{1, 2, \ldots\}$ of all positive integers is called an *infinite sequence*, or simply a *sequence*, and the values of f at $n = 1, 2, \ldots$

are called the *terms* of the sequence. More informally, a sequence is a rule or procedure assigning a number to every positive integer. We can write a sequence by listing some of its terms

$$f(1), f(2), \ldots, f(n), \ldots, \tag{4}$$

where $f(n)$ is called the *general term* of the sequence. The first set of dots in (4) means "and so on up to," while the second set means "and so on forever." It is customary to save a lot of parentheses by simply writing

$$f_1, f_2, \ldots, f_n, \ldots \tag{5}$$

instead of (4). A more concise way of specifying the sequence (5) is to write its general term inside curly brackets:

$$\{f_n\}.$$

Do not confuse $\{f_n\}$ in this context with the set whose only element is f_n. Note that the terms of a sequence are always listed in such a way that the integers $1, 2, \ldots, n, \ldots$ appear in their natural order, increasing from left to right, as in (4) and (5).

    b. Again there is nothing sacred about the letter f, and other letters will do just as well. Common choices are small Latin letters like a, b, c, s and x. Although, strictly speaking, x_n is the general term of the sequence $\{x_n\}$, we will often talk about "the sequence x_n." For example, suppose $\{x_n\}$ is the sequence such that $x_n = n^2$. Then we might talk about "the sequence $x_n = n^2$," or simply "the sequence n^2."

    c. The "law of formation" of a sequence is often given explicitly as a formula for its general term, as in the above example of the sequence $x_n = n^2$. A sequence may also be given *recursively*, that is, by showing how each term can be obtained from terms with lower subscripts. For example, suppose that

$$\begin{aligned} x_1 &= 1, \\ x_n &= x_{n-1} + n \quad \text{if} \quad n > 1 \end{aligned} \tag{6}$$

Then

$$x_1 = 1, \qquad x_2 = x_1 + 2 = 3, \qquad x_3 = x_2 + 3 = 6, \qquad x_4 = x_3 + 4 = 10, \ldots$$

This sequence can be written as $1, 3, 6, 10, \ldots$, but this does little more than suggest how the sequence was actually arrived at. A rule like (6) is called a *recursion formula*.

PROBLEMS

1.   What is the largest set of numbers x for which the following identities are true?

    (a) $\dfrac{x}{x} \equiv 1$;   (b) $x \equiv (\sqrt{x})^2$;   (c) $x \equiv \sqrt{x^2}$.

2.   Find values of a and b in the formula $f(x) = ax^2 + bx + 5$ such that

$$f(x + 1) - f(x) \equiv 8x + 3.$$

3.   Let f and g be two numerical functions with the same domain X. Then by the *sum* $f + g$ we mean the function with domain X whose value at every point $x \in X$ is just the sum of the value of f at x and the value of g at x. More concisely,

$$(f + g)(x) \equiv f(x) + g(x).$$

Other algebraic operations are defined similarly. Thus

$$fg(x) \equiv f(x)g(x)$$

$$f^n(x) \equiv \underbrace{f(x)f(x) \cdots f(x)}_{n \text{ factors}},$$

and so on.

Suppose that

$$f(x) = \frac{1}{1 + x}, \qquad g(x) = \sqrt{x - 1}.$$

Find the values of $f + g$, $f - g$, fg, f^3 and f/g all at $x = 5$.

4. Let

$$f(x) = \frac{1}{x}, \qquad g(x) = x^2.$$

Find $f(f(x))$, $f(g(x))$, $g(f(x))$ and $g(g(x))$.

5. Let

$$h(t) = \frac{1}{1 - t}.$$

Find $h(h(h(2)))$.

6. "In general, $f(g(x)) \equiv g(f(x))$." True or false?

7. Write the first five terms of the sequence $\{a_n\}$ with general term

(a) $a_n = \dfrac{n}{n + 1}$;   (b) $a_n = \dfrac{1}{n(n + 1)}$;   (c) $a_n = \dfrac{(-1)^{n-1}}{n}$;

(d) $a_n = \begin{cases} 1 \text{ for even } n, \\ \dfrac{1}{n} \text{ for odd } n. \end{cases}$

8. Let $\{x_n\}$ be the sequence $1, 3, 5, \ldots, 2n - 1, \ldots$ of all odd numbers written in increasing order, and let $s_1 = x_1$, $s_2 = x_1 + x_2, \ldots, s_n = x_1 + x_2 + \cdots + x_n, \ldots$ Write the first few terms of the sequence $\{s_n\}$, and find a simple expression for its general term.

*9. A sequence $\{x_n\}$ is specified by the following rule: Its first two terms equal 1, and the remaining terms are given by the recursion formula

$$x_n = x_{n-1} + x_{n-2} \qquad (n = 3, 4, \ldots).$$

Write the first eight terms of this sequence, known as the *Fibonacci sequence*.

*10. Find the terms a_1, a_3, a_4 and a_7 of the sequence $\{a_n\}$ determined by the formula $\sqrt{2} = 1.a_1 a_2 \ldots a_n \ldots$

*11. Let

$$f(x) = \begin{cases} -1 & \text{if} \quad -\infty < x < -1, \\ x & \text{if} \quad -1 \leqslant x \leqslant 1, \\ 1 & \text{if} \quad 1 < x < \infty, \end{cases} \qquad g(x) = \frac{1}{2}|x + 1| - \frac{1}{2}|x - 1|.$$

Prove that $f(x) \equiv g(x)$.

*12. Does $f(x)g(x) \equiv 0$ always imply $f(x) \equiv 0$ or $g(x) \equiv 0$?

*13.   Let

$$f(x, y) = \frac{1}{x^2 + y^2}, \qquad g(t) = t^2, \qquad h(t) = \sqrt{t}.$$

Find $f(g(2), h(2))$.

2.3   GRAPHS

2.31. a.   Let $F(x, y)$ be a numerical function of two real variables x and y. Then by the *solution set* of the equation

$$F(x, y) = 0 \tag{1}$$

we mean the set of all ordered pairs (x, y) for which (1) holds. For example, if

$$F(x, y) = xy,$$

then (1) becomes the equation

$$xy = 0, \tag{2}$$

which implies that either $x = 0$ or $y = 0$ (or both). Therefore the solution set of (2) is the set of all ordered pairs of the special form $(x, 0)$ or $(0, y)$, where x and y are arbitrary numbers.

Similarly, if

$$F(x, y) = x^2 + y^2,$$

we get the equation

$$x^2 + y^2 = 0. \tag{3}$$

But (3) implies that both $x = 0$ and $y = 0$. Therefore the solution set of (3) is the set whose only element is the pair $(0, 0)$.

b.   Let S be the solution set of equation (1). Then there is a simple way of "drawing a picture" of S. First we introduce a "system of rectangular coordinates," that is, we set up perpendicular axes Ox and Oy in the plane, as in Sec. 1.71. Next we plot all the elements of S as points in the xy-plane; since all the elements of S are ordered pairs of numbers, this can be done in the way described in Sec. 1.72. These points make up a "picture," called the *graph of S*, or, equivalently, the *graph of equation* (1). For example, the graph of equation (2) consists of the coordinate axes themselves, while the graph of equation (3) consists of a single point, namely the origin of coordinates.

c.   We can apply the same technique to a function

$$y = f(x) \tag{4}$$

of a single variable x. Let S be the set of all ordered pairs (x, y) for which (4) holds. Then, plotting all the elements of S as points in the xy-plane, we get a "picture," called the *graph of S*, or, equivalently, the *graph of the function* (4). For example, according to Sec. 1.9, the graph of the function

$$y = mx + b$$

is just the straight line with slope m and y-intercept b.

d.   The word "graph" will also be used as a verb, meaning "find the graph of."

Note that (4) is a special case of (1), corresponding to the choice $F(x, y) = y - f(x)$. Thus the graph of a function is the graph of a special kind of equation.

e.   The graph of an equation or function typically looks like a "curve," possibly made up of several "pieces." With the help of calculus methods, we will eventually become proficient at "curve sketching," learning how to draw the graph of a function without explicitly plotting more than a few points. The graph of an equation $F(x, y) = 0$ or of a function $y = f(x)$ is often simply called "the curve $F(x, y) = 0$" or "the curve $y = f(x)$."

2.32. Examples

a.   Graph the equation

$$x^2 + y^2 = 1. \tag{5}$$

SOLUTION.   Since $x^2 + y^2$ is the square of the distance between the point (x, y) and the origin O (Sec. 1.74), the point (x, y) belongs to the graph of (5) when the distance between (x, y) and O equals 1, and only then. Therefore the graph of (5) is the circle of radius 1 with its center at O, as shown in Figure 2.

b.   Graph the equation

$$x^2 - 6x + y^2 - 4y + 9 = 0. \tag{6}$$

SOLUTION.   First we "complete the squares" in (6), by noting that

$$x^2 - 6x + y^2 - 4y + 9 = (x^2 - 6x + 9) + (y^2 - 4y + 4) - 4$$
$$= (x - 3)^2 + (y - 2)^2 - 4,$$

so that (6) is equivalent to

$$(x - 3)^2 + (y - 2)^2 = 4.$$

But the expression on the left is just the square of the distance between the variable point (x, y) and the fixed point $(3, 2)$. Therefore the graph of (6) is the circle of radius $\sqrt{4} = 2$ with its center at the point $(3, 2)$, as shown in Figure 3. Note that the x-axis is tangent to the circle at the point $(3, 0)$.

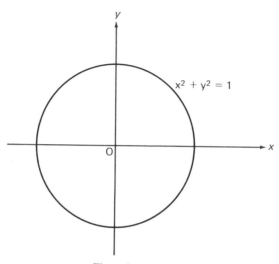

Figure 2.

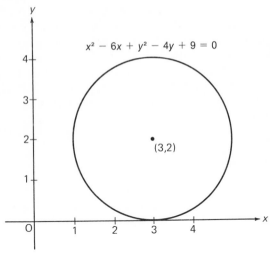

$$x^2 - 6x + y^2 - 4y + 9 = 0$$

$(3,2)$

Figure 3.

c. Graph the function

$$y = |x|. \tag{7}$$

SOLUTION.   If $x \geqslant 0$, then $|x| = x$ and (7) reduces to the straight line

$$y = x$$

with slope 1 going through the origin, while if $x < 0$, then $|x| = -x$ and (7) reduces to the straight line

$$y = -x$$

with slope -1 going through the origin. Therefore the graph of the function (7) is the curve shown in Figure 4, made up of "pieces" of the lines $y = x$ and $y = -x$. Note that the curve has a sharp "corner" at the origin.

A function like $y = |x|$, whose graph is made up of pieces of two or more straight lines, is said to be *piecewise linear*.

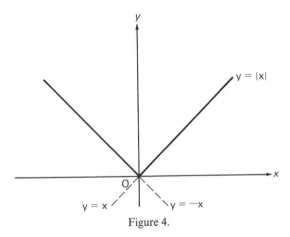

$y = |x|$

$y = x$ $y = -x$

Figure 4.

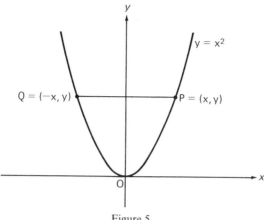

Figure 5.

d. Graph the function

$$y = x^2. \tag{8}$$

SOLUTION. The graph of (8) is the curve shown in Figure 5, known as a *parabola*. The x-axis seems to be tangent to the curve at the origin, and moreover the curve "opens upward" along its whole extent. These ideas will be made precise later, when we introduce the concepts of the "tangent to a curve" and "concavity."

The curve $y = x^2$ has another interesting property, namely it is *symmetric in the y-axis*. This simply means that for every point P of the curve on one side of the y-axis, there is another point Q of the curve on the other side such that the y-axis is the perpendicular bisector of the line segment PQ. To see that this is true, we merely note that changing x to $-x$ has no effect on the value of $y = x^2$, since $(-x)^2 = x^2$. But the points $P = (x, y)$ and $Q = (-x, y)$ clearly lie on opposite sides of the y-axis, and the y-axis is the perpendicular bisector of the horizontal segment PQ, as the figure makes apparent.

A function $f(x)$ is said to be *even* if $f(-x) \equiv f(x)$, where it is tacitly assumed that the domain of $f(x)$ contains $-x$ whenever it contains x. We have just shown that the function $f(x) = x^2$ is even and that its graph is symmetric in the y-axis. Clearly, the graph of every other even function has the same symmetry property. For example, the function $f(x) = |x|$ is even, since $|-x| \equiv |x|$, and hence the graph of $y = |x|$ is symmetric in the y-axis, as is apparent from Figure 4.

e. Graph the function

$$y = x^3. \tag{9}$$

SOLUTION. The graph of (9) is the curve shown in Figure 6, known as a *cubical parabola*. This curve seems to "open downward" to the left of the origin and "upward" to the right of the origin, while changing from "downward" to "upward" at the origin itself, which is accordingly called an "inflection point" of the curve. All these ideas will be made precise later, in connection with our discussion of "concavity."

The curve $y = x^3$ has another interesting property, namely it is *symmetric in the origin*. This simply means that for every point P of the curve, there is another point Q of the curve such that the origin is the midpoint of the line segment PQ

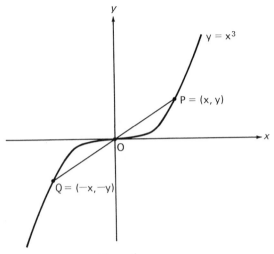

Figure 6.

(thus P and Q are, so to speak, on "opposite sides" of O). To see that this is true, we note that changing the sign of x also changes the sign of $y = x^3$, since $(-x)^3 = -x^3$. But the points $P = (x, y)$, $Q = (-x, -y)$ and $O = (0, 0)$ are collinear, as follows at once from the observation that the slope of the line through O and P has the same value y/x as the slope of the line through Q and O. Moreover, O is clearly the midpoint of the segment PQ, since

$$|OP| = |OQ| = \sqrt{x^2 + y^2}.$$

A function $f(x)$ is said to be *odd* if $f(-x) \equiv -f(x)$, where it is again assumed that the domain of $f(x)$ contains $-x$ whenever it contains x. We have just shown that the function $f(x) = x^3$ is odd and that its graph is symmetric in the origin. Clearly, the graph of every other odd function has the same symmetry property. For example, every line $y = mx$ through the origin has this property.

The problem of evenness versus oddness plays an important role in applied mathematics and physics. The question "What is the parity of $f(x)$?" simply means "Is $f(x)$ even or odd?"

2.33. Increasing and decreasing functions

a. Suppose the graph of a function $f(x)$ *rises* steadily as a variable point P on the graph moves from left to right, with its abscissa in some interval I. Then $f(x)$ is said to be *increasing in I*. For example, the functions $|x|$ and x^2 are both increasing in the interval $0 \leqslant x < \infty$, as we see at once from Figures 4 and 5, while the function x^3 is increasing on the whole real line, that is, in the whole interval $-\infty < x < \infty$, as we see from Figure 6. Similarly, if the graph of $f(x)$ *falls* steadily as P moves from left to right with its abscissa in an interval I, we say that $f(x)$ is *decreasing in I*. For example, the functions $|x|$ and x^2 are both decreasing in the interval $-\infty < x \leqslant 0$, as is again apparent from Figures 4 and 5.

b. The graph of the constant function $f(x) \equiv 1$ is simply the horizontal line $y = 1$, which neither rises nor falls. Therefore this function is neither increasing nor decreasing, in *every* interval. The same is true of any other constant function.

c. It is easy to give a purely algebraic definition of increasing and decreasing functions. Thus a function f defined in an interval I is said to be *increasing in I* if $f(x) < f(x')$ whenever x and x' are two points of I such that $x < x'$. Similarly, f is said to be *decreasing in I* if $f(x) > f(x')$ whenever x and x' are two points of I such that $x < x'$.

PROBLEMS

1. What is the graph of the equation $x^2 - y^2 = 0$?
2. What is the graph of the equation $x^2 + y^2 + 2x - 2y + 1 = 0$?
3. What is the equation of the circle of radius 2 with its center at the point $(-2, 3)$?
4. "No line parallel to the y-axis can intersect the graph of a *function* $y = f(x)$ in more than one point." True or false?
5. "No line parallel to the y-axis can intersect the graph of an *equation* $F(x, y) = 0$ in more than one point." True or false?
6. Neither of the graphs in Figures 2 and 3 is the graph of a function. Why not?
7. What is special about the graph of a *one-to-one* function $y = f(x)$?
8. Let G be the graph of the function $f(x)$. Describe the graphs of the functions $f(x) + c$ and $f(x + c)$.
9. Which of the following functions are even and which are odd?

   (a) $y \equiv 2$; (b) $y = x + 1$; (c) $y = \dfrac{1}{x}$; (d) $y = \dfrac{1}{|x|}$;

   (e) $y = \dfrac{x}{x^2 + 1}$; (f) $y = \dfrac{x^2}{x^2 + 1}$.

10. Show that the function

    $$y = x^n \qquad (n = 1, 2, \ldots) \tag{10}$$

    is even if n is even and odd if n is odd.

    Comment. By x^n we mean, of course, the *nth power* of x, that is,

    $$x^n = \underbrace{x \cdot x \cdots x}_{n \text{ factors}}.$$

11. Show that the product of two functions of the same parity is even, while the product of two functions of different parity is odd.
12. "If the function $f(x)$ is increasing in an interval I, then the function $-f(x)$ is decreasing in I, and conversely." True or false?
13. What is the equation of the circle circumscribed about the square with vertices $(0, 0)$, $(0, 1)$, $(1, 0)$ and $(1, 1)$?
14. Given the graph of a one-to-one function $y = f(x)$, how does one find the graph of the inverse function $x = f^{-1}(y)$?
15. Show that if a function f is increasing, then f is one-to-one, with an increasing inverse f^{-1}.
16. Show that if a function f is decreasing, then f is one-to-one, with a decreasing inverse f^{-1}
*17. Graph the function $y = |x + 1| + |x - 1|$. Is the function piecewise linear? Where is the function increasing and where decreasing? What happens in the interval $-1 \leqslant x \leqslant 1$?
*18. Graph the function $y = |x| + |x + 1| + |x + 2|$. Is the function piecewise linear? Where does the graph have corners? Where is the function increasing and where decreasing?

*19. Show that if $0 < x < x'$, then $0 < x^n < x'^n$ for all $n = 1, 2, \ldots$

*20. Show that if n is even, then the function (10) is increasing in the interval $0 \leqslant x < \infty$ and decreasing in the interval $-\infty < x \leqslant 0$, while if n is odd, then the function is increasing in the whole interval $(-\infty, \infty)$.

2.4 DERIVATIVES AND LIMITS

2.41. An instructive calculation

a. We now make a little calculation, leading us straight to the heart of our subject. Consider the function

$$Q(h) = \frac{(1 + h)^2 - 1}{h}, \tag{1}$$

defined for all nonzero values of h. Do not be disconcerted by our use of the "offbeat" letter h for the independent variable; this is a deliberate choice, made to avoid "tying up" the letter x, which will be needed a little later. We cannot allow $h = 0$ in (1), because $Q(h)$ would then reduce to the expression

$$Q(0) = \frac{(1 + 0)^2 - 1}{0} = \frac{1 - 1}{0} = \frac{0}{0},$$

which is meaningless. In fact, if $0/0$ is to make sense, it must mean the one and only number c such that $0 \cdot c = 0$. But *every* number c has this property! For this reason, the expression $0/0$ is often called an *indeterminate form*.

Despite the fact that $Q(0)$ itself is meaningless, the function $Q(h)$ is perfectly meaningful for values of h which are as close as we please to 0, *whether these values of h be positive or negative*. Thus it is only natural to ask: What happens to $Q(h)$ as h gets "closer and closer" to the forbidden value $h = 0$?

b. To answer this question, we first carry out the algebraic operations in the numerator of $Q(h)$, obtaining

$$Q(h) = \frac{1 + 2h + h^2 - 1}{h} = \frac{2h + h^2}{h} \quad (h \neq 0).$$

We then divide the numerator by the denominator h, which is permissible since $h \neq 0$. This gives the simple formula

$$Q(h) = 2 + h \quad (h \neq 0).$$

We now observe that as h gets "closer and closer" to 0, $Q(h)$ in turn gets "closer and closer" to 2. In fact, the distance between $Q(h)$ and 2, regarded as points of the real line, is just

$$|Q(h) - 2| = |(2 + h) - 2| = |h|$$

(Theorem 1.55), and $|h|$ is certainly very small whenever h is very near 0, for the simple reason that $|h|$ is just the distance between h and 0.

The fact that $Q(h)$ gets "closer and closer" to 2 as h gets "closer and closer" to 0 is summarized by writing

$$\lim_{h \to 0} Q(h) = 2. \tag{2}$$

In words, (2) says that "the limit of $Q(h)$ as h approaches zero equals 2." This is our first encounter with the concept of a *limit*, about which we will say much more later. An equivalent way of writing (2) is

$$Q(h) \to 2 \quad \text{as} \quad h \to 0,$$

which says that "$Q(h)$ approaches 2 as h approaches zero."

   c.   We have just killed two birds with one stone. Not only have we calculated the limit of $Q(h)$ as h approaches zero, but as we will see in a moment, we have also calculated something called "the derivative of the function $f(x) = x^2$ at the point $x = 1$." In fact, both the limit and the derivative equal 2.

2.42. The derivative concept

   a.   Let $f(x)$ be a function defined in some neighborhood of a point x_0. Then by the *difference quotient* of $f(x)$ at x_0, we mean the new function

$$Q(h) = \frac{f(x_0 + h) - f(x_0)}{h} \tag{3}$$

of the variable h. The letter Q stands for "quotient," and x_0 has a subscript zero to show that it is a *fixed* value of the argument x. For example, if $f(x) = x^2$ and $x_0 = 1$, then (3) reduces to

$$Q(h) = \frac{(1 + h)^2 - 1^2}{h},$$

which is nothing other than the expression (1).

   b.   Let $Q(h)$ be the difference quotient of the function $f(x)$ at the point x_0. Then by the *derivative of $f(x)$ at the point x_0*, denoted by $f'(x_0)$ and pronounced "f prime of x zero," we mean the limit

$$\lim_{h \to 0} Q(h), \tag{4}$$

provided that the limit "exists" (that is, makes sense). Here, as in formula (2), the expression (4) means the number, if any, which "$Q(h)$ approaches as h approaches zero." Combining (3) and (4), we find that

$$f'(x_0) = \lim_{h \to 0} \frac{f(x_0 + h) - f(x_0)}{h}. \tag{5}$$

It is important to note that $f'(x_0)$ is a number, not a function. Suppose once again that $f(x) = x^2$ and $x_0 = 1$. Then

$$f'(1) = \lim_{h \to 0} \frac{(1 + h)^2 - 1}{h} = \lim_{h \to 0} (2 + h) = 2.$$

Thus the derivative of the function $f(x) = x^2$ at the point $x = 1$ exists and equals 2. This justifies the claim made in Sec. 2.41c.

   c.   The derivative $f'(x_0)$ is also called the *rate of change of $y = f(x)$ with respect to x at the point x_0*. The reason for this designation is not hard to find. The numerator of the difference quotient (3) is just the change

$$f(x_0 + h) - f(x_0)$$

in the dependent variable $y = f(x)$ when the independent variable x is changed from x_0 to $x_0 + h$, while the denominator of (3) is just the change

$$(x_0 + h) - x_0 = h$$

in x itself. Therefore the difference quotient itself becomes

$$Q(h) = \frac{\text{Change in } y}{\text{Change in } x}, \tag{6}$$

and the derivative is the "limiting value" of this "change ratio" as the change in the independent variable x gets "smaller and smaller," that is, "approaches zero." As for the word "rate," which suggests something changing with respect to time, it is a metaphor borrowed from problems involving motion, where the independent variable is indeed time (usually denoted by t), and the dependent variable changes with respect to time at a certain "rate."

    d. In Sec. 1.12 we described calculus as the "mathematics of change" and formulated the two basic types of problems with which calculus deals. The first of these problems was stated in the following unsophisticated language:

> (1) Given a relationship between two changing quantities, what is the rate of change of one quantity with respect to the other?

We are now in a position to restate this problem in more precise language:

> (1') Given a function $y = f(x)$, what is the rate of change of y with respect to x?

The study of this problem is the province of a branch of calculus known as *differential calculus* and always involves the calculation of a derivative.

2.43. Examples

    a. Find the derivative of the function $f(x) = x^2$ at an arbitrary point x_0.

    SOLUTION.   For this function we have

$$f'(x_0) = \lim_{h \to 0} \frac{f(x_0 + h) - f(x_0)}{h} = \lim_{h \to 0} \frac{(x_0 + h)^2 - x_0^2}{h},$$

which becomes

$$f'(x_0) = \lim_{h \to 0} \frac{x_0^2 + 2x_0 h + h^2 - x_0^2}{h} = \lim_{h \to 0} \frac{2x_0 h + h^2}{h} = \lim_{h \to 0} (2x_0 + h),$$

after doing a little algebra. But as h gets "closer and closer" to 0, the quantity $2x_0 + h$ gets "closer and closer" to $2x_0$, for the simple reason that the distance between the points $2x_0 + h$ and $2x_0$ is just

$$|(2x_0 + h) - 2x_0| = |h|.$$

Therefore

$$\lim_{h \to 0} (2x_0 + h) = 2x_0,$$

so that finally

$$f'(x_0) = 2x_0.$$

Note that $f'(x_0) = 2$ when $x_0 = 1$, as we already know from the preliminary calculation made in Sec. 2.42b.

    b. Let

$$f(x) = mx + b, \tag{7}$$

where m and b are constants. Find the derivative of $f(x)$ at an arbitrary point x_0.

SOLUTION.   In this case,

$$f'(x_0) = \lim_{h \to 0} \frac{m(x_0 + h) + b - mx_0 - b}{h} = \lim_{h \to 0} \frac{mh}{h} = \lim_{h \to 0} m = m.$$

The last step calls for finding the number to which m gets "closer and closer" as $h \to 0$, but this can only be the number m itself, since m is a constant! Choosing $m = 0$ in (7), we find that *the derivative of any constant function $f(x) \equiv b$ equals 0 at every point x_0.* Choosing $m = 1$, $b = 0$ in (7), we find that *the derivative of the function $f(x) = x$ equals 1 at every point x_0.*

You will recognize (7) as the equation of the straight line with slope m and y-intercept b. Since the derivative of (7) equals m at every point x_0, you may begin to suspect that the derivative has something to do with slope. Indeed it has. In Sec. 2.52d we will see that the derivative $f'(x_0)$ is just the slope of the *tangent to the curve* $y = f(x)$ at the point with abscissa x_0. Of course, this will require that we first decide what is meant by the "tangent to a curve." Note that the word "tangent" as used here has nothing to do with the same word as used in trigonometry (Sec. 1.82b). In fact, the tangent to a curve is a line, while the tangent of an angle is a number.

2.44. Limits

a.   So far we have only considered limits of the form

$$\lim_{h \to 0} Q(h),$$

where the function $Q(h)$ is the difference quotient associated with another function $f(x)$. This, of course, is the special kind of limit leading to the notion of a derivative. More generally, we can consider the limit of an arbitrary function $f(x)$ as its argument x approaches an arbitrary point x_0, provided that $f(x)$ is defined in some *deleted* neighborhood of x_0 (Sec. 1.63a). Thus we say that a function $f(x)$ *approaches the limit A as x approaches x_0*, or that $f(x)$ *has the limit A at x_0*, if $f(x)$ gets "closer and closer" to A as x gets "closer and closer" to x_0 without ever actually coinciding with x_0. This fact is expressed by writing

$$\lim_{x \to x_0} f(x) = A \tag{8}$$

or

$$f(x) \to A \quad \text{as} \quad x \to x_0.$$

Put somewhat differently, (8) means that $f(x)$ is "arbitrarily near" A for all x which are "sufficiently near" x_0, or, equivalently, that $|f(x) - A|$ is "arbitrarily small" for all "sufficiently small" (but nonzero) values of $|x - x_0|$.

b.   Can this rather intuitive definition of a limit be made mathematically exact? Yes, it can, by resorting to the following procedure, invented by Cauchy in the early nineteenth century, which involves two positive numbers, traditionally called ε (the Greek letter epsilon) and δ (the Greek letter delta). What does it really mean to say that $|f(x) - A|$ is "arbitrarily small" for all "sufficiently small" $|x - x_0|$? Just this: Suppose somebody we call the "challenger" presents us with any positive number ε he pleases. Then we must be able to find another positive number δ such that $|f(x) - A| < \varepsilon$ for all x ($\neq x_0$) satisfying the inequality $|x - x_0| < \delta$. At this point, you may well ask: What has all this to do with the numbers $|f(x) - A|$ and $|x - x_0|$ being small? The answer is simply that we allow our challenger to present us with any positive number whatsoever, in particular, with a number which is as small as he pleases (that is, "arbitrarily small"). We must then find a corresponding number δ, which in general cannot be "too large" (and hence is "sufficiently small") such that $|f(x) - A| < \varepsilon$ whenever $|x - x_0| < \delta$.

Thus, once again, in more concise language, to say that $f(x)$ has the limit A at x_0 means that, *given any $\varepsilon > 0$, we can find a number $\delta > 0$ such that $|f(x) - A| < \varepsilon$ whenever $0 < |x - x_0| < \delta$.* Here the formula $0 < |x - x_0| < \delta$ is just a neat way of writing $|x - x_0| < \delta$ and $x \neq x_0$ at the same time, since $|x - x_0| > 0$ is equivalent to $x \neq x_0$.

You may find this definition a bit strange, but we urge you to master it anyway. It is a most valuable tool, which will help you keep many calculations brief and to the point (see Probs. 12–15, for example).

c. The fact that x is not allowed to take the value x_0 in the definition of the limit of $f(x)$ at x_0 is crucial. It shows that the limit (if any) of $f(x)$ at x_0 has nothing to do with the value of $f(x)$ at $x = x_0$, since this value does not even enter into the definition of the limit. In fact, a function can have a limit even at a point x_0 where it fails to be *defined*! For example, the limit of the function

$$Q(h) = \frac{f(x_0 + h) - f(x_0)}{h}$$

as $h \to 0$ is the derivative $f'(x_0)$, a fundamental concept of calculus, and yet $Q(h)$ is undefined at $h = 0$, where it reduces to the indeterminate form $0/0$.

d. It is often convenient to talk about having a limit without specifying what the limit is. Thus we say that a function $f(x)$ *has a limit at x_0* if there is some number A such that $f(x) \to A$ as $x \to x_0$.

2.45. Examples

a. Let x_0 be an arbitrary point. Then

$$\lim_{x \to x_0} x = x_0, \tag{9}$$

as we would certainly hope! This can be seen at once by using "ε, δ language." In fact, given $\varepsilon > 0$, we need only choose $\delta = \varepsilon$. It is then self-evident that $|x - x_0| < \varepsilon$ whenever $0 < |x - x_0| < \delta$.

b. The constant function $f(x) \equiv A$ approaches the limit A as x approaches an arbitrary point x_0. In fact, in this case $|f(x) - A| \equiv |A - A| \equiv 0$, so that, given any $\varepsilon > 0$, we have $|f(x) - A| < \varepsilon$ for *all* x, and in particular for all x such that $0 < |x - x_0| < \delta$, where $\delta > 0$ is arbitrary.

c. The function $f(x) = 3x$ approaches the limit 6 as $x \to 2$. In fact, given any $\varepsilon > 0$, choose $\delta = \varepsilon/3$. We then have

$$|f(x) - 6| = |3x - 6| = 3|x - 2| < 3\delta = \varepsilon$$

whenever $0 < |x - 2| < \delta$.

d. Show that

$$\lim_{x \to 2} x^2 = 4. \tag{10}$$

SOLUTION. It seems plausible enough that as a number gets "closer and closer" to 2, its square must get "closer and closer" to 4, and this is an acceptable, if somewhat crude, solution. A better solution is based on the use of "ε, δ language" and goes as follows: Given any $\varepsilon > 0$, let δ be the smaller of the two numbers 1 and $\varepsilon/5$. Then

$$|x^2 - 4| = |(x + 2)(x - 2)| = |x + 2|\,|x - 2| < 5 \cdot \frac{\varepsilon}{5} = \varepsilon$$

whenever $0 < |x - 2| < \delta$, since our choice of δ automatically forces x to satisfy the extra condition $|x + 2| < 5$, as well as $|x - 2| < \varepsilon/5$. To see this, note that $|x - 2| < \delta$ certainly implies $|x - 2| < 1$, or equivalently $1 < x < 3$, so that $3 < x + 2 < 5$, which in turn implies $|x + 2| < 5$.

Actually, even this solution is only a "stopgap measure." In Sec. 2.61 we will show that

$$\lim_{x \to 2} x^2 = \lim_{x \to 2} x \cdot \lim_{x \to 2} x = \left(\lim_{x \to 2} x \right)^2$$

and then (10) will be an immediate consequence of (9), with $x_0 = 2$.

e. Does the function

$$f(x) = \frac{|x|}{x} \qquad (x \neq 0)$$

have a limit at $x = 0$?

SOLUTION. No. If $x > 0$, then $|x| = x$ and $f(x) = 1$, while if $x < 0$, then $|x| = -x$ and $f(x) = -1$. Therefore $f(x)$ takes both values 1 and -1 in every deleted neighborhood of $x = 0$. But then $f(x)$ can hardly be "arbitrarily near" some number A for all x "sufficiently near" 0, even if we pick $A = 1$ or $A = -1$. This intuitive solution seems plausible, but its crudity is rather distressing. Again "ε, δ language" comes to the rescue, providing us with a solution which is both simple and perfectly sound. Suppose $f(x)$ has a limit A at $x = 0$. Then, choosing $\varepsilon = \frac{1}{2}$, we can find a number $\delta > 0$ such that $|f(x) - A| < \varepsilon = \frac{1}{2}$ whenever $0 < |x| < \delta$. Let $x_1 = \frac{1}{2}\delta$, $x_2 = -\frac{1}{2}\delta$, so that, in particular $0 < |x_1| < \delta, 0 < |x_2| < \delta$. Then, on the one hand,

$$f(x_1) = 1, \qquad f(x_2) = -1,$$

while, on the other hand,

$$|f(x_1) - A| < \frac{1}{2}, \qquad |f(x_2) - A| < \frac{1}{2}.$$

But these requirements are incompatible. In fact, using the triangle inequality (3), p. 14, we find that

$$|f(x_1) - f(x_2)| = |[f(x_1) - A] + [A - f(x_2)]|$$

$$\leqslant |f(x_1) - A| + |A - f(x_2)| < \frac{1}{2} + \frac{1}{2} = 1,$$

while, at the same time,

$$|f(x_1) - f(x_2)| = |1 - (-1)| = 2.$$

Thus the assumption that $f(x)$ has a limit at $x = 0$ has led to the absurd conclusion that $2 < 1$. Therefore $f(x)$ does not have a limit at $x = 0$.

f. Does the function $f(x) = |x|$ have a derivative at $x = 0$?

SOLUTION. No, since

$$f'(0) = \lim_{h \to 0} \frac{f(0 + h) - f(0)}{h} = \lim_{h \to 0} \frac{|0 + h| - |0|}{h} = \lim_{h \to 0} \frac{|h|}{h},$$

or

$$f'(0) = \lim_{x \to 0} \frac{|x|}{x},$$

if we denote the independent variable by x instead of by h, which is our privilege. But we have just shown that this limit fails to exist.

PROBLEMS

1. Let $f(x) = ax^2 + bx + c$, where a, b and c are constants. Verify that $f'(x_0) = 2ax_0 + b$ at every point x_0.
2. Let $f(x) = x^3$. Verify that $f'(x_0) = 3x_0^2$ at every point x_0.

    Comment. Problems 1 and 2 are just "warming up exercises." The technique of evaluating derivatives will be developed more systematically in Sec. 2.7.
3. Is the formula $f'(\alpha + \beta) = f'(\alpha) + f'(\beta)$ true for $f(x) = mx + b$? For $f(x) = x^2$?
4. Can $f'(x_0)$ ever equal $f(x_0)$?
5. Show that $f(x) \to A$ as $x \to x_0$ and $f(x) - A \to 0$ as $x \to x_0$ mean exactly the same thing.
6. Find the limit

    (a) $\lim\limits_{x \to 0} |x|$;   (b) $\lim\limits_{x \to 0} \dfrac{x}{x}$;   (c) $\lim\limits_{x \to 1} \dfrac{|x|}{x}$;   (d) $\lim\limits_{x \to 0} \dfrac{x^2}{x}$;   (e) $\lim\limits_{x \to 0} \dfrac{\sqrt{x^2}}{x}$.

7. Does the function

    $$f(x) = \begin{cases} x & \text{if } 0 \leqslant x \leqslant 1, \\ 2x & \text{if } 1 < x \leqslant 3 \end{cases}$$

    have a limit at the point $x = 1$? At the point $x = 2$?
8. Find the limit of the function

    $$f(x) = \begin{cases} x^2 & \text{if } x \neq 0, \\ 2 & \text{if } x = 0 \end{cases}$$

    at the point
    (a) $x = -1$;   (b) $x = 0$;   (c) $x = \sqrt{2}$.
9. To find the number δ in the "ε, δ language" must we know the number ε? Is δ a function of ε?
10. Show that if $f(x) \to 0$ as $x \to x_0$, then $|f(x)| \to 0$ as $x \to x_0$, and conversely.
*11. Show that if $f(x) \to A$ as $x \to x_0$, then $|f(x)| \to |A|$ as $x \to x_0$. Is the converse true?

The following problems are all easily solved with the help of "ε, δ language." Make sure that you understand what these problems mean, even if you don't work them out.

*12. Show that if $f(x)$ has a limit at x_0, then the limit is unique. In other words, show that $f(x)$ cannot have more than one limit at x_0.
*13. Show that changing the value of a function $f(x)$ at any point $x_1 \neq x_0$ has no effect on the limit (if any) of $f(x)$ at x_0.

    Comment. Thus only the values of $f(x)$ in the "immediate vicinity" of x_0 have any effect on the "limiting behavior" of $f(x)$ at x_0.
*14. Show that if $f(x) \to A$ as $x \to x_0$, then there is a deleted neighborhood of x_0 in which $|f(x)| < |A| + 1$.

    Comment. Thus a function cannot become "too large" in absolute value near a point where it has a limit.
*15. Show that if $f(x) \to A \neq 0$ as $x \to x_0$, then there is a deleted neighborhood of x_0 in which $f(x)$ has the same sign as A and $|f(x)| > \frac{1}{2}|A|$.

Comment. Thus a function cannot change sign or even become "too small" in absolute value near a point where it has a *nonzero* limit.

2.5 MORE ABOUT DERIVATIVES

2.51. Increment notation. The meaning of the derivative can be made even clearer by using a somewhat different notation. Given a function $y = f(x)$, let the change in the independent variable, namely the difference between the final and initial values of x, be denoted by Δx, instead of by h as in Sec. 2.42c. Here Δx, where Δ is the Greek capital letter delta, must be thought of as a *single entity*, pronounced "delta x," and *not* as the product of the separate symbols Δ and x. Let the corresponding change in the dependent variable, namely the difference between the final and initial values of y, be denoted by Δy ("delta y"), that is, let

$$\Delta y = f(x_0 + \Delta x) - f(x_0).$$

Then formula (6), p. 54, takes the particularly suggestive form

$$Q(\Delta x) = \frac{\Delta y}{\Delta x},$$

where the reason for the term "difference quotient" is now staring you in the face. We also call Δx the *increment of* x and Δy the *increment of* y. We will favor this "increment notation" from now on, because of the way it identifies the quantities which are actually being changed.

In terms of increment notation, formula (5), p. 53, defining the derivative of $f(x)$ at the point x_0 takes the form

$$f'(x_0) = \lim_{\Delta x \to 0} \frac{f(x_0 + \Delta x) - f(x_0)}{\Delta x}.$$

Bear in mind that in writing $\Delta x \to 0$, we impose no restriction on the sign of Δx, which is free to take both positive and negative values.

2.52. The tangent to a curve

a. In keeping with the remarks in Example 2.43b, we now decide what is meant by the "tangent to a curve." You already know how the tangent is defined in the case where the curve is a circle C. In fact, according to elementary geometry, the tangent to a circle C at a point P_0 is the line which intersects C in the point P_0 and *in no other point*. A moment's thought shows that this property of the tangent to a circle is useless for defining the tangent to a general curve. For example, in the case of the parabola $y = x^2$ graphed in Figure 7, there are *two* lines, namely the x-axis and the y-axis, intersecting the curve in the origin O and in no other point. But common sense rejects the idea of the y-axis being the tangent to the curve at O, although the x-axis seems a perfectly plausible candidate for the tangent to the curve at O.

As this example suggests, the key property of the tangent is that it "hugs the curve very closely at the point of tangency." For example, this seems to be true of the line T in Figure 7, which represents the tangent to the parabola at the point P_0. We now give a precise mathematical meaning to this qualitative idea. As you may suspect, the notions of limit and derivative will play a prominent role here.

b. Thus let $P_0 = (x_0, y_0)$ be a fixed point and $P = (x, y)$ a variable point of a given curve $y = f(x)$, and let S be the straight line going through the points P_0 and

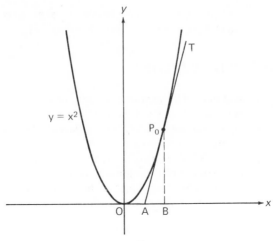

Figure 7.

P. Such a line is called a *secant* (*line*) of the curve $y = f(x)$. The slope of S is just

$$m_S = \frac{y - y_0}{x - x_0},$$

or equivalently

$$m_S = \frac{\Delta y}{\Delta x}, \tag{1}$$

in terms of the increments

$$\Delta x = x - x_0, \qquad \Delta y = y - y_0 = f(x_0 + \Delta x) - f(x_0).$$

The geometrical meaning of these increments is shown in Figure 8, which is drawn for the case where Δx and Δy are both positive. As an exercise, sketch similar figures for the other three choices of the signs of Δx and Δy. All of these figures are equiv-

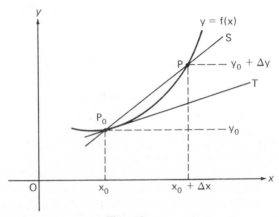

Figure 8.

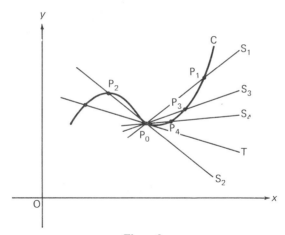

Figure 9.

alent from the standpoint of illustrating the construction of the tangent to the curve $y = f(x)$ at P_0.

    c.   We now vary the point P along the curve, making P move "closer and closer" to the fixed point P_0, and at the same time allowing P to move freely from one side of P_0 to the other. Then Δx approaches zero, and at the same time the secant through P_0 and P varies, taking first one position and then another. Suppose the limit

$$m = \lim_{\Delta x \to 0} m_S \tag{2}$$

exists. Then the straight line through P_0 with slope m is called the *tangent* (*line*) *to the curve* $y = f(x)$ *at the point* P_0. In other words, the tangent at P_0 is the "limiting position" of the secant through P_0 and P as the variable point P approaches the fixed point P_0, taking positions on both sides of P_0. This behavior is illustrated in Figure 9, where the secants go through a sequence of positions $S_1, S_2, S_3, S_4, \ldots$, getting "closer and closer" to the limiting tangent line T. This figure also shows that, unlike the case of a circle, the tangent to a general curve C may well intersect C in points other than the point of tangency P_0.

    d.   Finally, substituting (1) into (2), we find that

$$m = \lim_{\Delta x \to 0} \frac{\Delta y}{\Delta x} = \lim_{\Delta x \to 0} \frac{f(x_0 + \Delta x) - f(x_0)}{\Delta x},$$

where the limit on the right is, of course, just the derivative $f'(x_0)$ of the function $f(x)$ at the point x_0. Thus we have proved the following key result of differential calculus: *The curve* $y = f(x)$ *has a tangent* T *at the point* $P_0 = (x_0, f(x_0))$ *if the derivative* $f'(x_0)$ *exists, and only in this case. The slope of* T *is then equal to* $f'(x_0)$. This slope is often simply called the *slope of the curve at* P_0.

    According to Sec. 1.91a, the equation of the straight line with slope m going through the point (x_0, y_0) is just

$$y = m(x - x_0) + y_0.$$

Therefore the tangent to the curve $y = f(x)$ at the point $(x_0, y_0) = (x_0, f(x_0))$ has the equation

$$y = f'(x_0)(x - x_0) + f(x_0). \tag{3}$$

2.53. Examples

a. Find the tangent T to the parabola $y = x^2$ at the point $P_0 = (x_0, y_0)$.
SOLUTION.   Here $f(x) = x^2$, and hence

$$f'(x_0) = 2x_0, \tag{4}$$

as calculated in Example 2.43a. Substituting (4) into (3), we find that T has the equation

$$y = 2x_0(x - x_0) + x_0^2 = 2x_0x - x_0^2, \tag{5}$$

since $y_0 = x_0^2$. Setting $y = 0$ and solving for x, we see at once that the line T has the x-intercept $x_0/2$. Thus, to construct the tangent to the parabola $y = x^2$ at a point P_0 other than the origin, we need only drop the perpendicular P_0B to the x-axis, bisect the segment OB, and then draw the line T joining the midpoint A of OB to the point P_0, as shown in Figure 7. If P_0 is the origin, then $x_0 = 0$ and (5) reduces to the equation $y = 0$. Therefore in this case T is the line $y = 0$, namely the x-axis, as conjectured in Sec. 2.52a.
b. Find the tangent T to the curve $y = |x|$ at the point (x_0, y_0).
SOLUTION.   Here we have $f(x) = |x|$. If $x_0 > 0$, then

$$f'(x_0) = 1, \tag{6}$$

since $f(x) = x$ for all x in a suitable neighborhood of x_0. Substituting (6) into (3), we find that T has the equation

$$y = 1 \cdot (x - x_0) + x_0 = x,$$

since $y_0 = |x_0| = x_0$. Therefore in this case T is just the line $y = x$, as is geometrically apparent from Figure 10A. On the other hand, if $x_0 < 0$, we have

$$f'(x_0) = -1, \tag{7}$$

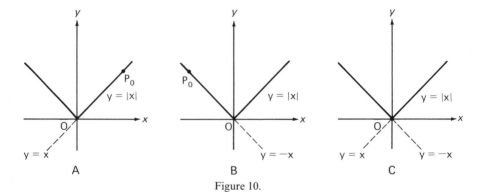

Figure 10.

instead of (6), since now $f(x) = -x$ for all x in a suitable neighborhood of x_0, and substitution of (7) into (3) gives

$$y = -1 \cdot (x - x_0) - x_0 = -x,$$

since in this case $y_0 = |x_0| = -x_0$. Therefore T is now just the line $y = -x$, as is geometrically apparent from Figure 10B.

If $x_0 = 0$, then $f'(x_0)$ fails to exist, as shown in Example 2.45f. Therefore the curve $y = |x|$ has no tangent at the origin O. The reason for this is geometrically evident from Figure 10C and is associated with the presence of the sharp "corner" of the graph of $y = |x|$ at the origin. The secant drawn through the origin O and a variable point P of the curve $y = |x|$ can hardly approach a "limiting position" as P approaches O, since the secant coincides with the line $y = x$ whenever P lies to the right of O (in the first quadrant) and with the perpendicular line $y = -x$ whenever P lies to the left of O (in the second quadrant). In fact, suppose P approaches O through a sequence of positions *on both sides of O*, in keeping with the definition of the tangent at O. Then the secant changes its inclination by a full 90° (in either the clockwise or the counterclockwise direction) every time P goes from one side of O to the other, and this "wild" behavior is clearly inconsistent with the secant approaching any tangent line at all as P approaches O.

2.54. Differentiation

a. The process leading from a function to its derivative is called *differentiation*, with respect to the independent variable. Another way of saying that a function $y = f(x)$ has a derivative at a point x_0 is to say that $f(x)$ is *differentiable at x_0*. If $f(x)$ is differentiable at every point of an interval I, we say that $f(x)$ is *differentiable in I*. For example, the function $f(x) = |x|$ is differentiable in both intervals $-\infty < x < 0$ and $0 < x < \infty$, although it fails to be differentiable at the point $x = 0$. Whenever we call a function *differentiable*, without further qualification, we always mean differentiable at some point or in some interval, where the context makes it clear just what is meant.

b. Suppose $f(x)$ is differentiable in an interval I. Then the derivative

$$f'(x_0) = \lim_{\Delta x \to 0} \frac{f(x_0 + \Delta x) - f(x_0)}{\Delta x} \tag{8}$$

exists for every x_0 in I. Hence there is a *new function* defined in I, whose value at every point x_0 is just $f'(x_0)$. This new function, which we denote by $(f(x))'$, or simply by $f'(x)$, is called the *derivative of $f(x)$*, with no mention of a point x_0. It will always be clear from the context whether the term "derivative" refers to a *function* or to a *number*, namely the value of the derivative function at some point. The results of Examples 2.43a and 2.43b can now be written more concisely as

$$(x^2)' = 2x, \qquad (mx + b)' = m.$$

The derivative of a function $y = f(x)$ is sometimes denoted simply by y'.

2.55. The differential

a. We will usually write (8) in the form

$$f'(x) = \lim_{\Delta x \to 0} \frac{f(x + \Delta x) - f(x)}{\Delta x}, \tag{9}$$

dropping the subscript zero in three places. This is done with the understanding that x is held *fixed* during the evaluation of the limit. The numerator of the difference quotient in (9) is often denoted by $\Delta f(x)$ instead of by Δy. This has the advantage of allowing us to make explicit the point x at which the increment $\Delta f(x) = f(x + \Delta x) - f(x)$ is taken. We call $\Delta f(x)$ the *increment of the function f at the point x*, where, as usual, Δf is to be thought of as a single entity. In terms of this notation, (9) takes the form

$$f'(x) = \lim_{\Delta x \to 0} \frac{\Delta f(x)}{\Delta x}. \tag{10}$$

Equation (10) can also be written as

$$\lim_{\Delta x \to 0} \left[\frac{\Delta f(x)}{\Delta x} - f'(x) \right] = 0$$

(Sec. 2.4, Prob. 5), or equivalently as

$$\lim_{\Delta x \to 0} \alpha(\Delta x) = 0, \tag{11}$$

where

$$\alpha(\Delta x) = \frac{\Delta f(x) - f'(x) \, \Delta x}{\Delta x}.$$

The numerator of $\alpha(\Delta x)$ is just the error made in replacing $\Delta f(x)$ by $f'(x) \, \Delta x$, and, according to (11), this error is small in absolute value compared to $|\Delta x|$ if $|\Delta x|$ is small. Thus it is often a good approximation to replace $\Delta f(x)$ by the quantity $f'(x) \, \Delta x$, called the *differential of the function f at the point x*. For the differential we introduce the special notation

$$df(x) = f'(x) \, \Delta x, \tag{12}$$

which stresses the connection between $df(x)$ and the increment $\Delta f(x)$. As in the case of Δf, the symbol df must be thought of as a *single entity*, pronounced "dee f," and *not* as the product of the separate symbols d and f. In the case of a function written as $y = f(x)$, we can write dy for $df(x)$, just as we write Δy for $\Delta f(x)$. Then (12) takes the form

$$dy = f'(x) \, \Delta x. \tag{13}$$

We will often drop arguments for brevity, writing f' for $f'(x)$, df for $df(x)$, and so on. Thus formulas like $df = f'(x) \, \Delta x$ or $dy = f' \, \Delta x$ shouldn't bother you a bit.

   b. Example. Find the increment Δy and the differential dy of the function $y = x^2$ for $x = 20$, $\Delta x = 0.1$. What is the error of the approximation $\Delta y \approx dy$? (The symbol $\approx$ means "is approximately equal to.")
   SOLUTION.

$$\Delta y = (x + \Delta x)^2 - x^2 = 2x \, \Delta x + (\Delta x)^2,$$
$$dy = (x^2)' \, \Delta x = 2x \, \Delta x,$$

and hence

$$\Delta y = 2(20)(0.1) + (0.1)^2 = 4.01,$$
$$dy = 2(20)(0.1) = 4.00$$

for $x = 20$, $\Delta x = 0.1$. Thus the error of the approximation $\Delta y \approx dy$ is only 0.01, about 0.25% of the actual value of Δy.

2.56. The *d* notation

a. For the function $y = f(x) = x$, we have $dy = dx$, $f'(x) = 1$, so that (13) reduces to

$$dx = 1 \cdot \Delta x = \Delta x.$$

In other words, the increment and the differential of the *independent* variable are equal. We can now write

$$df(x) = f'(x) \, dx, \tag{14}$$

instead of (12), and

$$dy = f'(x) \, dx, \tag{15}$$

instead of (13). These formulas lead at once to an important new way of writing derivatives, called the "*d* notation," which we will use freely from now on. Solving (14) for $f'(x)$, we get

$$f'(x) = \frac{df(x)}{dx},$$

or, without the arguments,

$$f' = \frac{df}{dx}.$$

This formula is read as "f prime equals dee f by dee x." Similarly, if $y = f(x)$, it follows from (15) and the last sentence of Sec. 2.54b that

$$y' = \frac{dy}{dx} = f'(x).$$

The advantage of the new notation, in which f' becomes df/dx and y' becomes dy/dx is that the independent variable is now indicated explicitly. Thus there is a distinction between

$$\frac{df}{dx}, \frac{df}{dt}, \frac{df}{du}, \ldots, \tag{16}$$

which is not so easily made in the old notation; here $x, t, u, \ldots$ indicate different symbols for the independent variable. To distinguish between the different derivatives (16) verbally, we call df/dx the derivative of f *with respect to x*, df/dt the derivative of f *with respect to t*, and so on. Such distinctions are often crucial.

b. Having learned enough about differentials to appreciate how the "*d* notation" arises, there is no further need to think of derivatives as ratios of differentials. Rather you should regard d/dx as a single entity, pronounced "dee by dee x" and called the *differentiation operator*, which has the effect of differentiating with respect to x any function written after it; for simplicity, the function is often written after the first letter d in d/dx rather than after the whole expression. Thus, for example,

$$\frac{d}{dx} f = \frac{df}{dx} = f', \qquad \frac{d}{dx}(f + g) = (f + g)',$$

where the prime denotes differentiation with respect to x. Similar remarks apply to differentiation operators like d/dt, d/du, and so on, where the independent variable is indicated by another letter.

PROBLEMS

1. Find the increment Δx of the independent variable and the corresponding increment Δy of the dependent variable for the function $y = 1/x^2$ if x is changed from 0.01 to 0.001.

2. Let $u = u(x)$ and $v = v(x)$ be two functions of x, where, for simplicity, we use the same letters to denote both the functions and the dependent variables. Show that $\Delta(u + v) = \Delta u + \Delta v$.

3. Verify that the tangent to the line $y = mx + b$ at every point of the line is just the line itself, as is to be expected.

4. Does the curve $y = x^2$ have two different tangents which are parallel? Does the curve $y = x^3$?

5. Does the curve $y = x^2$ have two different tangents which are perpendicular? Does the curve $y = x^3$?

6. Find the tangent to the curve $y = x^2$ going through the point $(2, 0)$. Note that $(2, 0)$ is not a point of the curve.

7. At what point of the curve $y = x^2$ is the tangent parallel to the secant drawn through the points with abscissas 1 and 3?

8. Find the value of x such that $df(x) = 0.8$ if $f(x) = x^2$ and $\Delta x = -0.2$.

9. Find the increment Δy and the differential dy of the function $y = x^3$ if $x = 1$ and
(a) $\Delta x = 1$;   (b) $\Delta x = 0.1$;   (c) $\Delta x = 0.01$;   (d) $\Delta x = 0.001$.
In each case, find the error $E = \Delta y - dy$ made in replacing Δy by dy, both as a number and as a percentage of Δy. What happens as Δx gets smaller?

10. When is $\Delta y \approx dy$ a bad approximation?

*11. Let $u = u(x)$ and $v = v(x)$ be the same as in Problem 2. Write two different expressions for $\Delta(uv)$ in terms of Δu and Δv.

*12. For what values of b and c does the curve $y = x^2 + bx + c$ have the line $y = x$ as a tangent at the point with abscissa 2?

*13. Where is the function $y = |x + 1| + |x - 1|$ differentiable? Where does the graph of the function fail to have a tangent?

*14. How much would the earth's surface area increase if its radius were increased by 1 foot?

*15. What is the geometrical meaning of the differential dy?

2.6   MORE ABOUT LIMITS

2.61. Algebraic operations on limits. The following basic theorem on limits is used time and again in calculus. It merely says that "the limit of a sum equals the sum of the limits," and similarly with the word "sum" replaced by "difference," "product" and "quotient."
    THEOREM. *If*

$$\lim_{x \to x_0} f(x) = A, \qquad \lim_{x \to x_0} g(x) = B, \qquad (1)$$

then

$$\lim_{x \to x_0} \left[f(x) + g(x) \right] = A + B, \tag{2}$$

$$\lim_{x \to x_0} \left[f(x) - g(x) \right] = A - B, \tag{3}$$

$$\lim_{x \to x_0} f(x)g(x) = AB, \tag{4}$$

$$\lim_{x \to x_0} \frac{f(x)}{g(x)} = \frac{A}{B}, \tag{5}$$

provided that $B \neq 0$ *in the last formula.*

Proof. As you might expect, "ε, δ language" is the thing to use here. Since $f(x) \to A$ as $x \to x_0$ and $g(x) \to B$ as $x \to x_0$, then, given any $\varepsilon > 0$, we can find numbers $\delta_1 > 0$ and $\delta_2 > 0$ such that $|f(x) - A| < \varepsilon/2$ whenever $0 < |x - x_0| < \delta_1$ and $|g(x) - B| < \varepsilon/2$ whenever $0 < |x - x_0| < \delta_2$. (Yes, we really mean $\varepsilon/2$ here, not ε.) Let δ be the smaller of the two numbers δ_1 and δ_2. Then, using our old standby, the triangle inequality (3), p. 14, we have

$$\left| \left[f(x) + g(x) \right] - (A + B) \right| = \left| \left[f(x) - A \right] + \left[g(x) - B \right] \right|$$

$$\leqslant |f(x) - A| + |g(x) - B| < \frac{\varepsilon}{2} + \frac{\varepsilon}{2} = \varepsilon$$

whenever $0 < |x - x_0| < \delta$. But this is just "ε, δ language" for the statement that $f(x) + g(x) \to A + B$ as $x \to x_0$. Thus we have proved (2). To prove (3), we need only note that

$$\left| \left[f(x) - g(x) \right] - (A - B) \right| = \left| \left[f(x) - A \right] + \left[B - g(x) \right] \right|$$

$$\leqslant |f(x) - A| + |B - g(x)| < \frac{\varepsilon}{2} + \frac{\varepsilon}{2} = \varepsilon$$

whenever $0 < |x - x_0| < \delta$.

To prove (4) and (5) we argue in the same way. For example to prove (4) we show that, given any $\varepsilon > 0$, there is a $\delta > 0$ such that $|f(x)g(x) - AB| < \varepsilon$ whenever $0 < |x - x_0| < \delta$, and similarly for (5). However, the details are not very instructive, and for that reason are left to Problem 15. You can work through this rather difficult problem if you have a special interest in mathematical technique. Otherwise, persuade yourself of the validity of formulas (4) and (5) by thinking of their intuitive meaning. □

2.62. Examples

a. Show that

$$\lim_{x \to 3} (x^2 + 2x) = 15.$$

SOLUTION. We already know from Example 2.45a that

$$\lim_{x \to 3} x = 3, \tag{6}$$

a result which is almost obvious. Therefore, using (4) twice, we have

$$\lim_{x \to 3} x^2 = \lim_{x \to 3} x \cdot \lim_{x \to 3} x = 3 \cdot 3 = 9$$

and

$$\lim_{x \to 3} 2x = \lim_{x \to 3} 2 \cdot \lim_{x \to 3} x = 2 \cdot 3 = 6$$

(by Example 2.45b, "the limit of a constant equals the constant itself"). It then follows from (2) that

$$\lim_{x \to 3} (x^2 + 2x) = \lim_{x \to 3} x^2 + \lim_{x \to 3} 2x = 9 + 6 = 15.$$

b. Evaluate

$$A = \lim_{x \to 3} \frac{x^2 - 5x + 6}{x^2 - 8x + 15}.$$

SOLUTION.   We have

$$A = \lim_{x \to 3} \frac{x^2 - 5x + 6}{x^2 - 8x + 15} = \lim_{x \to 3} \frac{(x - 3)(x - 2)}{(x - 3)(x - 5)} = \lim_{x \to 3} \frac{x - 2}{x - 5},$$

where in the last step we cancel the common factor $x - 3$ of the numerator and denominator, relying on the fact that x approaches 3 without ever being *equal* to 3, so that $x - 3$ is never zero. Using (3) and (6), we have the easy calculations

$$\lim_{x \to 3} (x - 2) = \lim_{x \to 3} x - \lim_{x \to 3} 2 = 3 - 2 = 1,$$

$$\lim_{x \to 3} (x - 5) = \lim_{x \to 3} x - \lim_{x \to 3} 5 = 3 - 5 = -2.$$

It then follows from (5) that

$$A = \lim_{x \to 3} \frac{x - 2}{x - 5} = \frac{\lim_{x \to 3} (x - 2)}{\lim_{x \to 3} (x - 5)} = \frac{1}{-2} = -\frac{1}{2}.$$

2.63. Continuous functions

a. Once again we stress that the *limit* of a function $f(x)$ at a point x_0 is something quite different from the *value* of $f(x)$ at x_0, and in fact $f(x)$ may not even be *defined* at x_0. There is a special name for a function "nice enough" to have a limit at x_0 which equals its value at x_0: Such a function is said to be *continuous at x_0*. In other words, we say that $f(x)$ is continuous at x_0 if

$$\lim_{x \to x_0} f(x) = f(x_0). \tag{7}$$

Of course, this presupposes that $f(x)$ is defined in some *nondeleted* neighborhood of x_0, so that both sides of (7) make sense. If formula (7) does not hold, we say that $f(x)$ is *discontinuous at x_0*.

b. If a function $f(x)$ is continuous at every point of an interval I, we say that $f(x)$ is *continuous in I*. When we call a function *continuous*, without further qualification, we always mean continuous at some point or in some interval, where the context makes it clear just what is meant. A function is said to be *discontinuous*, without further qualification, if it is discontinuous at one or more points. The property of being continuous is called *continuity*, and plays an important role in calculus.

c. Algebraic operations on continuous functions are governed by the following rule:

THEOREM.   *If the functions $f(x)$ and $g(x)$ are both continuous at x_0, then so are the sum $f(x) + g(x)$, the difference $f(x) - g(x)$, the product $f(x)g(x)$ and the quotient $f(x)/g(x)$, provided that $g(x_0) \neq 0$ in the case of the quotient.*

Proof. This is an immediate consequence of the corresponding theorem on limits. Instead of (1), we now have

$$\lim_{x \to x_0} f(x) = f(x_0), \qquad \lim_{x \to x_0} g(x) = g(x_0).$$

But then formulas (2) through (5) become

$$\lim_{x \to x_0} [f(x) + g(x)] = f(x_0) + g(x_0),$$

$$\lim_{x \to x_0} [f(x) - g(x)] = f(x_0) - g(x_0),$$

$$\lim_{x \to x_0} f(x)g(x) = f(x_0)g(x_0),$$

$$\lim_{x \to x_0} \frac{f(x)}{g(x)} = \frac{f(x_0)}{g(x_0)},$$

provided that $g(x_0) \neq 0$ in the last formula, and this is exactly what is meant by saying that the functions $f(x) + g(x)$, $f(x) - g(x)$, $f(x)g(x)$ and $f(x)/g(x)$ are continuous at x_0. $\square$

d. COROLLARY. *If the functions $f_1(x), f_2(x), \ldots, f_n(x)$ are all continuous at x_0, then so are the sum $f_1(x) + f_2(x) + \cdots + f_n(x)$ and the product $f_1(x)f_2(x) \cdots f_n(x)$*

Proof. For example, if

$$h(x) = f_1(x) + f_2(x) + f_3(x),$$

then $h(x) = g(x) + f_3(x)$, where $g(x) = f_1(x) + f_2(x)$. One application of the theorem just proved shows that $g(x)$ is continuous at x_0, being the sum of two functions $f_1(x)$ and $f_2(x)$ which are continuous at x_0, and then another application of the theorem shows that $h(x)$ is continuous at x_0, being the sum of two functions $g(x)$ and $f_3(x)$ which are continuous at x_0. The proof for products is virtually the same. $\square$

2.64. Examples

a. Any constant function is continuous *everywhere*, that is, at every point x_0. In fact, if $f(x) \equiv c$, then

$$\lim_{x \to x_0} f(x) = \lim_{x \to x_0} c = c = f(x_0).$$

Moreover, the function x is continuous everywhere, since, by Example 2.45a,

$$\lim_{x \to x_0} x = x_0$$

b. By a *polynomial* we mean a function of the form

$$P(x) = a_0 + a_1 x + a_2 x^2 + \cdots + a_n x^n,$$

where $a_0, a_1, a_2, \ldots, a_n$ are arbitrary constants and n is a positive integer, called the *degree* of the polynomial (if $a_n \neq 0$). Each term in the sum is continuous, by the corollary and the preceding example, and hence so is the sum itself, by the corollary again. Thus $P(x)$ is defined and continuous everywhere, that is, in the whole interval $(-\infty, \infty)$.

c. By a *rational function* we mean a quotient of two polynomials

$$R(x) = \frac{a_0 + a_1 x + a_2 x^2 + \cdots + a_n x^n}{b_0 + b_1 x + b_2 x^2 + \cdots + b_N x^N}, \tag{8}$$

where n and N are in general different. It follows from the preceding example and the theorem that $R(x)$ exists and is continuous except at the points (if any) where the denominator of (8) equals zero. For example, the rational function

$$f(x) = \frac{x}{1 + x^2}$$

is continuous in the whole interval $(-\infty, \infty)$, while

$$g(x) = \frac{x^2}{1 - x^2}$$

is continuous everywhere except at the two points $x = \pm 1$.

d. The function $|x|$ is continuous everywhere. In fact,

$$\lim_{x \to 0} |x| = 0 = |0|,$$

since $x \to 0$ and $|x| \to 0$ mean exactly the same thing, while

$$\lim_{x \to x_0} |x| = \lim_{x \to x_0} x = x_0 = |x_0|$$

if $x_0 > 0$, and

$$\lim_{x \to x_0} |x| = \lim_{x \to x_0} (-x) = -\lim_{x \to x_0} x = -x_0 = |x_0|$$

if $x_0 < 0$. In the last two calculations, we use the fact that x has the same sign as x_0 if x is "sufficiently close" to x_0.

e. The function

$$f(x) = \frac{|x|}{x} \qquad (x \neq 0),$$

which can also be written

$$f(x) = \begin{cases} -1 & \text{if} \quad x < 0, \\ 1 & \text{if} \quad x > 0, \end{cases} \tag{9}$$

is discontinuous at $x = 0$, since, as shown in Example 2.45e, it has no limit at $x = 0$. Moreover, there is no way of defining $f(x)$ at $x = 0$ which will make $f(x)$ continuous at $x = 0$, since the failure of $f(x)$ to have a limit at $x = 0$ has nothing to do with its value at $x = 0$ (Sec. 2.44c).

f. The situation is different for the function

$$g(x) = x \qquad (x \neq 0). \tag{10}$$

Here

$$\lim_{x \to 0} g(x) = 0,$$

and the function fails to be continuous at $x = 0$ only because the point $x = 0$ has been excluded, rather artificially, from the domain of $g(x)$. However, we can make $g(x)$ continuous at $x = 0$ by the simple expedient of setting $g(0)$ equal to 0, *by definition*. In this way, we can "remove the discontinuity" at $x = 0$, replacing the discontinuous function (10) by the function x which is continuous everywhere. On the other hand, if we set $g(0)$ equal to 1, say, instead of 0, we get a new function

$$h(x) = \begin{cases} x & \text{if} \quad x \neq 0, \\ 1 & \text{if} \quad x = 0, \end{cases} \tag{11}$$

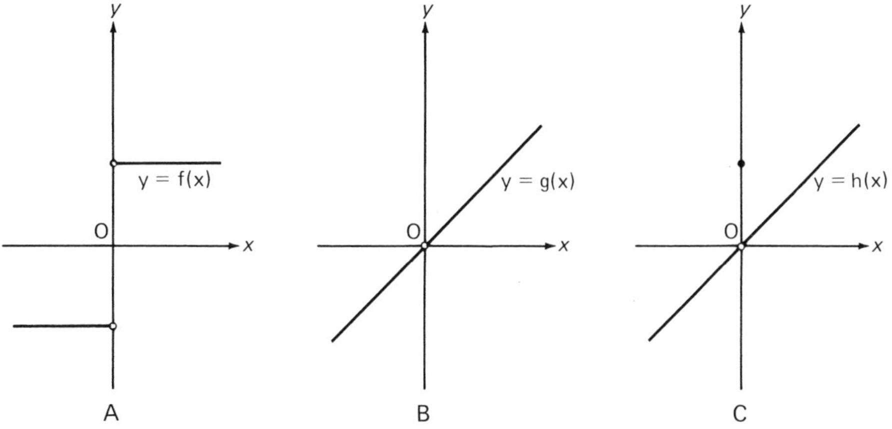

Figure 11.

which is still discontinuous at $x = 0$, since

$$\lim_{x \to 0} h(x) = 0 \neq 1 = h(0).$$

g. To get a better idea of the behavior of discontinuous functions, we graph the three functions (9), (10) and (11) in Figure 11, using hollow dots to indicate "missing points." The solid dot in Figure 11C indicates the "isolated point" $(0, 1)$ belonging to the graph of the function (11). There is something about all these graphs that shows us at a glance that each is the graph of a discontinuous function, namely each graph behaves "pathologically" at $x = 0$. Either the graph has no point at all with abscissa 0, as in Figures 11A and 11B, or there is such a point, as in Figure 11C, but it does not fall where it "ought to," namely at the origin. Moreover, the graph in Figure 11A has a "jump discontinuity" at $x = 0$, since a point moving along the graph of $f(x)$ from left to right has to make a "sudden jump" at $x = 0$ in order to get from the line $y = -1$ to the line $y = 1$. The graph of a function which is continuous in some interval cannot have "gaps" and "jumps" like these. Thus you can think of the graph of a continuous function as one which can be drawn without lifting pen from paper.

2.65. One-sided limits

a. Taking another look at Figure 11A, we are led at once to the idea of a *one-sided limit*, for we can hardly help noticing that the function $y = f(x)$ would have a limit at $x = 0$, equal to either -1 or 1, if we were to insist that x approach the origin O *from one side or the other*, either from the left of O, taking only negative values, or from the right of O, taking only positive values. In fact, under these circumstances, we could forget about the behavior of the function $y = f(x)$ on the other side of O, regarding $y = f(x)$ as either the constant function $y \equiv -1$ on the left of O or the constant function $y \equiv 1$ on the right of O. If x approaches a point x_0 from the left, we write $x \to x_0-$, while if x approaches x_0 from the right, we write $x \to x_0+$. Thus we have just observed that in the case of the function (9), $f(x) \to -1$ as $x \to 0-$ and $f(x) \to 1$ as $x \to 0+$, or equivalently

$$\lim_{x \to 0-} f(x) = -1, \qquad \lim_{x \to 0+} f(x) = 1, \tag{12}$$

where the first limit is called the *left-hand limit* of $f(x)$ at $x = 0$ and the second is called the *right-hand limit* of $f(x)$ at $x = 0$.

b.   As you may have already guessed, there is also a kind of continuity involving one-sided limits. In fact, if

$$\lim_{x \to x_0 -} f(x) = f(x_0),$$

we say that $f(x)$ is *continuous from the left at* x_0, while if

$$\lim_{x \to x_0 +} f(x) = f(x_0),$$

we say that $f(x)$ is *continuous from the right at* x_0. For example, the function $f(x)$ defined by (9) has the one-sided limits (12) at the point $x = 0$, so that $f(x)$ can be made continuous from the left at $x = 0$ by setting $f(0) = -1$ or continuous from the right by setting $f(0) = 1$ (remember that $f(0)$ was not defined originally). However, there is clearly no way to make this function continuous both from the left and from the right at $x = 0$, since the one-sided limits (12) are *different*.

c.   We now make a small improvement in our definition of a function which is continuous in an interval I. In Sec. 2.63b we insisted that such a function be continuous at every point of I. We now relax this requirement a bit, but *only* at the end points of I. Suppose I contains its *left* end point, call it a. Then we require only that $f(x)$ be continuous *from the right* at a. This makes sense, since a point which stays inside I can only approach a from the right. Similarly, if I contains its *right* end point, call it b, we now require only that $f(x)$ be continuous *from the left* at b. Again this makes sense, since a point inside I can only approach b from the left.

For example, the function

$$f(x) = \begin{cases} -1 & \text{if } x < 0, \\ 1 & \text{if } x \geqslant 0, \end{cases}$$

differing from (9) only by having $x \geqslant 0$ instead of $x > 0$, is regarded as continuous in the closed interval $0 \leqslant x \leqslant 1$, even though it is not continuous at $x = 0$, because it is continuous at every point $x > 0$ and continuous *from the right* at $x = 0$.

2.66.   We have already encountered a function $|x|$ which fails to have a derivative at a point (namely $x = 0$) where it is continuous. On the other hand, *a function is automatically continuous at any point where it has a derivative.* In fact, suppose $f(x)$ has a derivative $f'(x_0)$ at a point x_0. Then

$$\lim_{\Delta x \to 0} [f(x_0 + \Delta x) - f(x_0)] = \lim_{\Delta x \to 0} \frac{f(x_0 + \Delta x) - f(x_0)}{\Delta x} \Delta x$$

$$= \lim_{\Delta x \to 0} \frac{f(x_0 + \Delta x) - f(x_0)}{\Delta x} \lim_{\Delta x \to 0} \Delta x$$

$$= f'(x_0) \lim_{\Delta x \to 0} \Delta x,$$

with the help of Theorem 2.61 and the definition of the derivative $f'(x_0)$. But

$$\lim_{\Delta x \to 0} \Delta x = 0,$$

and therefore

$$\lim_{\Delta x \to 0} [f(x_0 + \Delta x) - f(x_0)] = 0,$$

or equivalently

$$\lim_{\Delta x \to 0} f(x_0 + \Delta x) = f(x_0),$$

since $f(x_0)$ is a constant. But this is just another way of writing the formula

$$\lim_{x \to x_0} f(x) = f(x_0),$$

expressing the continuity of $f(x)$ at x_0 (let $x = x_0 + \Delta x$, $\Delta x = x - x_0$).

PROBLEMS

1. Deduce from (4) that if $f(x) \to A$ as $x \to x_0$, then $cf(x) \to cA$ as $x \to x_0$, where c is an arbitrary constant.

2. Show that if $f_1(x) \to A_1$, $f_2(x) \to A_2, \ldots, f_n(x) \to A_n$ as $x \to x_0$, then $f_1(x) + f_2(x) + \cdots + f_n(x) \to A_1 + A_2 + \cdots + A_n$ and $f_1(x)f_2(x)\cdots f_n(x) \to A_1 A_2 \cdots A_n$ as $x \to x_0$.

3. What goes wrong in Example 2.62b if we try to evaluate the limit A directly by using Theorem 2.61, without making the preliminary factorization?

4. Evaluate

    (a) $\lim_{x \to 2} \dfrac{x^2 - 5x + 6}{x^2 - 8x + 15}$;   (b) $\lim_{x \to 0} \dfrac{x^2 - 2x - 3}{x^2 + 2x + 1}$;

    (c) $\lim_{x \to 0} \dfrac{x^3 - 3x + 2}{x^4 - 4x + 3}$;   (d) $\lim_{x \to 1} \dfrac{(1 + x)(1 + 2x)(1 + 3x) - 1}{x}$.

5. At what points does the function

    $$f(x) = \frac{x^2 - 1}{x^2 - 3x + 2}$$

    fail to be continuous?

6. What choice of $f(0)$ makes the function

    $$f(x) = \frac{5x^2 - 3x}{2x} \qquad (x \neq 0)$$

    continuous at $x = 0$?

7. Find the one-sided limits at $x = 2$ of the function

    $$f(x) = \begin{cases} x^2 - 1 & \text{if } 1 \leqslant x < 2, \\ 2x + 1 & \text{if } 2 \leqslant x \leqslant 3. \end{cases}$$

8. Graph the function $f(x) = [x]$, where $[x]$ is the integral part of x (Sec. 1.4, Prob. 10). At what points is $f(x)$ discontinuous?

9. Verify that the function $f(x) = [x]$ is continuous from the right at every point of the real line.

10. Is the function

    $$f(x) = \begin{cases} 2x & \text{if } 0 \leqslant x \leqslant 1, \\ 2 - x & \text{if } 1 < x \leqslant 2 \end{cases}$$

    continuous in the interval $0 \leqslant x \leqslant 2$? In the interval $0 \leqslant x \leqslant 1$? In $0 < x < 2$?

11. Show that the ordinary limit

    $$\lim_{x \to x_0} f(x)$$

exists if both one-sided limits

$$\lim_{x \to x_0 -} f(x), \qquad \lim_{x \to x_0 +} f(x)$$

exist and are equal, and conversely. Show that if the ordinary limit exists, then

$$\lim_{x \to x_0} f(x) = \lim_{x \to x_0 -} f(x) = \lim_{x \to x_0 +} f(x).$$

12. Show that $f(x)$ is continuous at x_0 if $f(x)$ is continuous both from the left and from the right at x_0, and conversely.

13. Do the considerations of Sec. 2.65c apply to *open* intervals?

*14. Show that if $f(x)$ is continuous at x_0, then so is $|f(x)|$.

*15. Prove formulas (4) and (5) with the help of Sec. 2.4, Problems 14 and 15.

2.7 DIFFERENTIATION TECHNIQUE

So far we have only calculated the derivatives of a few functions, resorting each time to the definition of the derivative as a limit. This is, of course, very inefficient, and what we really want are ways to evaluate derivatives simply and methodically, without the need to always go back to first principles. To this end, we now prove a number of easy theorems, each establishing an important differentiation rule. As in Sec. 2.55a, we will avoid the use of subscripts to keep the notation as simple as possible. For the same reason, we will often leave out arguments of functions, writing f instead of $f(x)$, F instead of $F(x)$, and so on.

2.71. a. Theorem (Derivative of a sum or difference). *If the functions f and g are both differentiable at x, then so is the sum $F = f + g$ and the difference $G = f - g$. The derivatives of F and G at x are given by*

$$F'(x) = f'(x) + g'(x), \tag{1}$$

$$G'(x) = f'(x) - g'(x), \tag{2}$$

or equivalently by

$$\frac{dF(x)}{dx} = \frac{df(x)}{dx} + \frac{dg(x)}{dx},$$

$$\frac{dG(x)}{dx} = \frac{df(x)}{dx} - \frac{dg(x)}{dx}$$

in the "d notation."

Proof. By the definition of a derivative, we have

$$F'(x) = \lim_{\Delta x \to 0} \frac{F(x + \Delta x) - F(x)}{\Delta x} = \lim_{\Delta x \to 0} \frac{f(x + \Delta x) + g(x + \Delta x) - f(x) - g(x)}{\Delta x}$$

$$= \lim_{\Delta x \to 0} \left[\frac{f(x + \Delta x) - f(x)}{\Delta x} + \frac{g(x + \Delta x) - g(x)}{\Delta x} \right].$$

But "the limit of a sum is the sum of the limits" (Theorem 2.61), and therefore

$$F'(x) = \lim_{\Delta x \to 0} \frac{f(x + \Delta x) - f(x)}{\Delta x} + \lim_{\Delta x \to 0} \frac{g(x + \Delta x) - g(x)}{\Delta x} = f'(x) + g'(x),$$

which proves (1). The proof of (2) is virtually the same. $\square$

b. By an *algebraic sum* we mean a sum whose terms can have either sign.

COROLLARY. *If the functions $f_1, f_2, \ldots, f_n$ are all differentiable at x, then so is the algebraic sum $F = f_1 \pm f_2 \pm \cdots \pm f_n$. The derivative of F at x is given by*

$$F'(x) = f'_1(x) \pm f'_2(x) \pm \cdots \pm f'_n(x), \tag{3}$$

or equivalently by

$$\frac{dF(x)}{dx} = \frac{df_1(x)}{dx} \pm \frac{df_2(x)}{dx} \pm \cdots \pm \frac{df_n(x)}{dx}.$$

Proof. Here we can choose any combination of pluses and minuses in $F = f_1 \pm f_2 \pm \cdots \pm f_n$, just as long as we pick the same combination in (3). To prove (3), we merely apply Theorem 2.71a repeatedly. For example, if $F = f_1 + f_2 - f_3$, then $F = g - f_3$, where $g = f_1 + f_2$, and therefore

$$F'(x) = g'(x) - f'_3(x), \qquad g'(x) = f'_1(x) + f'_2(x),$$

which together imply

$$F'(x) = f'_1(x) + f'_2(x) - f'_3(x),$$

and similarly for other combinations. $\square$

In other words, to calculate the derivative of the algebraic sum of two or more differentiable functions, we differentiate the sum "term by term." This fact can be expressed even more simply by writing

$$(f_1 \pm f_2 \pm \cdots \pm f_n)' = f'_1 \pm f'_2 \pm \cdots \pm f'_n.$$

2.72. a. THEOREM **(Derivative of a product).** *If the functions f and g are both differentiable at x, then so is the product $F = fg$. The derivative of F at x is given by*

$$F'(x) = f'(x)g(x) + f(x)g'(x), \tag{4}$$

or equivalently by

$$\frac{dF(x)}{dx} = \frac{df(x)}{dx} g(x) + f(x) \frac{dg(x)}{dx}.$$

Proof. Again, by definition,

$$F'(x) = \lim_{\Delta x \to 0} \frac{F(x + \Delta x) - F(x)}{\Delta x} = \lim_{\Delta x \to 0} \frac{f(x + \Delta x)g(x + \Delta x) - f(x)g(x)}{\Delta x}$$

$$= \lim_{\Delta x \to 0} \frac{f(x + \Delta x)g(x + \Delta x) - f(x)g(x + \Delta x) + f(x)g(x + \Delta x) - f(x)g(x)}{\Delta x}$$

$$= \lim_{\Delta x \to 0} \frac{f(x + \Delta x) - f(x)}{\Delta x} g(x + \Delta x) + \lim_{\Delta x \to 0} f(x) \frac{g(x + \Delta x) - g(x)}{\Delta x}$$

$$= \lim_{\Delta x \to 0} \frac{f(x + \Delta x) - f(x)}{\Delta x} \lim_{\Delta x \to 0} g(x + \Delta x) + \lim_{\Delta x \to 0} f(x) \lim_{\Delta x \to 0} \frac{g(x + \Delta x) - g(x)}{\Delta x},$$

where we repeatedly use Theorem 2.61. It follows that

$$F'(x) = f'(x) \lim_{\Delta x \to 0} g(x + \Delta x) + f(x)g'(x), \tag{5}$$

where we use the definitions of the derivatives $f'(x)$ and $g'(x)$, as well as the fact that $f(x)$ is a constant in this calculation. But g is continuous at x, by Sec. 2.66, and therefore

$$\lim_{\Delta x \to 0} g(x + \Delta x) = g(x). \tag{6}$$

Substituting (6) into (5), we get the desired formula (4). $\square$

b. COROLLARY.   *If the functions $f_1, f_2, \ldots, f_n$ are all differentiable at x, then so is the product $F = f_1 f_2 \cdots f_n$. The derivative of F is given by*

$$F'(x) = f'_1(x)f_2(x) \cdots f_n(x) + f_1(x)f'_2(x) \cdots f_n(x) + \cdots + f_1(x)f_2(x) \cdots f'_n(x),$$

or equivalently by

$$\frac{dF(x)}{dx} = \frac{df_1(x)}{dx} f_2(x) \cdots f_n(x) + f_1(x) \frac{df_2(x)}{dx} \cdots f_n(x) + \cdots + f_1(x)f_2(x) \cdots \frac{df_n(x)}{dx}.$$

Proof.   This time we apply Theorem 2.72a repeatedly. For example, if $F = f_1 f_2 f_3$, then $F = gf_3$, where $g = f_1 f_2$. Therefore, by two applications of the theorem,

$$
\begin{aligned}
F'(x) &= g'(x)f_3(x) + g(x)f'_3(x) \\
&= [f'_1(x)f_2(x) + f_1(x)f'_2(x)]f_3(x) + f_1(x)f_2(x)f'_3(x) \\
&= f'_1(x)f_2(x)f_3(x) + f_1(x)f'_2(x)f_3(x) + f_1(x)f_2(x)f'_3(x). \quad \square
\end{aligned}
$$

In other words, to calculate the derivative of the product of two or more differentiable functions, we add the result of differentiating the first factor and leaving the other factors alone to the result of differentiating the second factor and leaving the others alone, and then if necessary we add this sum to the result of differentiating the third factor and leaving the others alone, and so on until all the factors have been differentiated.

2.73. THEOREM **(Derivative of a quotient).**   *If the functions f and g are both differentiable at x, then so is the quotient $F = f/g$, provided that $g(x) \neq 0$. The derivative of F at x is given by*

$$F'(x) = \frac{f'(x)g(x) - f(x)g'(x)}{g^2(x)}, \tag{7}$$

or equivalently by

$$\frac{dF(x)}{dx} = \frac{\dfrac{df(x)}{dx} g(x) - f(x) \dfrac{dg(x)}{dx}}{g^2(x)}.$$

Proof.   This time we have

$$
F'(x) = \lim_{\Delta x \to 0} \frac{F(x + \Delta x) - F(x)}{\Delta x} = \lim_{\Delta x \to 0} \frac{\dfrac{f(x + \Delta x)}{g(x + \Delta x)} - \dfrac{f(x)}{g(x)}}{\Delta x}
$$

$$
= \lim_{\Delta x \to 0} \frac{f(x + \Delta x)g(x) - f(x)g(x) + f(x)g(x) - f(x)g(x + \Delta x)}{\Delta x g(x)g(x + \Delta x)}
$$

$$
\doteq \lim_{\Delta x \to 0} \frac{\dfrac{f(x + \Delta x) - f(x)}{\Delta x} g(x) - f(x) \dfrac{g(x + \Delta x) - g(x)}{\Delta x}}{g(x)g(x + \Delta x)}
$$

$$
= \frac{\displaystyle\lim_{\Delta x \to 0} \frac{f(x + \Delta x) - f(x)}{\Delta x} g(x) - \lim_{\Delta x \to 0} f(x) \frac{g(x + \Delta x) - g(x)}{\Delta x}}{\displaystyle\lim_{\Delta x \to 0} g(x)g(x + \Delta x)}, \tag{8}
$$

where Theorem 2.61 has been used twice. Taking the limits called for in (8) and using formula (6) again, we get the desired formula (7), where, of course, $g^2(x)$ is shorthand for $[g(x)]^2$. □

2.74. Examples

a.   We have already shown in Examples 2.43a and 2.43b that

$$\frac{dc}{dx} = 0, \qquad \frac{dx}{dx} = 1, \qquad \frac{dx^2}{dx} = 2x, \tag{9}$$

where c is an arbitrary constant. Of course, each of these formulas is an almost effortless application of the definition of the derivative as a limit.

b.   If c is an arbitrary constant, then

$$\frac{d}{dx} cf(x) = c\frac{df(x)}{dx}. \tag{10}$$

SOLUTION.   By Theorem 2.72a,

$$\frac{d}{dx} cf(x) = \frac{dc}{dx} f(x) + c\frac{df(x)}{dx}.$$

But this immediately implies (10), since $dc/dx = 0$, by the first of the formulas (9).

c.   Show that

$$\frac{d}{dx} x^n = nx^{n-1} \qquad (n = 1, 2, \dots), \tag{11}$$

where $x^0 = 1$, by definition.

SOLUTION.   Formula (11) holds for $n = 1$ and $n = 2$. In fact, for these values it reduces to the last two of the formulas (9). Suppose (11) holds for $n = k$, so that

$$\frac{d}{dx} x^k = kx^{k-1}.$$

Then, by Theorem 2.72a,

$$\frac{d}{dx} x^{k+1} = \frac{d}{dx}(x^k \cdot x) = \frac{dx^k}{dx} x + x^k \frac{dx}{dx} = kx^{k-1} \cdot x + x^k \cdot 1$$

$$= kx^k + x^k = (k + 1)x^k.$$

Thus if formula (11) holds for $n = k$, it also holds for $n = k + 1$. But the formula holds for $n = 1$ (or, for that matter, for $n = 2$), and therefore it holds for all $n = 1, 2, \dots$, by mathematical induction (Sec. 1.37).

d.   Differentiate the polynomial

$$P(x) = a_0 + a_1 x + a_2 x^2 + \cdots + a_n x^n.$$

SOLUTION.   Using the corollary to Theorem 2.71a, together with formulas (10) and (11), we get

$$\frac{dP(x)}{dx} = a_1 + 2a_2 x + \cdots + na_n x^{n-1}.$$

This is, of course, another polynomial, whose degree is one less than that of the original polynomial $P(x)$.

e. Show that

$$\frac{d}{dx}\frac{1}{x^n} = -\frac{n}{x^{n+1}} \qquad (x \neq 0).$$ (12)

SOLUTION. Using Theorem 2.73, we have

$$\frac{d}{dx}\frac{1}{x^n} = \frac{x^n \dfrac{d}{dx} 1 - 1 \cdot \dfrac{d}{dx} x^n}{x^{2n}} = -\frac{nx^{n-1}}{x^{2n}} = -\frac{n}{x^{n+1}},$$

with the help of (11). Suppose we set

$$x^{-n} = \frac{1}{x^n} \qquad (n = 1, 2, \ldots),$$ (13)

in accordance with the usual definition of negative powers. Then (12) can be written in the form

$$\frac{d}{dx} x^{-n} = -nx^{-n-1} \qquad (x \neq 0).$$

But, apart from the necessary stipulation that $x \neq 0$, this is just formula (11) with $-n$ instead of n. Thus we see that (11) remains valid for *negative* integers. Note that (11) also holds for $n = 0$ (and $x \neq 0$), since it then reduces to the formula

$$\frac{d}{dx} x^0 = \frac{d}{dx} 1 = 0 \cdot x^{-1} = 0,$$

which merely expresses the fact that the derivative of the constant 1 equals 0.

We have just shown that formula (11) is valid for any integer n, positive, negative or zero. Remarkably enough, it can be shown that (11) remains valid even when n is an arbitrary real number. This is worth writing down as a separate formula:

$$\frac{d}{dx} x^r = rx^{r-1} \qquad (r \text{ real}).$$ (14)

Here, of course, we assume that x^r and x^{r-1} are both defined. If r is irrational, we must require that $x > 0$, but if r is rational, x^r and x^{r-1} may well be defined for all x or for all $x \neq 0$ (see Probs. 5–7). We will use formula (14) freely from now on, and it should be committed to memory. The proof of (14), and of formulas (15) and (16) below, as well as the reason for the requirement $x > 0$ if r is irrational, will be given in Sec. 4.45, where we will decide just what is meant by x^r in the first place! Formula (14) is used in conjunction with the natural extension of (13) to the case of an arbitrary real number r:

$$x^{-r} = \frac{1}{x^r} \qquad (r \text{ real}).$$ (15)

f. Differentiate $\sqrt{x}$.

SOLUTION. First we note that $\sqrt{x} = x^{1/2}$. To see this, we use the formula

$$(x^r)^s = x^{rs},$$ (16)

valid for arbitrary real numbers r and s. Choosing $r = \frac{1}{2}$, $s = 2$ in (16), we find that

$$(x^{1/2})^2 = x,$$

which implies $x^{1/2} = \sqrt{x}$.

Applying formula (14), with $r = \frac{1}{2}$, we get

$$\frac{d}{dx} \sqrt{x} = \frac{d}{dx} x^{1/2} = \frac{1}{2} x^{-1/2} = \frac{1}{2} \frac{1}{x^{1/2}},$$

where in the last step we use (15). Thus, finally,

$$\frac{d}{dx} \sqrt{x} = \frac{1}{2\sqrt{x}} \qquad (x > 0).$$

2.75. Higher derivatives

a. Let $f(x)$ be differentiable in an interval I, with derivative $f'(x)$, and suppose $f'(x)$ is itself differentiable in I. Then the function

$$\frac{df'(x)}{dx} = (f'(x))'$$

is called the *second derivative* of $f(x)$, written $f^{(2)}(x)$ or $f''(x)$. Similarly, if $f''(x)$ is differentiable in I, the function

$$\frac{df''(x)}{dx} = (f''(x))'$$

is called the *third derivative* of $f(x)$, written $f^{(3)}(x)$ or $f'''(x)$. More generally, by the *derivative of order n of $f(x)$*, or briefly the *nth derivative of $f(x)$*, denoted by $f^{(n)}(x)$, we mean the function

$$\frac{df^{(n-1)}(x)}{dx} = (f^{(n-1)}(x))',$$

assuming that the derivative $f^{(n-1)}(x)$ of order $n - 1$ exists and is itself differentiable in I. We also write

$$f(x) = f^{(0)}(x),$$

that is, $f(x)$ is the result of not differentiating $f(x)$ at all!

b. In terms of the "d notation," $f^{(n)}(x)$ is written as

$$f^{(n)}(x) = \frac{d^n f(x)}{dx^n} = \frac{d^n}{dx^n} f(x).$$

Note that in the numerator the exponent n is attached to the symbol d, while in the denominator it is attached to the independent variable x. The expression d^n/dx^n should be thought of as a single entity calling for n-fold differentiation of any function written after it. Similarly, $d^n f(x)/dx^n$ should be regarded as just another way of writing $f^{(n)}(x)$, without attempting to ascribe separate meaning to the different symbols making up the expression. Higher derivatives of the dependent variable are defined in the natural way. Thus, if $y = f(x)$, we have

$$y' = \frac{dy}{dx} = f'(x), \qquad y'' = \frac{d^2 y}{dx^2} = f''(x), \ldots, \qquad y^{(n)} = \frac{d^n y}{dx^n} = f^{(n)}(x).$$

c. Example. If $y = x^4$, then

$$\frac{dy}{dx} = 4x^3, \quad \frac{d^2y}{dx^2} = \frac{d}{dx} 4x^3 = 12x^2, \quad \frac{d^3y}{dx^3} = \frac{d}{dx} 12x^2 = 24x,$$

$$\frac{d^4y}{dx^4} = \frac{d}{dx} 24x = 24, \quad \frac{d^5y}{dx^5} = \frac{d}{dx} 24 = 0.$$

It is clear that in this case all derivatives of order $n > 5$ also equal zero. Note that the fourth derivative of x^4 equals $4 \cdot 3 \cdot 2 \cdot 1 = 24$. More generally,

$$\frac{d^n}{dx^n} x^n = \frac{d^{n-1}}{dx^{n-1}} nx^{n-1} = \frac{d^{n-2}}{dx^{n-2}} n(n-1)x^{n-2}$$

$$= \cdots = \frac{d}{dx} n(n-1) \cdots 2x = n(n-1) \cdots 2 \cdot 1 = n!,$$

where we use the symbol $n!$, pronounced "n factorial," as shorthand for the product of the first n positive integers.

PROBLEMS

1. Differentiate

   (a) $x^4 + 3x^2 - 6$;   (b) $2ax^3 - bx^2 + c$;   (c) $x - \dfrac{1}{x}$;

   (d) $\dfrac{1}{x} + \dfrac{2}{x^2} + \dfrac{3}{x^3}$.

2. Differentiate

   (a) $(x - a)(x - b)$;   (b) $x(x - a)(x - b)$;   (c) $(1 + 4x^2)(1 + 2x^2)$;
   (d) $(2x - 1)(x^2 - 6x + 3)$.

3. Differentiate

   (a) $\dfrac{x - a}{x + a}$;   (b) $\dfrac{2x}{1 - x^2}$;   (c) $\dfrac{x^2 - 5x}{x^3 + 3}$;   (d) $\dfrac{x^3 + 1}{x^2 - x - 2}$.

4. "The derivative of a rational function is also a rational function." True or false?

5. Given any positive integer n, by the *nth root* of x, denoted by $\sqrt[n]{x}$ or $x^{1/n}$ (with the conventions $\sqrt[1]{x} = x$, $\sqrt[2]{x} = \sqrt{x}$), we mean either the unique *nonnegative* number whose nth power equals x if n is even, or the unique number (possibly negative) whose nth power equals x if n is odd. Show that this definition is in keeping with formula (16). Show that if n is odd, then $\sqrt[n]{x}$ is defined for all x and is an odd function, while if n is even, then $\sqrt[n]{x}$ is defined only for $x \geqslant 0$.

6. Given any positive integers m and n, where the fraction m/n is in lowest terms, let

$$x^{m/n} = \sqrt[n]{x^m} = (x^m)^{1/n},$$

by *definition*. Show that this is in keeping with formula (16). Show that if n is odd, then $x^{m/n}$ is defined for all x and is an even function if m is even and an odd function if m is odd, while if n is even, then $x^{m/n}$ is defined only for $x \geqslant 0$.

7. Let $x^{m/n}$ be the same as in the preceding problem, and let

$$x^{-m/n} = \frac{1}{x^{m/n}},$$

by definition. Show that this is in keeping with formula (15). Show that if n is odd, then $x^{-m/n}$ is defined for all $x \neq 0$ and is an even function if m is even and an odd function if m is odd, while if n is even, then $x^{-m/n}$ is defined only for $x > 0$.

8. Differentiate

   (a) $\sqrt[3]{x}$;   (b) $\dfrac{1}{\sqrt[3]{x}}$;   (c) $\sqrt{x^3}$;   (d) $\sqrt[3]{x^2}$.

9. "The nth derivative of a polynomial of degree n is a nonzero constant." True or false?

10. Find the first n derivatives of the function $y = 1/x$.

11. Given two functions f and g with third derivatives, evaluate $(fg)'''$.

12. Let $y = x(2x - 1)^2(x + 3)^3$. Find $y^{(6)}$ and $y^{(7)}$ with as little work as possible.

*13. Show that the segment of any tangent to the curve $y = 1/x$ cut off by the coordinate axes is bisected by the point of tangency.

*14. Why are the denominators in (8) all nonzero, as required?

2.8 FURTHER DIFFERENTIATION TECHNIQUE

2.81. a.   The concept of an inverse function was introduced in Sec. 2.16. The next rule shows how to express the derivative of an inverse function in terms of the derivative of the original function.

THEOREM.   *Let f be a one-to-one function with inverse $g = f^{-1}$. Suppose f is differentiable at x, with derivative $f'(x) \neq 0$, and suppose g is continuous at $y = f(x)$. Then g is differentiable at y, with derivative*

$$g'(y) = \frac{1}{f'(x)}. \tag{1}$$

Proof. If

$$y = f(x), \qquad y + \Delta y = f(x + \Delta x),$$

then

$$x = g(y), \qquad x + \Delta x = g(y + \Delta y),$$

so that, in particular,

$$\Delta x = g(y + \Delta y) - g(y), \qquad \Delta y = f(x + \Delta x) - f(x)$$

$$\frac{g(y + \Delta y) - g(y)}{\Delta y} = \frac{\Delta x}{f(x + \Delta x) - f(x)}.$$

Since g is continuous at y, $\Delta y \to 0$ implies $\Delta x \to 0$. But then

$$g'(y) = \lim_{\Delta y \to 0} \frac{g(y + \Delta y) - g(y)}{\Delta y} = \lim_{\Delta x \to 0} \frac{\Delta x}{f(x + \Delta x) - f(x)}$$

where the denominator $f(x + \Delta x) - f(x)$ cannot vanish since f is one-to-one. Therefore

$$g'(y) = \lim_{\Delta x \to 0} \frac{1}{\dfrac{f(x + \Delta x) - f(x)}{\Delta x}} = \frac{1}{\displaystyle\lim_{\Delta x \to 0} \frac{f(x + \Delta x) - f(x)}{\Delta x}} = \frac{1}{f'(x)}. \qquad \square$$

b. In the "d notation," (1) becomes

$$\frac{dg(y)}{dy} = \frac{df^{-1}(y)}{dy} = \frac{1}{\dfrac{df(x)}{dx}}.$$

More concisely, we have

$$\frac{dx}{dy} = \frac{1}{\dfrac{dy}{dx}},$$

or equivalently

$$\frac{dy}{dx} = \frac{1}{\dfrac{dx}{dy}}, \qquad \frac{dy}{dx}\frac{dx}{dy} = 1,$$

in terms of the variables $y = f(x)$ and $x = g(y)$. All three formulas resemble algebraic identities, but they do not, of course, constitute a proof of our theorem. They do show, however, that the "d notation" is so apt that it tends to suggest true theorems!

c. In order to use Theorem 2.81a, we must somehow know that the inverse function $g = f^{-1}$ is continuous at x, so that $\Delta y \to 0$ will imply $\Delta x \to 0$. In every case of interest, this will follow from the following fact, a complete proof of which is beyond the scope of this book (for the easy part of the proof, see Sec. 2.3, Probs. 15 and 16): *If f is continuous and one-to-one in a closed interval $[a, b]$, then there are only two possibilities*:

(1) f *is increasing in $[a, b]$ and its inverse function f^{-1} is increasing and continuous in the closed interval $[f(a), f(b)]$*;
(2) f *is decreasing in $[a, b]$ and its inverse function f^{-1} is decreasing and continuous in the closed interval $[f(b), f(a)]$.*

The meaning of this assertion is illustrated by Figure 12A for the case of increasing f (and f^{-1}) and by Figure 12B for the case of decreasing f (and f^{-1}).

    d. Example. The function

$$y = f(x) = x^2$$

is one-to-one and continuous in every closed interval $a \leqslant x \leqslant b$, where $a \geqslant 0$, by Examples 2.16c and 2.64b, and we already know that $f(x)$ is increasing in $a \leqslant x \leqslant b$, since $0 \leqslant x_1 < x_2$ implies $x_1^2 < x_2^2$ (why?). It follows from the italicized assertion that the inverse function

$$x = f^{-1}(y) = \sqrt{y}$$

is increasing and continuous in the interval $a^2 \leqslant y \leqslant b^2$. In particular, f^{-1} is continuous at every point $y \geqslant 0$, since every such point belongs to an interval of the type $a^2 \leqslant y \leqslant b^2$, where $a \geqslant 0$.

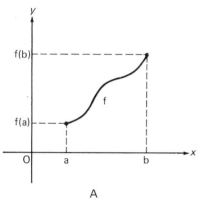

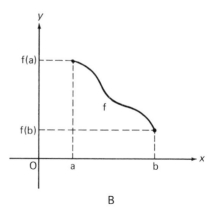

A   B

Figure 12.

Having proved the continuity of $x = \sqrt{y}$, we can now use Theorem 2.81a to differentiate $\sqrt{y}$, without recourse to formula (14), p. 78. In fact,

$$\frac{d}{dy}\sqrt{y} = \frac{dx}{dy} = \frac{1}{\dfrac{dy}{dx}} = \frac{1}{2x} = \frac{1}{2\sqrt{y}},$$

which is the same as the result of Example 2.74f, except that the roles of x and y have been reversed (why is this?).

2.82. a.   The concept of a composite function was introduced in Sec. 2.22. The next rule, one of the most important in calculus, shows how to express the derivative of a composite function in terms of the derivatives of its "constituent functions."

THEOREM.   *Let f and g be two functions such that f is differentiable at x and g is differentiable at $f(x)$. Then the composite function F, defined by $F(x) \equiv g(f(x))$, is differentiable at x, with derivative*

$$F'(x) = g'(f(x))f'(x). \tag{2}$$

Proof.   Let $y = f(x)$ and $z = g(y) = F(x)$. Since f is differentiable at x and g is differentiable at $y = f(x)$, both limits

$$f'(x) = \lim_{\Delta x \to 0} \frac{f(x + \Delta x) - f(x)}{\Delta x},$$

$$g'(y) = \lim_{\Delta y \to 0} \frac{g(y + \Delta y) - g(y)}{\Delta y}$$

exist. But then

$$\lim_{\Delta x \to 0} \left[\frac{f(x + \Delta x) - f(x)}{\Delta x} - f'(x) \right] = 0,$$

$$\lim_{\Delta y \to 0} \left[\frac{g(y + \Delta y) - g(y)}{\Delta y} - g'(y) \right] = 0,$$

or equivalently

$$\lim_{\Delta x \to 0} \alpha(\Delta x) = 0, \qquad \lim_{\Delta y \to 0} \beta(\Delta y) = 0,$$

where

$$\alpha(\Delta x) = \frac{f(x + \Delta x) - f(x)}{\Delta x} - f'(x),$$

$$\beta(\Delta y) = \frac{g(y + \Delta y) - g(y)}{\Delta y} - g'(y)$$

(3)

(recall Sec. 2.55a). Using (3) to write the increments of y and z in terms of the increments of x and y, we get

$$\Delta y = f(x + \Delta x) - f(x) = [f'(x) + \alpha(\Delta x)]\,\Delta x,$$
$$\Delta z = g(y + \Delta y) - g(y) = [g'(y) + \beta(\Delta y)]\,\Delta y.$$

The trick now is to substitute the expression for Δy into the formula for Δz. This gives

$$\Delta z = [g'(y) + \beta(\Delta y)][f'(x) + \alpha(\Delta x)]\,\Delta x,$$

(4)

which is beginning to look a little like (2). It follows from the expression for Δy (or from the continuity of f at x) that $\Delta x \to 0$ implies $\Delta y \to 0$, so that $\Delta x \to 0$ implies both $\alpha(\Delta x) \to 0$ and $\beta(\Delta y) \to 0$. Therefore, dividing (4) by Δx and taking the limit as $\Delta x \to 0$, we find that

$$\lim_{\Delta x \to 0} \frac{\Delta z}{\Delta x} = \lim_{\Delta x \to 0} [g'(y) + \beta(\Delta y)] \lim_{\Delta x \to 0} [f'(x) + \alpha(\Delta x)]$$
$$= g'(y)f'(x) = g'(f(x))f'(x).$$

(5)

At the same time,

$$\Delta z = g(y + \Delta y) - g(y) = g(f(x) + f(x + \Delta x) - f(x)) - g(f(x))$$
$$= g(f(x + \Delta x)) - g(f(x)) = F(x + \Delta x) - F(x),$$

which implies

$$\lim_{\Delta x \to 0} \frac{\Delta z}{\Delta x} = \lim_{\Delta x \to 0} \frac{F(x + \Delta x) - F(x)}{\Delta x} = F'(x).$$

(6)

Comparing (5) and (6), we immediately get (2). □

b. Thus, to differentiate the composite function $g(f(x))$, we multiply the result of differentiating g with respect to *its* argument $f(x)$ by the result of differentiating f with respect to *its* argument x. Roughly speaking, we "peel off" the layers of parentheses one by one, differentiating each function encountered on the way. This procedure applies equally well to more than two functions. For example, if $F(x) \equiv h(g(f(x)))$, then

$$F'(x) = h'(g(f(x)))\frac{d}{dx}g(f(x)) = h'(g(f(x)))g'(f(x))f'(x)$$

(7)

if f, g and h are differentiable at x, $f(x)$ and $g(f(x))$, respectively.

Theorem 2.82a is called the *chain rule*, a term suggesting the process of differentiation just described. The term is even more suggestive in the case of functions of several variables (see Sec. 6.3).

c. In the "d notation," (2) becomes

$$\frac{dz}{dx} = \frac{dz}{dy}\frac{dy}{dx},$$

in terms of the variables $y = f(x)$, $z = g(y)$. Similarly, introducing variables $y = f(x)$, $z = g(y)$, $u = h(z)$, we can write (7) in the form

$$\frac{du}{dx} = \frac{du}{dz}\frac{dz}{dy}\frac{dy}{dx}.$$

Do not make the mistake of regarding these formulas as trivial algebraic calculations, involving nothing more than cancelling dy and dz from the numerators and denominators. We have not done away with the need for proving the chain rule, but have merely written it in a very suggestive way, which, in particular, makes it very easy to remember.

   d. In connection with composite functions, it should be noted that "a continuous function of a continuous function is continuous." More exactly, *if f and g are two functions such that f is continuous at x_0 and g is continuous at $f(x_0)$, then the composite function F, defined by $F(x) \equiv g(f(x))$, is continuous at x_0.* This is easily shown with the help of "ε, δ language." Since g is continuous at $f(x_0)$, given any $\varepsilon > 0$, we can find a number $\delta_1 > 0$ such that

$$|F(x) - F(x_0)| = |g(f(x)) - g(f(x_0))| < \varepsilon \tag{8}$$

whenever $|f(x) - f(x_0)| < \delta_1$. (Note that there is now no need to require that $f(x) \neq f(x_0)$ or $x \neq x_0$, since $g(f(x_0))$ and $f(x_0)$ are defined.) But since f is continuous at x_0, we can also find a number $\delta > 0$ such that $|f(x) - f(x_0)| < \delta_1$ whenever $|x - x_0| < \delta$. Therefore (8) holds whenever $|x - x_0| < \delta$, that is, $F(x) \to F(x_0)$ as $x \to x_0$. In other words, F is continuous at x_0, as asserted.

   Thus, to prove the continuity of the function

$$F(x) = \sqrt{1 + x^2},$$

we use the continuity of $g(x) = \sqrt{x}$, established in Example 2.81d, and the continuity of $f(x) = 1 + x^2$, established in Example 2.64b, together with the observation that $F(x) \equiv g(f(x))$. In fact, $F(x)$ is continuous in the whole interval $(-\infty, \infty)$, since $1 + x^2 \geq 1$ for all x, while $\sqrt{x}$ is continuous for all $x \geq 0$.

2.83. Examples

   a. Differentiate

$$F(x) = \left(1 + \frac{1}{x^2}\right)^{100} \tag{9}$$

SOLUTION. Here $F(x) \equiv g(f(x))$, where $g(x) = x^{100}$, $f(x) = 1 + x^{-2}$. Therefore

$$g'(x) = 100x^{99}, \qquad f'(x) = -2x^{-3},$$

by Examples 2.74c and 2.74e, or, if you prefer, by formula (14), p. 78. The chain rule then gives

$$F'(x) = g'(f(x))f'(x) = 100(1 + x^{-2})^{99}(-2x^{-3}) = -\frac{200}{x^3}\left(1 + \frac{1}{x^2}\right)^{99}$$

It would be the height of folly to actually calculate the right side of (9) explicitly and then differentiate!

b. Differentiate $\sqrt{x + \sqrt{x}}$.

SOLUTION. By the chain rule,

$$\frac{d}{dx}\sqrt{x + \sqrt{x}} = \frac{1}{2\sqrt{x + \sqrt{x}}}\frac{d}{dx}(x + \sqrt{x})$$

$$= \frac{1}{2\sqrt{x + \sqrt{x}}}\left(1 + \frac{1}{2\sqrt{x}}\right) = \frac{2\sqrt{x} + 1}{4\sqrt{x}\sqrt{x + \sqrt{x}}},$$

where we use Example 2.74f twice.

c. Given a function $y = f(x)$, find the derivative of y^n.

SOLUTION. By the chain rule,

$$\frac{dy^n}{dx} = \frac{dy^n}{dy}\frac{dy}{dx} = ny^{n-1}\frac{dy}{dx},$$

or, more concisely,

$$(y^n)' = ny^{n-1}y'. \tag{10}$$

d. If

$$x^2 - xy + y^3 = 1, \tag{11}$$

find y'.

SOLUTION. Rather than solve (11) for y as a function of x and then differentiate y, we differentiate (11) with respect to x and then solve for y'. Thus

$$\frac{d}{dx}(x^2 - xy + y^3) = \frac{d}{dx}1 = 0$$

or

$$2x - y - xy' + 3y^2y' = 0,$$

with the help of (10). Solving for y', we find that

$$y' = \frac{2x - y}{x - 3y^2}. \tag{12}$$

Since we make no attempt to express y explicitly as a function of x, this process is called *implicit differentiation*. In the present case, direct calculation of y from the cubic equation (11), followed by differentiation of the resulting expression for y, would lead at once to a mass of tedious and completely unnecessary calculations.

2.84. Two remarks must be made in connection with implicit differentiation:

a. The method cannot be used blindly, since it gives a formal answer for y' even in cases where y (and hence y') fails to exist! For example, the solution set (Sec. 2.31a) of the equation

$$x^2 + y^2 = a$$

is empty if $a < 0$, and yet implicit differentiation of this equation gives

$$2x + 2yy' = 0,$$

and hence

$$y' = -\frac{x}{y},$$

regardless of the sign of a.

b. Extra work is required to evaluate the derivative y' at a particular point $x = x_0$. For example, to evaluate (12) at $x = 1$, we need the value of y at $x = 1$. Substituting $x = 1$ into (11), we get

$$1 - y + y^3 = 1$$

or

$$y^3 = y,$$

which has three solutions $y = 0$ and $y = \pm 1$. The corresponding values of y' are

$$y'|_{x=1, y=0} = \frac{2 \cdot 1 - 0}{1 - 3 \cdot 0} = 2,$$

$$y'|_{x=1, y=1} = \frac{2 \cdot 1 - 1}{1 - 3 \cdot 1} = -\frac{1}{2},$$

$$y'|_{x=1, y=-1} = \frac{2 \cdot 1 + 1}{1 - 3 \cdot 1} = -\frac{3}{2}.$$

Here, of course, $y'|_{x=x_0, y=y_0}$ stands for the value of y' corresponding to $x = x_0$, $y = y_0$. We will often find this kind of "single vertical bar notation" useful.

PROBLEMS

1. Use Theorem 2.81a to differentiate $\sqrt[3]{x}$. Why is this function continuous everywhere? Check the result by using formula (14), p. 78.
2. Which is larger, $\sqrt{3} + \sqrt{5}$ or $\sqrt{2} + \sqrt{6}$? Justify your answer.
3. Differentiate

    (a) $(x + 1)(x + 2)^2(x + 3)^3$;   (b) $\dfrac{(x + 4)^2}{x + 3}$;   (c) $\dfrac{(2 - x)(3 - x)}{(1 - x)^2}$.

4. Let $y = (2x + 3)^{100}$. Find $y'|_{x=0}$ with as little work as possible.
5. Show that "differentiation changes parity," which means that the derivative of an even function is odd, while the derivative of an odd function is even.
6. Use the chain rule and the rule for differentiating a product to deduce the rule for differentiating a quotient.
7. Differentiate

    (a) $\sqrt{x^2 + a^2}$;   (b) $\sqrt{\dfrac{1 + x}{1 - x}}$;   (c) $\dfrac{x}{\sqrt{a^2 - x^2}}$.

8. Differentiate

$$\frac{(1 - x)^r}{(1 + x)^s},$$

   where r and s are arbitrary real numbers.
9. Verify that

$$\frac{d}{dx} \frac{1}{\sqrt{1 + x^2}(x + \sqrt{1 + x^2})} = -\frac{1}{(1 + x^2)^{3/2}}.$$

10. Verify that

$$\frac{d^n}{dx^n} \frac{1}{x(1-x)} = n! \left[\frac{(-1)^n}{x^{n+1}} + \frac{1}{(1-x)^{n+1}} \right].$$

11. Where is the function $\sqrt{x + \sqrt{x}}$ continuous? How about the function $\sqrt[3]{x} - \sqrt[3]{x}$?

12. If

$$x^2 + xy + y^3 = 1, \qquad (13)$$

find y'. Evaluate $y'|_{x=1}$.

13. If

$$x^2 - xy + y^2 = 1, \qquad (14)$$

find y'. Evaluate $y'|_{x=1}$.

14. Use implicit differentiation to find y'' if $x^m y^n = 1$, where m and n are nonzero integers.

15. We have already assumed the validity of the formula

$$\frac{d}{dx} x^r = rx^{r-1}, \qquad (15)$$

where r is an arbitrary real number (this will be proved in Sec. 4.45). Use the chain rule to verify (15) for the case where r is an arbitrary *rational* number m/n, starting from the fact that $y = x^{m/n}$ is equivalent to $y = t^m$, where $t = x^{1/n}$.

16. Use implicit differentiation to verify (15) for rational r, this time starting from the fact that $y = x^{m/n}$ is equivalent to $y^n = x^m$.

*17. If $x^2 + y^2 = 25$, find the values of y', y'' and y''' at the point $(3, 4)$.

*18. Solve Problem 13 by first finding an explicit formula for y as a function of x.

*19. "If

$$x^4 + y^4 = x^2 y^2, \qquad (16)$$

find y'." Why is this an impossible assignment?

* 20. Heeding the warning in Sec. 2.84a, verify the existence of the derivatives in Sec. 2.84b and Problem 12.

2.9 OTHER KINDS OF LIMITS

2.91. Limits involving infinity

a. The graph of the function

$$y = f(x) = \frac{1}{x} \qquad (x \neq 0) \qquad (1)$$

is shown in Figure 13A. Examining this graph, we see that $f(x)$ has a number of interesting "limiting properties" of a kind not yet encountered:

(a) As x takes "smaller and smaller" positive values, y takes "larger and larger" positive values;

(b) As x takes "smaller and smaller" negative values, y takes "larger and larger" negative values;

(c) As x takes "larger and larger" positive values, y takes "smaller and smaller" positive values;

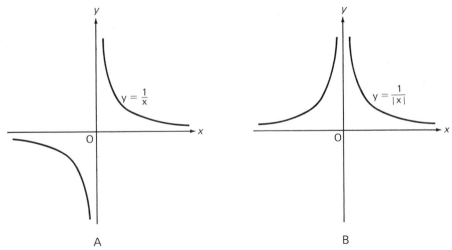

Figure 13.

(d) As x takes "larger and larger" negative values, y takes "smaller and smaller" negative values.

By a "small" or "large" negative number, we mean, of course, a negative number of "small" or "large" absolute value.

These properties of $f(x)$ all express a kind of limiting behavior in which "large-ness" plays a role, as well as "smallness." How do we modify the language of limits to cover situations of this type? Very simply. If a variable, say x, takes "larger and larger" positive values, we say that "x approaches (plus) infinity" and write $x \to \infty$, while if x takes "larger and larger" negative values, we say that "x approaches minus infinity" and write $x \to -\infty$. This is in keeping with the use of the symbols ∞ and $-\infty$ in writing infinite intervals (Sec. 1.64). Once again, we emphasize that ∞ and $-\infty$ are not numbers, so that we can never have $x = \infty$ or $x = -\infty$.

We can now express the four listed properties of the function (1) much more concisely:

(a) As $x \to 0+$, $y \to \infty$, or

$$\lim_{x \to 0+} f(x) = \infty;$$

(b) As $x \to 0-$, $y \to -\infty$, or

$$\lim_{x \to 0-} f(x) = -\infty;$$

(c) As $x \to \infty$, $y \to 0$ (more exactly, $y \to 0+$), or

$$\lim_{x \to \infty} f(x) = 0;$$

(d) As $x \to -\infty$, $y \to 0$ (more exactly, $y \to 0-$), or

$$\lim_{x \to -\infty} f(x) = 0.$$

In (a) and (b) we have "infinite limits," and in (c) and (d) we have "limits at infinity," as opposed to the "finite limits"

$$\lim_{x \to x_0} f(x) = A, \qquad \lim_{x \to x_0+} f(x) = A, \qquad \lim_{x \to x_0-} f(x) = A,$$

considered previously, where A and x_0 are both *numbers*, rather than one of the symbols ∞, $-\infty$, a fact often emphasized by calling A and x_0 "finite." There are also ordinary, "two-sided" infinite limits. For example, it is clear from Figure 13B that

$$\lim_{x \to 0} \frac{1}{|x|} = \infty.$$

We can also have infinite limits at infinity. For example, x^2 takes "arbitrarily large" positive values when x takes "arbitrarily large" values of either sign (see Figure 5, p. 49), and therefore

$$\lim_{x \to \infty} x^2 = \infty, \qquad \lim_{x \to -\infty} x^2 = \infty.$$

Similarly,

$$\lim_{x \to \infty} x^3 = \infty, \qquad \lim_{x \to -\infty} x^3 = -\infty$$

(see Figure 6, p. 50).

We will sometimes say that a variable "becomes infinite." This simply means that it approaches either (plus) infinity or minus infinity. A function $f(x)$ is said to become infinite at a point x_0 if $y = f(x)$ becomes infinite as x approaches x_0.

b. All this can be made mathematically exact by using a version of the "ε, δ language" in which letters other than ε and δ are used for numbers that are typically *large*, since ε and δ have a built-in connotation of *smallness*. For example, $f(x) \to \infty$ as $x \to x_0$ means that, given any $M > 0$, *no matter how large*, we can find a number $\delta > 0$ such that $f(x) > M$ whenever $0 < |x - x_0| < \delta$, $f(x) \to \infty$ as $x \to \infty$ means that, given any $M > 0$, we can find a *suitably large* number $L > 0$ such that $f(x) > M$ whenever $x > L$, $f(x) \to A$ as $x \to -\infty$ means that, given any number $\varepsilon > 0$, we can find a number $L > 0$ such that $|f(x) - A| < \varepsilon$ whenever $x < -L$, and so on.

c. Every problem involving infinite limits or limits at infinity can be reduced to an analogous problem involving a finite limit at a finite point. To see this, we observe that if

$$x = \frac{1}{t}, \tag{2}$$

or equivalently

$$t = \frac{1}{x},$$

then $x \to \infty$ is equivalent to $t \to 0+$, while $x \to -\infty$ is equivalent to $t \to 0-$. In fact, if x takes "larger and larger" positive values, then its reciprocal t takes "smaller and smaller" positive values, and conversely, while if x takes "larger and larger" negative values, t takes "smaller and smaller" negative values, and conversely. It follows that $f(x) \to A$ as $x \to \infty$ is equivalent to $f(1/t) \to A$ as $t \to 0+$, while $f(x) \to A$ as $x \to -\infty$ is equivalent to $f(1/t) \to A$ as $t \to 0-$. By virtually the same argument, $f(x) \to \infty$ as $x \to x_0$ is equivalent to $1/f(x) \to 0+$ as $x \to x_0$, while $f(x) \to -\infty$ as $x \to x_0$ is equivalent to $1/f(x) \to 0-$ as $x \to x_0$.

2.92. Examples

a. If

$$f(x) = \frac{x^2 + 2}{x^2 + 1},$$

find

$$\lim_{x \to \infty} f(x), \qquad \lim_{x \to -\infty} f(x).$$

SOLUTION. We make the substitution (2) and investigate the behavior of the resulting function of t as $t \to 0\pm$. Thus

$$\lim_{x \to \infty} f(x) = \lim_{t \to 0+} f\left(\frac{1}{t}\right) = \lim_{t \to 0+} \frac{\left(\frac{1}{t}\right)^2 + 2}{\left(\frac{1}{t}\right)^2 + 1} = \lim_{t \to 0+} \frac{1 + 2t^2}{1 + t^2} = 1,$$

and similarly

$$\lim_{x \to -\infty} f(x) = \lim_{t \to 0-} f\left(\frac{1}{t}\right) = \lim_{t \to 0-} \frac{1 + 2t^2}{1 + t^2} = 1.$$

This behavior as $x \to \pm\infty$ is apparent from the graph of $f(x)$, shown in Figure 14.

b. If

$$f(x) = \frac{1}{x^2 - 1},$$

find

$$\lim_{x \to 1+} f(x), \qquad \lim_{x \to 1-} f(x).$$

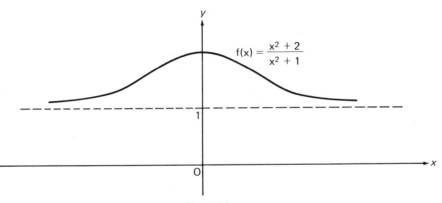

Figure 14.

SOLUTION.   Noting that $f(x)$ is undefined at $x = 1$, we go over to the function

$$g(x) = \frac{1}{f(x)} = x^2 - 1,$$

which is perfectly well-behaved at $x = 1$. Clearly $x^2 - 1 \to 0$ as $x \to 1+$, and moreover $x^2 - 1 > 0$ if $x > 1$ (why?). Therefore $g(x) = x^2 - 1 \to 0+$ as $x \to 1+$. It follows from the last sentence of Sec. 2.91c that

$$\lim_{x \to 1+} f(x) = \infty.$$

In virtually the same way, we see that $g(x) \to 0-$ as $x \to 1-$, and hence

$$\lim_{x \to 1-} f(x) = -\infty.$$

This behavior as $x \to 1\pm$ is apparent from the graph of $f(x)$, shown in Figure 15. From the graph we also deduce at a glance that

$$\lim_{x \to \infty} f(x) = \lim_{x \to -\infty} f(x) = 0, \qquad \lim_{x \to (-1)+} f(x) = -\infty, \qquad \lim_{x \to (-1)-} f(x) = \infty.$$

As this example illustrates, and as is quite generally true, a rational function approaches infinity at precisely those points where its denominator equals zero, provided, of course, that all common factors of the numerator and denominator have been cancelled out.

2.93. Asymptotes

a.   Suppose a function $f(x)$ becomes infinite at certain points, or is defined in an infinite interval, so that the argument x becomes infinite. The graph of $f(x)$ then consists of one or more parts, called "infinite branches," which "extend out to infinity" in one direction or another. For example, the function graphed in Figure 15 has three such branches, namely the part of the graph to the left of the line $x = -1$, the

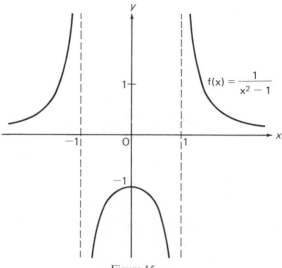

$$f(x) = \frac{1}{x^2 - 1}$$

Figure 15.

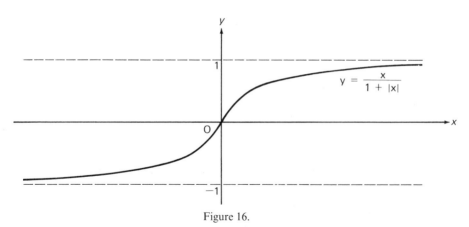

Figure 16.

part of the graph between the lines $x = -1$ and $x = 1$, and the part of the graph to the right of the line $x = 1$.

Now suppose an infinite branch of $f(x)$ approaches a straight line L (without touching it) as x approaches infinity in one or both directions, or as x approaches certain "exceptional points" from one or both sides. Then L is called an *asymptote* of $f(x)$, and the function $f(x)$, or its graph, is said to approach L *asymptotically*.

b. Horizontal asymptotes. If the horizontal line $y = y_0$ is an asymptote of $f(x)$, then the distance between the point $(x, f(x))$ and the line $y = y_0$ approaches 0 as $x \to \infty$ or as $x \to -\infty$. But this distance is just $|f(x) - y_0|$, and therefore at least one and possibly both of the formulas

$$\lim_{x \to \infty} f(x) = y_0, \qquad \lim_{x \to -\infty} f(x) = y_0$$

must hold. Thus, to find the horizontal asymptotes (if any) of $f(x)$, we need only examine the limiting behavior of $f(x)$ as $x \to \pm\infty$. For example, the line $y = 1$ is a horizontal asymptote of the function $f(x)$ graphed in Figure 14, while the function

$$y = \frac{x}{1 + |x|},$$

whose graph is the "S-shaped" curve shown in Figure 16, has two horizontal asymptotes, namely the lines $y = \pm 1$. (How can this also be seen without drawing a graph?) It is clear that a function can have no more than two horizontal asymptotes.

c. Vertical asymptotes. If the vertical line $x = x_0$ is an asymptote of $f(x)$, then the distance between the points $(x, f(x))$ and the line $x = x_0$ approaches 0 as $x \to x_0 +$ or as $x \to x_0 -$. This is automatically true for any function $f(x)$, but in the case of an asymptote, $f(x)$ must at the same time become infinite, since an asymptote is defined only for an infinite branch. Therefore, excluding the case (of no practical interest) where $f(x)$ does not stay of fixed sign as $f(x)$ approaches its asymptote, we see that at least one and possibly two of the formulas

$$\lim_{x \to x_0+} f(x) = \infty, \qquad \lim_{x \to x_0+} f(x) = -\infty, \qquad \lim_{x \to x_0-} f(x) = \infty, \qquad \lim_{x \to x_0-} f(x) = -\infty$$

must hold. Thus, in looking for the vertical asymptotes of $f(x)$, we can confine our attention to the points (if any) at which $f(x)$ becomes infinite. Note that $f(x)$ is necessarily undefined at any such point. For example, the lines $x = 1$ and $x = -1$ are vertical asymptotes of the function $f(x)$ graphed in Figure 15.

An example of a function with an asymptote which is neither horizontal nor vertical is given in Problem 20.

2.94. The limit of a sequence

a.   We say that a sequence $\{x_n\}$ *approaches (or has) a limit A as n approaches infinity* if the general term x_n gets "closer and closer" to A as n gets "larger and larger." This fact is expressed by writing

$$\lim_{n \to \infty} x_n = A, \tag{3}$$

or $x_n \to A$ as $n \to \infty$. Put somewhat differently, (3) means that $|x_n - A|$ is "arbitrarily small" for all "sufficiently large" n. Better still, in the natural analogue of the "ε, δ language," (3) means that, given any $\varepsilon > 0$, *no matter how small*, we can find an integer n_0 such that $|x_n - A| < \varepsilon$ whenever $n \geqslant n_0$, that is, for all n starting from n_0. Clearly, this also means that every ε-neighborhood of A, namely every open interval of the form $(A - \varepsilon, A + \varepsilon)$, contains all the terms of the sequence x_n starting from some value of n, where this value, of course, depends on the choice of ε. Choosing $\varepsilon = 1$, we find that all the terms of the sequence x_n starting from some value of n fall in the interval $(A - 1, A + 1)$. This fact will be used in a moment.

b.   A sequence is said to be *convergent* if it has a finite limit as $n \to \infty$ and *divergent* otherwise. If a sequence is convergent, with limit A, we also say that the sequence *converges to A*. A sequence x_n is said to be *bounded* if there is some number $M > 0$ such that $|x_n| < M$ for all $n = 1, 2, \ldots$ and *unbounded* if no such number exists. (For emphasis, we sometimes write "for all $n = 1, 2, \ldots$" instead of the equivalent phrase "for all n.") For example, the sequence $x_n = 1/n$ is bounded, since $0 < x_n \leqslant 1$ for all n, while the sequence $x_n = n$ is unbounded, since there is clearly no number $M > 0$ such that $|x_n| = n < M$ for all n.

c.   *A convergent sequence is necessarily bounded.* In fact, if $\{x_n\}$ is a convergent sequence, with limit A, then there is an integer n_0 such that all the terms $x_{n_0}, x_{n_0+1}, x_{n_0+2}, \ldots$, that is, all the terms x_n starting from n_0, lie in the interval $(A - 1, A + 1)$. By choosing $M > 0$ large enough we can see to it that the interval $(-M, M)$, with its midpoint at the origin, contains the interval $(A - 1, A + 1)$, together with the remaining terms $x_1, x_2, \ldots, x_{n_0-1}$, some or all of which may not lie in $(A - 1, A + 1)$. But then $|x_n| < M$ for all $n = 1, 2, \ldots$, so that the sequence is indeed bounded, as claimed.

Since a convergent sequence is necessarily bounded, *an unbounded sequence is necessarily divergent.*

d.   A sequence x_n is said to be *increasing* if $x_n < x_{n+1}$ for all n and *decreasing* if $x_n > x_{n+1}$ for all n. By a *monotonic sequence* we mean either an increasing sequence or a decreasing sequence. An important tool in the study of sequences is the following key proposition, whose proof lies beyond the scope of this book: *A bounded monotonic sequence is necessarily convergent.*

e.   Algebraic operations on convergent sequences obey the same rules as algebraic operations on limits of functions (why?). For example, if $x_n \to A$ and $y_n \to B$ as $n \to \infty$, then $x_n + y_n \to A + B$ and $x_n y_n \to AB$ as $n \to \infty$.

2.95. Examples

a.   The sequence

$$x_n = \frac{1}{n}$$

is convergent, with limit 0. In fact, given any $\varepsilon > 0$, let n_0 be any integer greater than $1/\varepsilon$. Then $|x_n - 0| = |x_n| = 1/n < \varepsilon$ for all $n \geq n_0$, since $1/n \leq 1/n_0 < \varepsilon$ for such n (use Theorem 1.46 twice). Note that this sequence is bounded and decreasing, so that its convergence follows from the proposition in Sec. 2.94d. However, the proposition does not tell us how to *find* the limit.

    b.   The sequence

$$x_n = n! = n(n - 1) \cdots 2 \cdot 1$$

is unbounded and hence divergent. In fact, to make $|x_n|$ larger than any given positive number M, we need only choose $n > M$, since then $|x_n| = n! > n > M$.

    c.   A bounded sequence need not be convergent. For example, the sequence

$$x_n = (-1)^n, \tag{4}$$

which looks like

$$-1, \quad 1, \quad -1, \quad 1, \quad -1, \quad 1, \ldots,$$

is obviously bounded, since $|x_n| = 1$ for all n. On the other hand, the sequence is divergent. To see this, take any proposed limit A, and make ε so small that the interval $I = (A - \varepsilon, A + \varepsilon)$ fails to contain at least one of the points 1 and -1. Clearly this can always be done, even if $A = 1$ or $A = -1$. Then all the terms of (4) with even n lie outside I if I fails to contain the point $x = 1$, while all the terms of (4) with odd n lie outside I if I fails to contain the point $x = -1$. Thus, in any event, the sequence (4) cannot be convergent.

    d.   The sequence

$$x_n = a^n \tag{5}$$

is convergent for $-1 < a < 1$. To see this, suppose first that $0 < a < 1$. Then the sequence is decreasing, since

$$x_{n+1} = a^{n+1} = ax_n < x_n$$

for all n. Moreover, the sequence is bounded, since

$$0 < x_n < x_1 = a$$

for all n. It follows from the proposition in Sec. 2.94d that the sequence is convergent. Let the limit of the sequence as $n \to \infty$ be A. To find A, we note that

$$A = \lim_{n \to \infty} x_{n+1} = \lim_{n \to \infty} ax_n = a \lim_{n \to \infty} x_n = aA.$$

But since $a \neq 1$, this is possible only if $A = 0$. Therefore

$$\lim_{n \to \infty} a^n = 0 \qquad (0 < a < 1). \tag{6}$$

    Next, if $-1 < a < 0$, we have $0 < |a| < 1$, and therefore $|a|^n$ approaches 0 as $n \to \infty$, by formula (6) with $|a|$ instead of a. But then, since $|a^n| = |a|^n$, it follows that a^n also approaches 0 as $n \to \infty$ (explain further), that is

$$\lim_{n \to \infty} a^n = 0 \qquad (-1 < a < 0).$$

Combining this with (6) and the obvious fact that the sequence a^n converges to 0 if $a = 0$, since all its terms then equal 0, we get

$$\lim_{n \to \infty} a^n = 0 \qquad (-1 < a < 1). \tag{7}$$

Finally, we note that the sequence a^n converges to 1 if $a = 1$, since all its terms then equal 1.

e. The sequence (5) is divergent for $a = -1$ and $|a| > 1$. For $a = -1$ the sequence reduces to the sequence (4), which has already been shown to be divergent. If $|a| > 1$, we first write

$$|a| = 1 + (|a| - 1).$$

Then

$$|a|^n = [1 + (|a| - 1)]^n \geqslant 1 + n(|a| - 1) > n(|a| - 1),$$

with the help of the inequality

$$(1 + x)^n \geqslant 1 + nx \qquad (x > -1),$$

proved in Problem 16. Since $|a| - 1$ is positive, the product $n(|a| - 1)$ is greater than any given number $M > 0$ for all n greater than

$$\frac{M}{|a| - 1}.$$

It follows that the sequence (5) is unbounded and hence divergent if $|a| > 1$, as claimed.

2.96. The sum of an infinite series

a. Summation notation. First we introduce a concise way of writing sums, involving the symbol $\sum$ (capital Greek sigma). Let p and q be nonnegative integers such that $p \leqslant q$, and let $f(n)$ be a function defined for all integers n from p to q. Then

$$\sum_{n=p}^{q} f(n) \tag{8}$$

is shorthand for the sum

$$f(p) + f(p + 1) + \cdots + f(q).$$

The symbol n is a "dummy index" (of summation), in the sense that it can be replaced by any other symbol without changing the meaning of (8). For example,

$$\sum_{n=0}^{3} 2^n = \sum_{k=0}^{3} 2^k = \sum_{\alpha=0}^{3} 2^\alpha = 2^0 + 2^1 + 2^2 + 2^3 = 1 + 2 + 4 + 8 = 15.$$

If $p = q$, the sum (8) reduces to the single term $f(p)$.

b. Given a sequence $\{x_n\}$, the expression

$$\sum_{n=1}^{\infty} x_n = x_1 + x_2 + \cdots + x_n + \cdots, \tag{9}$$

involving the terms of the sequence, is called an *infinite series*, or simply a *series*, with *terms* $x_1, x_2, \ldots, x_n, \ldots$ The symbol ∞ on top of the summation sign $\sum$ means that the sum on the right "goes on forever." This is also expressed by the second set of dots $\cdots$ on the right. Suppose the sequence

$$s_1 = x_1, \qquad s_2 = x_1 + x_2, \ldots, \qquad s_n = x_1 + x_2 + \cdots + x_n, \ldots \qquad (10)$$

of *partial sums* of the terms of the series (9) is convergent, with limit s. Then we say that the series (9) is *convergent*, with *sum* s. By the same token, if the sequence of partial sums (10) is divergent, we call the series (9) *divergent* and assign it no sum at all.

    c. Example. Investigate the convergence of the *geometric series*

$$\sum_{n=0}^{\infty} a^n = 1 + a + a^2 + \cdots + a^n + \cdots. \qquad (11)$$

    SOLUTION. In other words, we are asked to find the values of a for which the series (11) is convergent and the values for which it is divergent. Here the sum of the first n terms of the series is just

$$s_n = 1 + a + \cdots + a^{n-1}. \qquad (12)$$

Multiplying s_n by a, we get

$$as_n = a + a^2 + \cdots + a^n.$$

Subtracting as_n from (12), we find that all but two terms cancel out, leaving

$$s_n - as_n = (1 - a)s_n = 1 - a^n.$$

Therefore

$$s_n = \frac{1 - a^n}{1 - a}, \qquad (13)$$

provided that $a \neq 1$. It follows from Examples 2.95d and 2.95e that s_n is convergent with limit

$$\lim_{n \to \infty} \frac{1 - a^n}{1 - a} = \frac{1}{1 - a} \lim_{n \to \infty} (1 - a^n) = \frac{1}{1 - a} (1 - 0) = \frac{1}{1 - a}$$

if $-1 < a < 1$ and divergent if $a = -1$ or if $|a| > 1$. If $a = 1$, formula (13) breaks down (why?), but in this case the series (11) reduces to simply

$$1 + 1 + \cdots + 1 + \cdots, \qquad (14)$$

so that $s_n = n$. Being unbounded, the sequence s_n is divergent, and hence so is the series (14).

    Thus, to summarize, the geometric series (11) is convergent, with sum

$$\sum_{n=0}^{\infty} a^n = \frac{1}{1 - a}$$

if $-1 < a < 1$ and divergent otherwise. For example,

$$\sum_{n=0}^{\infty} \left(\frac{1}{2}\right)^n = 1 + \frac{1}{2} + \frac{1}{4} + \cdots + \frac{1}{2^n} + \cdots = \frac{1}{1 - \frac{1}{2}} = \frac{1}{\frac{1}{2}} = 2,$$

while

$$\sum_{n=0}^{\infty} \left(-\frac{1}{2}\right)^n = 1 - \frac{1}{2} + \frac{1}{4} - \cdots + \frac{(-1)^n}{2^n} + \cdots = \frac{1}{1 - (-\frac{1}{2})} = \frac{1}{\frac{3}{2}} = \frac{2}{3}.$$

PROBLEMS

1. Evaluate

    (a) $\displaystyle\lim_{x\to\infty}\frac{\sqrt{x^2+1}}{x+1}$;   (b) $\displaystyle\lim_{x\to-\infty}\frac{\sqrt{x^2+1}}{x+1}$;   (c) $\displaystyle\lim_{x\to\pm\infty}\frac{2x^2-5x+1}{5x^2+x-1}$.

2. Evaluate

    (a) $\displaystyle\lim_{x\to3-}\frac{x+3}{x^2-9}$;   (b) $\displaystyle\lim_{x\to3+}\frac{x^2+9}{x^2-9}$;   (c) $\displaystyle\lim_{x\to0+}\frac{1}{\sqrt{x}}$.

3. If

$$f(x)=\frac{1}{x(x-1)(x-2)},$$

    find all the one-sided limits of $f(x)$ at 0, 1 and 2.

4. Verify that

$$\lim_{x\to\infty}\frac{(x-1)(x-2)(x-3)(x-4)(x-5)}{(5x-1)^5}=\frac{1}{5^5}.$$

5. Evaluate

$$\lim_{x\to-\infty}\frac{(2x-3)^{20}(3x+2)^{30}}{(2x+1)^{50}}.$$

6. If

$$f(x)=\frac{x-1}{x+2},$$

    verify that $f(x)\to1$ as $x\to\infty$. Find all positive x such that $|f(x)-1|<0.001$.

7. If

$$f(x)=\frac{x-2}{2x},$$

    verify that $f(x)\to\frac{1}{2}$ as $x\to-\infty$. Find all negative x such that $|f(x)-\frac{1}{2}|<0.001$.

8. If

$$f(x)=\frac{x}{x-3},$$

    verify that $f(x)\to\infty$ as $x\to3+$ and $f(x)\to-\infty$ as $x\to3-$. Find all x such that $f(x)>1000$ and all x such that $f(x)<-1000$.

9. Find all asymptotes of the function

    (a) $y=\dfrac{x-4}{2x+4}$;   (b) $y=\dfrac{ax+b}{cx+d}$;   (c) $y=\dfrac{x^2}{x^2-4}$.

10. Given any positive integer n, find a function $f(x)$ with n vertical asymptotes.

11. Give an example of a function approaching a horizontal asymptote from one direction only.

12. Give an example of a function approaching a vertical asymptote from one side only.

13. Find the limit of the sequence

(a) $0, \dfrac{2}{3}, \dfrac{8}{9}, \ldots, \dfrac{3^n - 1}{3^n}, \ldots;$ (b) $0, 1, 0, \dfrac{1}{2}, \ldots, \dfrac{1 + (-1)^n}{n}, \ldots;$

(c) $0.2, 0.22, 0.222, \ldots, 0.\underbrace{222 \ldots 2}_{n \text{ times}}, \ldots;$ (d) $0, \dfrac{3}{2}, -\dfrac{2}{3}, \ldots, (-1)^n + \dfrac{1}{n}, \ldots$

14. Which of the following sequences are convergent?

(a) $x_n = (-1)^n n;$ (b) $x_n = n^{(-1)^n};$ (c) $x_n = n - (-1)^n;$

(d) $x_n = \begin{cases} 1 \text{ for even } n, \\ \dfrac{1}{n} \text{ for odd } n. \end{cases}$

15. Starting from what value of n are the terms of the sequence $(-\tfrac{1}{2})^n$ within 10^{-6} of its limit?

16. Use mathematical induction to verify that

$$(1 + x)^n \geqslant 1 + nx$$

for all $n = 1, 2, \ldots$ if $x > -1$.

17. Evaluate

$$\lim_{n \to \infty} \frac{a^n}{1 + a^n} \qquad (a \neq -1).$$

18. Write the following expressions out in full, and then calculate their numerical values:

(a) $\displaystyle\sum_{n=1}^{5} \frac{1}{n};$ (b) $\displaystyle\sum_{n=1}^{6} n!;$ (c) $\displaystyle\sum_{n=2}^{5} n^{n-1}.$

19. Find the sum of the series

(a) $\displaystyle\sum_{n=1}^{\infty} \frac{1}{2^{n-1}};$ (b) $\displaystyle\sum_{n=1}^{\infty} \left(\frac{1}{\sqrt{n}} - \frac{1}{\sqrt{n + 1}} \right);$ (c) $\displaystyle\sum_{n=1}^{\infty} \frac{1}{n(n + 1)}.$

20. Verify that the function

$$f(x) = \frac{x^3}{2x^2 + 1}$$

has neither horizontal nor vertical asymptotes. Convince yourself that $f(x)$ has the line $y = x/2$ as an asymptote.

*21. Give an example of a convergent sequence of rational numbers with an irrational limit.

*22. Verify that the *harmonic series*

$$1 + \frac{1}{2} + \frac{1}{3} + \cdots + \frac{1}{n} + \cdots$$

is divergent.

*23. Show that if the series (9) is convergent, then $x_n \to 0$ as $n \to \infty$. Is the converse true?

Chapter 3

DIFFERENTIATION
AS A TOOL

3.1 VELOCITY AND ACCELERATION

3.11. By a *particle* we mean an object whose actual size can be ignored in a given problem, and which can therefore be idealized as a point. There are problems in which the earth itself can be regarded as a particle, just as there are problems in which a pinhead is a complicated structure made up of vast numbers of tiny particles.

Consider the motion of a particle along a straight line. Let s be the particle's distance at time t from some fixed reference point, where s is positive if measured in a given direction along the line and negative if measured in the opposite direction. Then the particle's motion is described by some *distance function*

$$s = s(t). \tag{1}$$

Here, for simplicity, we denote the dependent variable and the function by the same letter, a common practice. In the language of physics, (1) is the *equation of motion* of the particle.

We now ask a key question: How fast is the particle going? There are two answers, depending on whether we ask about a given *interval* of time or about a given *instant* of time. In the first case, we get the *average velocity*, which is a difference quotient. In the second case, we get the *instantaneous velocity*, which is a derivative.

3.12. Average velocity

a. By the *average velocity* of the particle with equation of motion (1), *over the interval from t to* $t + \Delta t$, we mean the function of *two* variables

$$v_{av}(t, \Delta t) = \frac{s(t + \Delta t) - s(t)}{\Delta t}.$$

It is meaningless to ask for the average velocity at a given instant t without specifying the *averaging time*.

b. To be useful, an average velocity should not be too "crude," that is, Δt should not be too large. Consider, for example, a particle whose equation of motion is described (in part) by the table

t (in seconds)	0	1	2	3	4
s (in feet)	10	0	12	2	14

The particle is actually moving back and forth rather dramatically, but you would never know it calculating the "two-second averages"

$$v_{av}(0, 2) = \frac{12 - 10}{2} = 1, \qquad v_{av}(2, 2) = \frac{14 - 12}{2} = 1,$$

which seems to suggest that the particle is moving slowly and steadily in the positive direction with a velocity of 1 foot per second! Choosing a shorter averaging time $\Delta t = 1$, we get

$$v_{av}(0, 1) = -10, \qquad v_{av}(1, 1) = 12, \qquad v_{av}(2, 1) = -10, \qquad v_{av}(3, 1) = 12.$$

This gives a better picture of the particle's motion. At least, it shows that the direction of the particle's motion changes. But how do we know that it's an accurate picture? After all, everything depends on what the particle is doing between the times of measurement.

3.13. Instantaneous velocity

a. By the *instantaneous velocity* (or "true velocity") of the particle with equation of motion (1), *at the time t*, we mean the function of *one* variable

$$v(t) = \lim_{\Delta t \to 0} v_{av}(t, \Delta t) = \lim_{\Delta t \to 0} \frac{s(t + \Delta t) - s(t)}{\Delta t},$$

obtained by taking the limit of the average velocity as the averaging time Δt "goes to zero." This is, of course, just the derivative

$$v(t) = s'(t) = \frac{ds(t)}{dt}$$

of the distance function $s(t)$ with respect to the time t. Since our averaging time is now "infinitesimal," we can rest assured that no details of the particle's motion have been overlooked in calculating $v(t)$.

From now on, when we talk about "velocity" without further qualification, we mean *instantaneous* velocity. The quantity called "speed" in common parlance is just the absolute value of the velocity.

b. Suppose a stone is dropped from a high tower, and let distance be measured vertically downward from the initial position of the stone. Then, as in Sec. 2.11, the stone, regarded as a particle, has the equation of motion

$$s = s(t) = 16t^2, \tag{2}$$

where s is measured in feet and t in seconds, provided that the stone has not yet hit the ground. The stone's velocity at time t is just

$$v = v(t) = \frac{ds(t)}{dt} = \frac{d}{dt} 16t^2 = 32t. \tag{3}$$

Note that v is an increasing function of time (Sec. 2.33), and is in fact directly proportional to t.

3.14. Acceleration

a. Suppose we differentiate the velocity function $v(t)$ itself. This gives a new function

$$a = a(t) = v'(t) = \frac{dv(t)}{dt},$$

called the *acceleration* of the particle at the time t. Since $v(t) = s'(t)$ is the first derivative (that is, the ordinary derivative) of the distance function $s(t)$, the acceleration is just the second derivative of $s(t)$:

$$a(t) = \frac{dv(t)}{dt} = \frac{d}{dt}\frac{ds(t)}{dt} = \frac{d^2 s(t)}{dt^2}.$$

Both velocity and acceleration are, of course, *rates of change* (Sec. 2.42c), the first the rate of change of the distance with respect to time, the second the rate of change of the velocity with respect to time. Negative acceleration is often called *deceleration*.

b.   The acceleration corresponding to the velocity (3), and hence in turn to the distance function (2), is just

$$a = a(t) = \frac{dv(t)}{dt} = \frac{d}{dt} 32t = 32.$$

Thus the acceleration of the falling stone has the constant value of 32 feet per second per second (more concisely, 32 ft/sec²), the so-called "acceleration due to gravity."

3.15. Example.   As will be shown in Sec. 5.32c, a stone thrown vertically upward from ground level with an initial velocity of v_0 ft/sec (at the time $t = 0$) has the equation of motion

$$s = s(t) = v_0 t - 16t^2, \tag{4}$$

where, as usual, t is the time in seconds, and s is now the height of the stone above the ground, in feet. Suppose $v_0 = 96$ ft/sec. At what time does the stone stop rising and begin to fall? What is the maximum height reached by the stone?

SOLUTION.   Substituting $v_0 = 96$ in (4), we get

$$s = s(t) = 96t - 16t^2. \tag{5}$$

Differentiating with respect to t, we then find that

$$v = v(t) = \frac{d}{dt}(96t - 16t^2) = 96 - 32t. \tag{6}$$

The stone stops rising and begins to fall when its velocity changes from positive to negative values. This change occurs at the precise instant when the velocity equals zero. Setting $v = 0$ in (6) and solving for t, we find that this happens at the time

$$t = \frac{96}{32} = 3.$$

Thus the stone rises for 3 seconds, comes to rest instantaneously, and then falls down for 3 more seconds, finally hitting the ground 6 seconds after being thrown upward (note that $s(6) = 0$).

To find the maximum height achieved by the stone, we make the substitution $t = 3$ in equation (5). This gives

$$s = 96 \cdot 3 - 16 \cdot 9 = 144.$$

Thus the stone rises to a height of 144 feet before beginning to fall back to the ground. To get the stone's acceleration, we differentiate (6) once again, obtaining

$$a = \frac{d}{dt}(96 - 32t) = -32.$$

Thus the acceleration has the constant value of -32 ft/sec².

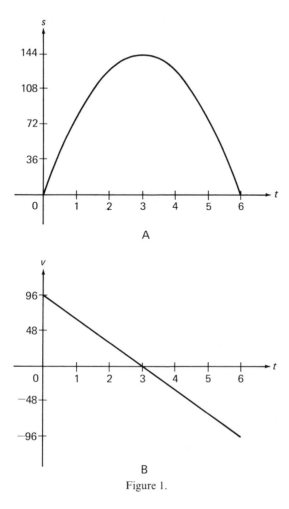

A

B

Figure 1.

We graph the distance function (5) in Figure 1A and the velocity function (6) in Figure 1B. There is no reason to make the horizontal and vertical units the same in these graphs, and we have not done so. The curve in Figure 1A is an upside-down version of the curve in Figure 5, p. 49, with a shift and a scale change, and is again called a *parabola*. Do not make the mistake of confusing this curve with the stone's trajectory! In fact, the stone's trajectory is just the vertical line segment $0 \leqslant s \leqslant 144$, traversed once in the upward direction and once in the downward direction. It is clear from the figure that the function $s(t)$ is increasing for the first 3 seconds of the stone's motion and decreasing for the next 3 seconds, while $v(t)$ is decreasing for the whole 6 seconds. Note that the stone hits the ground at the same *speed* as its initial speed.

As we will see in Sec. 3.3, the fact that the derivative $v(t) = s'(t)$ equals zero for the value of t at which the function $s(t)$ achieves its maximum is not just a special feature of this problem, but rather reflects a general property of differentiable functions.

PROBLEMS

1. Suppose a particle moving along a straight line has the equation of motion
$$s = 10t + 5t^2,$$
where s is measured in feet and t in seconds. Find the average velocity of the particle over the interval from 20 to $20 + \Delta t$ for $\Delta t = 1, 0.1$ and 0.01. What is the particle's instantaneous velocity at the time $t = 20$?

2. Suppose a particle moving along a straight line has the equation of motion
$$s = \frac{1}{3}t^3 - 2t^2 + 3t$$
(units unspecified). Find the particle's velocity v and acceleration a at the time t. When does the direction of motion of the particle change? When does the particle return to its initial position (at $t = 0$)?

3. What can be said about the motion of a particle whose equation of motion contains powers of t greater than 2?

4. A stone is thrown vertically upward with an initial velocity of 32 ft/sec by a man standing at the edge of a roof 48 feet above the ground. Find the time when the stone hits the ground, assuming that it misses the roof on the way down. How fast is the stone going when it hits the ground?

5. How high should the roof be in the preceding problem if the stone is to hit the ground 4 seconds later? 5 seconds later?

6. The equation of motion of a car starting from rest is
$$s = \frac{1}{2}kt^2, \tag{7}$$
where s is measured in feet and t in seconds. Interpret k. Find k if the car reaches a speed of 60 mi/hr in 10 seconds flat. How long do you expect equation (7) to be valid?

7. A car is going v_0 mi/hr when its brakes are suddenly applied. Suppose its subsequent motion is described by the equation
$$s = v_0 t - \frac{1}{2}kt^2. \tag{8}$$
Interpret k. Find k if $v_0 = 60$ mi/hr and the car brakes to a complete stop in 22 seconds.

*8. Show that the distance travelled by a car after its brakes are suddenly applied is proportional to the square of its speed v_0.

*9. The graph of $s(t)$ in Figure 1A is symmetric in the line $t = 3$, that is, reflection in this line does not change the graph. What does this mean physically?

*10. After t seconds a braked flywheel rotates through an angle of
$$\theta = \theta(t) = a + bt - ct^2$$
degrees, where a, b and c are positive constants. Suitably define and then determine the flywheel's angular velocity and angular acceleration. When does the flywheel stop rotating?

3.2 RELATED RATES AND BUSINESS APPLICATIONS

3.21. First we consider a class of problems involving *related rates*. In such problems we are given the rate of change of one quantity (usually with respect to time), and we are asked to find the rate of change of another related quantity.

a. Example. A large spherical balloon is losing air at the rate of one tenth of a cubic foot per second (more concisely, 0.1 ft^3/sec). How fast is the radius of the balloon decreasing when its diameter is 6 feet?

SOLUTION. Let R be the radius and V the volume of the balloon. Then

$$V = \frac{4}{3}\pi R^3, \tag{1}$$

by elementary geometry. Since the size of the balloon is changing, both V and R are functions of time. We could express this fact by writing $V = V(t)$, $R = R(t)$, but it is better to just bear in mind that V and R depend on time. Differentiating (1) with respect to time (identical functions have identical derivatives), we get

$$\frac{dV}{dt} = \frac{4}{3}\pi \cdot 3R^2 \frac{dR}{dt} = 4\pi R^2 \frac{dR}{dt},$$

with the help of Example 2.83c. We then solve this equation for dR/dt, obtaining

$$\frac{dR}{dt} = \frac{1}{4\pi R^2}\frac{dV}{dt},$$

or

$$\frac{dR}{dt} = -\frac{0.1}{4\pi R^2}\text{ ft}^3/\text{sec}, \tag{2}$$

since $dV/dt = -0.1$ ft^3/sec, according to the statement of the problem. Note that dV/dt is negative because air is being *lost*. When the balloon's diameter is 6 feet, its radius is 3 feet. Substituting $R = 3$ into (2), we find that at that moment

$$\frac{dR}{dt} = -\frac{0.1}{4\pi \cdot 9} = -\frac{1}{360\pi}\text{ ft/sec},$$

or equivalently

$$\frac{dR}{dt} = -\frac{12 \cdot 60}{360\pi} = -\frac{2}{\pi} \approx -0.64 \text{ in/min}$$

(inches per minute). Thus R is decreasing at the rate of about 0.64 in/min, a rather slow leak for a large balloon. Note that dR/dt is itself a function of the radius. In fact, the smaller the balloon, the larger dR/dt, as shown by (2).

b. Example. A ladder 20 feet long is leaning against a wall. Suppose the bottom of the ladder is pulled away from the wall at a constant rate of 6 ft/min. How fast is the top of the ladder moving down the wall when

(a) The bottom of the ladder is 12 feet from the wall;
(b) The top of the ladder is 12 feet from the ground?

SOLUTION. Idealizing the ladder as a straight line segment, we introduce rectangular coordinates as shown in Figure 2, where x is the distance between the wall and the bottom of the ladder, and y is the height above ground of the top of the ladder. By the Pythagorean theorem,

$$x^2 + y^2 = 20^2 = 400. \tag{3}$$

Since the position of the ladder is changing, both x and y are functions of the time t, a fact that could be emphasized by writing $x = x(t)$, $y = y(t)$. To find dy/dt, we use the technique of implicit differentiation. Thus we differentiate (3) with respect to t,

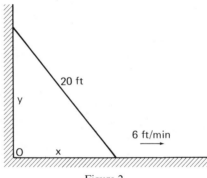

Figure 2.

obtaining the equation

$$2x\frac{dx}{dt} + 2y\frac{dy}{dt} = 0,$$

which we then solve for dy/dt. The result is

$$\frac{dy}{dt} = -\frac{x}{y}\frac{dx}{dt}$$

or

$$\frac{dy}{dt} = -\frac{6x}{y} \text{ ft/min},$$

since $dx/dt = 6$ ft/min, according to the statement of the problem. Note that dx/dt is positive because the bottom of the ladder is moving *away from* the wall, and dy/dt is negative because the top of the ladder is moving *down* the wall. Therefore the top of the ladder moves down the wall at the rate

$$\left|\frac{dy}{dt}\right| = \frac{6 \cdot 12}{\sqrt{400 - 12^2}} = \frac{72}{16} = 4.5 \text{ ft/min}$$

when the bottom of the ladder is 12 feet from the wall, and at the rate of

$$\left|\frac{dy}{dt}\right| = \frac{6\sqrt{400 - 12^2}}{12} = \frac{6 \cdot 16}{12} = 8 \text{ ft/min}$$

when the top of the ladder is 12 feet from the ground.

3.22. a.   The word "marginal" is encountered repeatedly in business and economics, in expressions like "marginal cost," "marginal revenue," "marginal profit," etc. The second word in each expression is always some function, and the word "marginal" merely calls for taking the derivative of this function, with respect to the independent variable. For example, the *total cost* to a firm of producing a quantity Q of some commodity is some function of Q, called the *cost function* and denoted by $C(Q)$. The derivative of the cost function, namely

$$C'(Q) = \frac{dC(Q)}{dQ}, \tag{4}$$

is called the *marginal cost*, denoted by $MC(Q)$. Here we follow the convention, standard in economic theory, of denoting certain functions by pairs of capital letters, like MC for "marginal cost," AR for "average revenue," and so on. (Do not think of these pairs as products!) In this notation, (4) takes the form

$$MC(Q) = \frac{dC(Q)}{dQ}. \tag{5}$$

   b. In writing $C(Q)$ and its derivative $MC(Q)$, we tacitly assume that the units of Q are such that $C(Q)$ is defined for arbitrary $Q \geqslant 0$, and not just for the integers $Q = 0, 1, 2, \ldots$ This assumption is certainly appropriate for oil, measured in pints or gallons, or for salt, measured in pounds or tons, but it is absurd for aircraft carriers. For TV sets the assumption makes sense if the output is large and if we are not too literal-minded. Thus if the answer to a production problem is "Make 31.5 TV sets a day," we can either make 63 sets in 2 days or else settle for making 31 or 32 sets a day. The same remark applies to the application of calculus methods to a host of other problems involving objects that come one at a time, like members of an animal population.
   c. Example. The cost function $C(Q)$ is typically the sum of a constant term, representing certain fixed costs, called the *overhead*, which are independent of the output Q, and a variable term which depends on the actual value of Q. Prove that the marginal cost is independent of the overhead. Find the marginal cost $MC(Q)$ corresponding to the commonly used model of a *cubic* cost function

$$C(Q) = aQ^3 + bQ^2 + cQ + d, \tag{6}$$

where a, b, c and d are constants. What can be said about the constant d?
   SOLUTION.   If $C(Q) = f(Q) + k$, where k is the overhead and hence a constant, then clearly

$$MC(Q) = \frac{d}{dQ}\left[f(Q) + k\right] = \frac{df(Q)}{dQ} + \frac{dk}{dQ} = \frac{df(Q)}{dQ},$$

so that $MC(Q)$ is independent of the overhead. For the cost function (6), the constant d is the overhead and therefore must be positive. Differentiating (6), we get the corresponding marginal cost

$$MC(Q) = 3aQ^2 + 2bQ + c.$$

   d. Example. The function

$$AC(Q) = \frac{C(Q)}{Q} \tag{7}$$

is called the *average cost*. Express the marginal cost in terms of the average cost. Show that the derivative of the average cost equals zero when the marginal cost equals the average cost, and only then.
   SOLUTION.   Combining (5) and (7), we get the formula

$$MC(Q) = \frac{d}{dQ}\, QAC(Q),$$

expressing the marginal cost in terms of the average cost. The derivative of the average cost equals

$$\frac{d}{dQ}\, AC(Q) = \frac{d}{dQ}\frac{C(Q)}{Q} = \frac{C'(Q)Q - C(Q)}{Q^2},$$

by the rule for differentiating a quotient, and this equals zero when

$$C'(Q)Q - C(Q) = 0, \qquad (8)$$

and only then. But $C'(Q) = MC(Q)$, so that (8) can be written as

$$MC(Q)Q - C(Q) = 0,$$

or equivalently

$$MC(Q) = \frac{C(Q)}{Q} = AC(Q).$$

This proves the assertion made in the statement of the example.

PROBLEMS

1. Air is being pumped into a large spherical balloon at the rate of 10 ft³/min. How fast is the radius of the balloon increasing when its diameter is 4 feet?

2. Two ships A and B sail away from a point P along perpendicular routes. Ship A is going 15 mi/hr, while ship B is going 20 mi/hr. Suppose that at a certain time A is 5 miles from P and B is 10 miles from P. How fast are the ships moving apart 1 hour later?

3. The radius of a circle is increasing at a constant rate. Is the same true of its area? Of its circumference?

4. A point moves away from the origin in the first quadrant along the curve $y = \frac{1}{48}x^3$. Which coordinate, x or y, is increasing faster?

5. The length of one side of a rectangle increases at 2 in/sec, while the length of the other side decreases at 3 in/sec. At a certain moment the first side is 20 inches long and the second side is 50 inches long. Is the area of the rectangle increasing or decreasing at this moment? How fast?

6. A man 6 feet tall walks at a speed of 4 ft/sec toward a street light 18 feet above the ground. How fast is the length of the man's shadow decreasing? Does the answer depend on his distance from the light?

7. Let $R(Q)$ be the *total revenue* received by a firm from the sale of a quantity Q of some commodity. Then the derivative $R'(Q)$ is called the *marginal revenue*, denoted by $MR(Q)$, and the function

$$AR(Q) = \frac{R(Q)}{Q}$$

is called the *average revenue*. Express the marginal revenue in terms of the average revenue.

8. Suppose the curve of average cost is a straight line

$$AC(Q) = a - mQ \qquad (a > 0, \ m > 0),$$

with negative slope (average cost typically decreases as output increases). Find the curve of marginal cost.

9. Suppose the demand for a commodity produced by a monopolistic firm is described by the function $Q = Q(P)$, where Q is the quantity demanded at the price P. What does the truism "The greater the price, the less the demand" tell us about the function $Q(P)$?

10. Let $Q(P)$ be the same as in the preceding problem. Then the firm's total revenue is clearly $R(Q) = PQ(P)$, so that its *profit* is just

$$\Pi\,(Q) = R(Q) - C(Q) = PQ(P) - C(Q),$$

where $C(Q)$ is the firm's total cost function. Show that $Q = Q(P)$ has an inverse function $P = P(Q)$, allowing us to also write the profit as

$$\Pi\,(Q) = QP(Q) - C(Q).$$

*11. How fast is the surface area of the balloon in Problem 1 increasing when its diameter is 8 feet?

*12. In the ladder problem of Example 3.21b, find the acceleration of the top of the ladder when the bottom of the ladder is 16 feet from the wall.

3.3 PROPERTIES OF CONTINUOUS FUNCTIONS

So far, one of our chief concerns has been to acquire the technique of differentiation, so that we can find the derivative f' of a given function f. But what do we do with f' once we have found it? As we will see later in this chapter, knowledge of f' can tell us a great deal about the behavior of the original function f. Knowledge of the second derivative f'' also turns out to be a great asset in many cases, because of the further light it sheds on the behavior of f.

Remarkably enough, there is also much that can be deduced from the mere fact that a function is *continuous* in an interval, especially in a *closed* interval, as we now show.

3.31. The continuous image of a closed interval

a. The properties of a function continuous in a *closed* interval (Sec. 2.65c) all stem from the following key proposition, whose proof is a bit too hard for a first course in calculus:

THEOREM. *If f is continuous in a closed interval $I = [a, b]$ and if f is not a constant function, then the range of f, namely the set of all values taken by f at the points of I, is itself a closed interval.*

This fact is expressed by saying that "a continuous function maps a closed interval into a closed interval," or that "the continuous image of a closed interval is a closed interval."

b. We must insist that f be nonconstant, since the function $f(x) \equiv C$ maps I into the single point C, hardly a closed interval! A function which is continuous in an *open* interval can map the interval into any other kind of interval, open, closed or half-open (see Prob. 10). Thus it is crucial to the validity of the theorem that the interval I be *closed*. The set of values taken by f at the points of I will be denoted by $f(I)$. Do not think of $f(I)$ as the value of f at I, which is meaningless. In fact, $f(I)$ is not a number, but rather the set $\{f(x): x \in I\}$.

c. What this means geometrically is shown in Figure 3, where the solid curve is the graph of a function f continuous in a closed interval $I = [a, b]$. Suppose we drop perpendiculars from all the points of the curve $y = f(x)$ onto the y-axis. Then, according to our theorem, the resulting points completely "fill up" some closed interval $[m, M]$ on the y-axis, as shown in the figure. Note that in general several points of $[a, b]$ correspond to the same point of $[m, M]$. For example, the points r and s shown in the figure are both "mapped" by f into the same point $l \in [m, M]$.

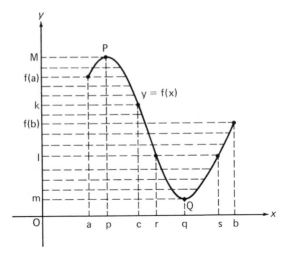

Figure 3.

3.32. Maxima and minima

a. Let f be a function defined in an interval I, and suppose there is a point $p \in I$ such that $f(x) \leqslant f(p)$ for all $x \in I$. Then $f(p)$ is called the *maximum* of f in I, and we say that f has this maximum at the point p. Similarly, suppose there is a point $q \in I$ such that $f(x) \geqslant f(q)$ for all $x \in I$. Then $f(q)$ is called the *minimum* of f in I, and we say that f has this minimum at the point q. We also say that "f takes its maximum value at p and its minimum value at q." Note that f may well take its maximum (or minimum) value, provided there is one, at more than one point of I.

The word *extremum* refers to either a maximum or a minimum, and the phrase *extreme value* refers to either a maximum value or a minimum value. The words "maximum," "minimum" and "extremum" have Latin plurals, namely "maxima," "minima" and "extrema." Extrema of a different kind will be introduced in Sec. 3.51, and will be known as *local* extrema, as opposed to the kind of extrema considered here, which are often called *global* extrema.

b. For example, if I is the closed interval $[-1, 1]$, the function x^2 has its maximum in I, equal to 1, at both points $x = \pm 1$, and its minimum in I, equal to 0, at the point $x = 0$. In the same interval, the function x^3 has its maximum, equal to 1, at the point $x = 1$, and its minimum, equal to -1, at the point $x = -1$. A constant function, say $f(x) \equiv k$, has a maximum and a minimum, both equal to k, at *every* point of any interval. On the other hand, the function x has neither a maximum nor a minimum in the *open* interval $(0, 1)$, since there is no largest number less than 1 and no smallest number greater than 0 (Sec. 1.4, Prob. 12), and the same is true of the functions x^2 and x^3 in this interval. All three functions x, x^2 and x^3 have a minimum, equal to 0, in the half-open interval $[0, 1)$, at the point $x = 0$, but again none of them has a maximum in $[0, 1)$. As we now see, this failure to have one or both extrema cannot occur if the function is continuous in a *closed* interval.

c. THEOREM. *If f is continuous in a closed interval $I = [a, b]$, then I contains points p and q such that*

$$f(q) \leqslant f(x) \leqslant f(p) \tag{1}$$

for all $x \in I$. *In other words, f has both a maximum and a minimum in I, at the points p and q, respectively.*

Proof. We can immediately exclude the case of a constant function, for which the theorem is trivially true. Thus, suppose f is not a constant function. Then, by Theorem 3.31a, the range of f, namely the set $f(I)$, is a closed interval. Let this interval be $[m, M]$. Then, for all $x \in I$, we have $f(x) \in [m, M]$, or equivalently

$$m \leqslant f(x) \leqslant M, \tag{2}$$

by the very meaning of $f(I)$. Let p be any point of I such that $f(p) = M$ and q any point of I such that $f(q) = m$. Again, such points p and q exist by the very meaning of $f(I)$. We can then write (2) in the form (1). $\square$

d. Naturally, M is the maximum of f in I, and m is the minimum of f in I. Interpreted geometrically, the theorem means that the graph of a function f continuous in a closed interval I must have both a "highest point" $P = (p, M)$ and a "lowest point" $Q = (q, m)$, as illustrated by Figure 3. It is important to note that f may take one or both of its extreme values at end points of I (give examples), although this is not the case for the function shown in Figure 3.

3.33. The intermediate value theorem

a. Let k be any number between $f(a)$ and $f(b)$. Then, since both $f(a)$ and $f(b)$ belong to the interval $f(I) = [m, M]$, so does k. Therefore k is a value of f taken at some point c in the interval $[a, b]$. But c cannot be one of the end points a and b, since k lies between $f(a)$ and $f(b)$, and therefore cannot coincide with either $f(a)$ and $f(b)$. The situation is illustrated in Figure 3.

b. This key property of continuous functions deserves a name of its own:

THEOREM **(Intermediate value theorem).** *If f is continuous in an interval I, which need not be closed, and if f takes different values $f(\alpha)$ and $f(\beta)$ at two points α and β of I, then f takes every value between $f(\alpha)$ and $f(\beta)$ at some point between α and β.*

Proof. In view of the preceding remarks, we need only show that f is continuous in $[\alpha, \beta]$ if $\alpha < \beta$, or in $[\beta, \alpha]$ if $\beta < \alpha$. But, by hypothesis, f is continuous in an interval I containing α and β. Therefore f is certainly continuous in the closed interval with end points α and β, since this is the smallest interval containing both α and β. $\square$

PROBLEMS

1. Find the extrema, if any, of the function $f(x) = 1/x$ in the interval
    (a) $(0, 1)$;   (b) $(0, 1]$;   (c) $[1, 2]$;   (d) $(0, \infty)$;   (e) $(-\infty, -1]$.

2. Find the extrema, if any, of the function $f(x) = [x]$, where $[x]$ is the integral part of x (Sec. 1.4, Prob. 10), in the interval
    (a) $(0, 1)$;   (b) $(0, 1]$;   (c) $(-1, 0]$;   (d) $(0, \infty)$;   (e) $(-\infty, \infty)$.

3. Verify that if a function f is increasing in a closed interval $I = [a, b]$, then f has global extrema in I, even if f is discontinuous. Where does f take its extreme values? How about the case of decreasing f?

4. "If f is continuous in an interval I and if f takes values $f(\alpha)$ and $f(\beta)$ with opposite signs at two points α and β of I, then f equals zero at some point between α and β." True or false? Why?

5. Is the function

$$f(x) = \begin{cases} x + 1 & \text{if} & -1 \leqslant x < 0, \\ x - 1 & \text{if} & 0 \leqslant x \leqslant 1 \end{cases} \tag{3}$$

continuous? Does it map the closed interval $-1 \leqslant x \leqslant 1$ into a closed interval?

6. Does the function (3) map the closed interval $-\frac{1}{2} \leqslant x \leqslant \frac{1}{2}$ into a closed interval?

7. Investigate the global extrema of the function (3) in the interval $-1 \leqslant x \leqslant 1$.

*8. Give an example of a continuous function mapping a finite interval into an infinite interval.

*9. Give an example of a continuous function mapping an infinite interval into a finite interval.

*10. Give an example of a continuous function mapping an open interval into

(a) An open interval;   (b) A half-open interval;   (c) A closed interval.

3.4 PROPERTIES OF DIFFERENTIABLE FUNCTIONS

3.41. a. Having investigated the properties of continuous functions, we now return to the study of differentiable functions. We begin by establishing the following interesting fact: *If f is differentiable at a point p, with derivative f'(p), and if f'(p) is **positive**, then f is **increasing** in some neighborhood of p.* To show this, we observe that just as in Sec. 2.55a

$$\lim_{\Delta x \to 0} \left[\frac{f(p + \Delta x) - f(p)}{\Delta x} - f'(p) \right] = 0,$$

by the definition of the derivative $f'(p)$, or equivalently

$$\lim_{\Delta x \to 0} \left[Q(\Delta x) - f'(p) \right] = 0,$$

in terms of the difference quotient

$$Q(\Delta x) = \frac{f(p + \Delta x) - f(p)}{\Delta x} = \frac{\Delta f(p)}{\Delta x}.$$

In "ε, δ language" this means that, given any $\varepsilon > 0$, there is a $\delta > 0$ such that

$$\left| Q(\Delta x) - f'(p) \right| < \varepsilon$$

whenever $0 < |\Delta x| < \delta$. Let $\varepsilon = f'(p) > 0$. Then there is a $\delta > 0$ such that

$$\left| Q(\Delta x) - f'(p) \right| < f'(p),$$

or equivalently

$$-f'(p) < Q(\Delta x) - f'(p) < f'(p),$$

whenever $0 < |\Delta x| < \delta$. But then

$$0 < Q(\Delta x) < 2f'(p) \tag{1}$$

whenever $0 < |\Delta x| < \delta$. Therefore the difference quotient $Q(\Delta x) = \Delta f(p)/\Delta x$ is *positive* whenever $0 < \Delta x < \delta$ or $-\delta < \Delta x < 0$, which means that *the increments $\Delta f(p)$ and Δx have the same sign* under these conditions. In other words,

$$f(p + \Delta x) - f(p) > 0 \tag{2}$$

if $0 < \Delta x < \delta$, while

$$f(p + \Delta x) - f(p) < 0 \tag{3}$$

if $-\delta < \Delta x < 0$. Changing the sign of Δx in (3), we find that

$$f(p - \Delta x) - f(p) < 0 \tag{4}$$

if $0 < \Delta x < \delta$. Combining (2) and (4), we finally get

$$f(p - \Delta x) < f(p) < f(p + \Delta x)$$

if $0 < \Delta x < \delta$, which shows at once that f is *increasing* in some neighborhood of the point p, namely in the neighborhood $(p - \delta, p + \delta)$ or in any smaller neighborhood.

    b. In virtually the same way, we can show that *if f is differentiable at a point p, with derivative $f'(p)$, and if $f'(p)$ is **negative**, then f is **decreasing** in some neighborhood of p.* This time we choose $\varepsilon = -f'(p) > 0$, obtaining first

$$f'(p) < Q(\Delta x) - f'(p) < -f'(p)$$

and then

$$2f'(p) < Q(\Delta x) < 0,$$

instead of (1), whenever $0 < |\Delta x| < \delta$. Therefore $Q(\Delta x)$ is *negative* whenever $0 < \Delta x < \delta$ or $-\delta < \Delta x < 0$, which means that *the increments $\Delta f(p)$ and Δx have opposite signs* under these conditions. In other words,

$$f(p + \Delta x) - f(p) < 0 \tag{5}$$

if $0 < \Delta x < \delta$, while

$$f(p + \Delta x) - f(p) > 0 \tag{6}$$

if $-\delta < \Delta x < 0$. Changing the sign of Δx in (6), we find that

$$f(p - \Delta x) - f(p) > 0 \tag{7}$$

if $0 < \Delta x < \delta$. Combining (5) and (7), we finally get

$$f(p - \Delta x) > f(p) > f(p + \Delta x)$$

if $0 < \Delta x < \delta$, which shows at once that f is *decreasing* in some neighborhood of the point p, namely in the neighborhood $(p - \delta, p + \delta)$ or in any smaller neighborhood.

    c. By an *interior point* of an interval I (open, closed or half-open), we mean any point of I other than its end points. A function f is said to *vanish* at a point c if $f(c) = 0$. In other words, "to vanish" means the same thing as "to equal zero." If a function f vanishes at every point of an interval I, we say that f *vanishes identically* in I. This extra vocabulary will come in handy time and again.

3.42. Rolle's theorem

    a. Our next result is a stepping stone on the way to another, more important result, called the "mean value theorem," but it is of considerable interest in its own right.

    THEOREM (**Rolle's theorem**). *Let f be continuous in the closed interval $[a, b]$ and differentiable, with derivative f', in the open interval (a, b). Suppose that $f(a) = f(b) = k$. Then there is a point $c \in (a, b)$ such that $f'(c) = 0$.*

    Proof. Do not be put off by the formal language. A more informal way of stating the theorem is the following: Let f be continuous in a closed interval $I = [a, b]$ and differentiable at every interior point of I. Suppose f takes the same value at both end points of I. Then the derivative f' vanishes at some interior point of I.

To prove the theorem, we first observe that f has both a maximum M and a minimum m in $I = [a, b]$, by Theorem 3.32c, where $m \leqslant k \leqslant M$, since f equals k at the points a and b. If $m = k = M$, then f reduces to the constant function $f(x) \equiv k$, whose derivative vanishes at *every* interior point of I, and the theorem is proved. Otherwise, we have either $m < k$ or $M > k$. Suppose $M > k$, and let c be a point of I such that $f(c) = M$. Then $c \in (a, b)$, that is, c is an interior point of I, so that the derivative $f'(c)$ exists. If $f'(c) \neq 0$, then $f'(c)$ is either positive or negative. In the first case, f is increasing in some neighborhood of c, by Sec. 3.41a, while in the second case, f is decreasing in some neighborhood of c, by Sec. 3.41b. In either case, the neighborhood contains values of f larger than M, so that M cannot be the maximum of f. It follows that $f'(c) = 0$. The proof for $m < k$ is almost identical (give the details). □

b. Rolle's theorem has a simple geometrical interpretation. It merely says that if the end points of the curve

$$y = f(x) \qquad (a \leqslant x \leqslant b) \tag{8}$$

have the same ordinate, so that $f(a) = f(b)$, then the slope of the tangent to the curve vanishes and hence is *horizontal* at some "intermediate point," that is, at some point of the curve other than its end points. This situation is illustrated by Figure 4, which shows that the curve can actually have horizontal tangents at more than one intermediate point, in particular, at points other than those with the maximum and minimum ordinates M and m.

3.43. The mean value theorem

a. If $f(a) \neq f(b)$, we can no longer assert that the curve (8) has a horizontal tangent at some intermediate point. However, we can now assert that at some intermediate point the curve has a tangent with the same slope as the chord joining its end points $A = (a, f(a))$ and $B = (b, f(b))$, that is, a tangent with slope

$$\frac{f(b) - f(a)}{b - a},$$

as illustrated by Figure 5.

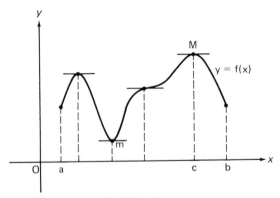

Figure 4.

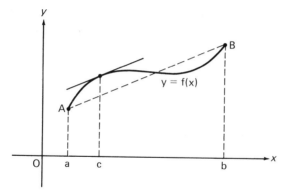

Figure 5.

THEOREM **(Mean value theorem).** *Let f be continuous in the closed interval* $[a, b]$ *and differentiable, with derivative* f', *in the open interval* (a, b). *Then there is a point* $c \in (a, b)$ *such that*

$$f'(c) = \frac{f(b) - f(a)}{b - a}, \tag{9}$$

or equivalently

$$f(b) - f(a) = f'(c)(b - a). \tag{10}$$

Proof. We introduce a new function

$$g(x) = f(x) - kx,$$

choosing the constant k in such a way that $g(x)$ has the same value at both end points a and b. Then k must satisfy the equation

$$g(a) = f(a) - ka = f(b) - kb = g(b),$$

with solution

$$k = \frac{f(b) - f(a)}{b - a}.$$

With this choice of k, $g(x)$ satisfies all the conditions of Rolle's theorem. But then there is a point $c \in (a, b)$ such that

$$g'(c) = f'(c) - k = 0,$$

that is, such that

$$f'(c) = k = \frac{f(b) - f(a)}{b - a}. \quad \square$$

b. COROLLARY **(Mean value theorem in increment form).** *If f is differentiable in an interval I containing the points x and x + Δx, then the increment of f at x can be written in the form*

$$\Delta f(x) = f(x + \Delta x) - f(x) = f'(x + \alpha \Delta x)\, \Delta x, \tag{11}$$

where $0 < \alpha < 1$.

Proof. Here it is not necessary to state explicitly that f is continuous in I, since this follows automatically from the assumption that f is differentiable in I (Sec. 2.66). Choosing $a = x$, $b = x + \Delta x$ in (10), we obtain

$$\Delta f(x) = f'(c) \, \Delta x, \tag{12}$$

where c lies between x and $x + \Delta x$, regardless of the sign of Δx. Therefore the number

$$\alpha = \frac{c - x}{\Delta x}$$

is always positive and lies in the interval $(0, 1)$, or equivalently

$$c = x + \alpha \Delta x \qquad (0 < \alpha < 1). \tag{13}$$

Comparing (12) and (13), we get (11). $\square$

PROBLEMS

1. Check the validity of Rolle's theorem for the function

$$f(x) = (x - 1)(x - 2)(x - 3).$$

   In other words, verify that the derivative f' vanishes at a point in the interval $(1, 2)$ and at a point in the interval $(2, 3)$.

2. The function

$$f(x) = |x| \qquad (-a \leqslant x \leqslant a)$$

   takes the same value $|a|$ at both points $x = \pm a$, but the derivative f' does not vanish at any point $c \in (-a, a)$. Why doesn't this contradict Rolle's theorem?

3. Show that the mean value theorem (10) remains valid even if $a > b$.

4. At what points of the curve $y = x^3$ is the tangent parallel to the chord joining the points $A = (-3, -9)$ and $B = (3, 9)$?

5. According to formula (10),

$$f(2) - f(1) = f'(c) \qquad (1 < c < 2).$$

   Find c if $f(x) = 1/x$.

6. According to formula (11),

$$f(1 + \Delta x) - f(1) = f'(1 + \alpha \Delta x) \, \Delta x \qquad (0 < \alpha < 1).$$

   Find α if $f(x) = x^3$, $\Delta x = -1$.

7. Justify the following "kinematic interpretation" of the mean value theorem: If a train traverses the distance between two stations at an average velocity v_{av}, then there is a moment when the train's instantaneous velocity equals v_{av}.

*8. Let

$$f(x) = \begin{cases} \dfrac{1}{2}(3 - x^2) & \text{if } x < 1, \\[2mm] \dfrac{1}{x} & \text{if } x \geqslant 1. \end{cases}$$

   Find two points c satisfying formula (10) for $a = 0$, $b = 2$.

*9. Show that the square roots of any two consecutive integers exceeding 24 differ by less than 0.1.

3.5 APPLICATIONS OF THE MEAN VALUE THEOREM

3.51.   We already know that the derivative of a constant function vanishes everywhere. More concisely, if $f(x) \equiv$ constant, then $f'(x) \equiv 0$. But, conversely, does $f'(x) \equiv 0$ imply $f(x) \equiv$ constant? Yes, if the domain of f is an interval, as we now show.

THEOREM.   *If f is differentiable in an interval I, and if the derivative f' vanishes identically in I, then $f(x) \equiv$ constant in I, that is, f has the same value at every point of I.*

Proof.   Let x_0 be a fixed point of I, and let x be any other point of I. Then, by the mean value theorem,

$$f(x) - f(x_0) = f'(c)(x - x_0),$$

where c lies between x_0 and x. But $f'(c) = 0$, since c belongs to I, and therefore $f(x) - f(x_0) = 0$ or $f(x) = f(x_0)$. Since x is an arbitrary point of I, it follows that $f(x) = f(x_0)$ for every $x \in I$. □

3.52. Antiderivatives

a.   Given a function $f(x)$ defined in an interval I, suppose $F(x)$ is another function defined in I such that

$$\frac{dF(x)}{dx} = F'(x) = f(x)$$

for every $x \in I$. Then $F(x)$ is said to be an *antiderivative* of $f(x)$, in the interval I. For example, $\frac{1}{3}x^3$ is an antiderivative of x^2 in $(-\infty, \infty)$, since

$$\frac{d}{dx}\frac{1}{3}x^3 = x^2$$

for every $x \in (-\infty, \infty)$. If $F(x)$ is an antiderivative of $f(x)$ in I, then so is

$$G(x) \equiv F(x) + C, \tag{1}$$

where C is an arbitrary constant, since

$$\frac{dG(x)}{dx} = \frac{d}{dx}[F(x) + C] = \frac{dF(x)}{dx} + \frac{dC}{dx} = \frac{dF(x)}{dx} = f(x).$$

The next proposition shows that there are no other antiderivatives of $f(x)$.

b. THEOREM.   *Let $F(x)$ be any antiderivative of $f(x)$ in an interval I. Then every other antiderivative of $f(x)$ in I is of the form* (1).

Proof.   Let $G(x)$ be any other antiderivative of $f(x)$ in I, and let $H(x) = G(x) - F(x)$. Then

$$H'(x) = G'(x) - F'(x) = f(x) - f(x) = 0$$

for every $x \in I$, that is, the derivative H' vanishes identically in I. It follows from the preceding theorem that H has the same value, call it C, at every point of I. In other words,

$$H(x) = G(x) - F(x) \equiv C,$$

which is equivalent to (1). □

3.53. The indefinite integral

a. Let $F(x)$ be an antiderivative of $f(x)$ in I, so that $F'(x) = f(x)$ for every $x \in I$. Then, as just shown, the "general antiderivative" of $f(x)$ in I is of the form

$$F(x) + C,$$

where C is an arbitrary constant. This expression is also called the *indefinite integral* of $f(x)$, and is denoted by

$$\int f(x)\, dx \tag{2}$$

Thus

$$\int f(x)\, dx = F(x) + C, \tag{3}$$

by definition, so that the indefinite integral is defined only to within an arbitrary "additive constant." The symbol $\int$ is called the *integral sign*. The operation leading from the function $f(x)$, called the *integrand*, to the expression (2) is called (*indefinite*) *integration*, with respect to x, the argument x is called the *variable of integration*, and the constant C is called the *constant of integration*. Note that the expression behind the integral sign in (2) is the product of the integrand $f(x)$ and the differential dx of the variable of integration. Recalling Sec. 2.55a, we recognize this product as the differential

$$dF(x) = F'(x)\, dx = f(x)\, dx$$

of the antiderivative $F(x)$, so that (3) can also be written as

$$\int dF(x) = F(x) + C.$$

In writing (3), it is tacitly assumed that the formula is an identity for all x in some underlying interval I in which $f(x)$ and $F(x)$ are both defined; however, I is usually left unspecified. The convention is to give f and F the same argument, in keeping with the formula $F'(x) = f(x)$, but we could just as well write

$$\int f(t)\, dt = F(t) + C,$$

say, replacing x by some other symbol (here t) in both sides of (3). Differentiating (3), we get

$$\frac{d}{dx} \int f(x)\, dx = \frac{d}{dx} [F(x) + C] = \frac{dF(x)}{dx} = F'(x),$$

so that

$$\frac{d}{dx} \int f(x)\, dx = f(x). \tag{4}$$

b. Since a function $f(x)$ is obviously an antiderivative of its own derivative $f'(x)$, we have

$$\int f'(x)\, dx = f(x) + C.$$

This formula can be used to derive an integration formula from any differentiation formula. For example, the formula

$$\frac{d}{dx} \frac{x^{r+1}}{r+1} = \frac{(r+1)x^r}{r+1} = x^r$$

(Sec. 2.74e), valid for arbitrary real $r \neq -1$, leads at once to the formula

$$\int x^r \, dx = \frac{x^{r+1}}{r+1} + C \tag{5}$$

if $r \neq -1$. Choosing $r = -\frac{1}{2}$ in (5), we get the formula

$$\int \frac{dx}{\sqrt{x}} = 2\sqrt{x} + C,$$

valid in the interval $(0, \infty)$, but in no larger interval, while the choice $r = \frac{1}{3}$ gives the formula

$$\int x^{1/3} \, dx = \frac{3}{4} x^{4/3} + C,$$

valid in the whole interval $(-\infty, \infty)$.

    c. THEOREM. *Suppose $f(x)$ and $g(x)$ have indefinite integrals in the same interval I. Then*

$$\int [af(x) + bg(x)] \, dx = a \int f(x) \, dx + b \int g(x) \, dx, \tag{6}$$

where a and b are arbitrary constants.

    Proof. It follows from (4), applied to the function $af(x) + bg(x)$, that

$$\frac{d}{dx} \int [af(x) + bg(x)] \, dx = af(x) + bg(x).$$

On the other hand, by the usual rules of differentiation,

$$\frac{d}{dx} \left[a \int f(x) \, dx + b \int g(x) \, dx \right] = a \frac{d}{dx} \int f(x) \, dx + b \frac{d}{dx} \int g(x) \, dx = af(x) + bg(x),$$

where we use (4) twice more. Thus the two sides of (6) have the same derivative in I, and hence can differ only by an arbitrary constant, by the same argument as in the proof of Theorem 3.52b. But then (6) holds, since the indefinite integral on the left is defined only to within an arbitrary "additive constant." $\square$

    By virtually the same argument, you can easily convince yourself of the validity of the more general formula

$$\int [c_1 f_1(x) + c_2 f_2(x) + \cdots + c_n f_n(x)] \, dx$$

$$= c_1 \int f_1(x) \, dx + c_2 \int f_2(x) \, dx + \cdots + c_n \int f_n(x) \, dx, \tag{7}$$

where $f_1(x), f_2(x), \ldots, f_n(x)$ are functions which have indefinite integrals in the same interval, and $c_1, c_2, \ldots, c_n$ are arbitrary constants.

    d. Example. Evaluate

$$\int \left(5x^4 - 6x^2 + \frac{2}{x^2} \right) dx.$$

    SOLUTION. By (7), we have

$$\int \left(5x^4 - 6x^2 + \frac{2}{x^2} \right) dx = 5 \int x^4 \, dx - 6 \int x^2 \, dx + 2 \int \frac{dx}{x^2} = x^5 - 2x^3 - \frac{2}{x} + C,$$

with the help of (5). Note that the constants of integration contributed by each of the three integrals separately can be combined into a single constant of integration C.

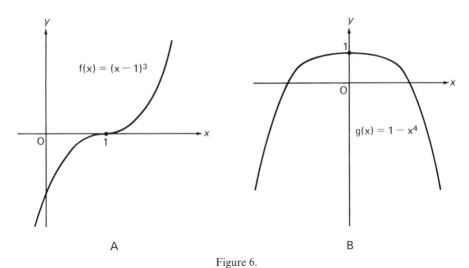

Figure 6.

3.54. In Sec. 3.41c we showed that if f is differentiable at a point x_0, with derivative $f'(x_0)$, and if $f'(x_0)$ is positive, then f is increasing in some neighborhood of x_0. We now use the mean value theorem to establish the following closely related result: *If f is differentiable in an interval I, with derivative f', and if f' is **positive** at every interior point of I, then f is **increasing** in I.* To see this, let x_1 and x_2 be any two points of I such that $x_1 < x_2$. Then, by the mean value theorem,

$$f(x_2) - f(x_1) = f'(c)(x_2 - x_1) \tag{8}$$

for some point c between x_1 and x_2. Therefore c is an interior point of I, so that $f'(c) > 0$. The right side of (8) is positive, being the product of two positive numbers, and hence $f(x_2) - f(x_1)$ is also positive. In other words, $x_1 < x_2$ implies $f(x_1) < f(x_2)$, which means that f is increasing in I, as claimed.

Virtually the same argument shows that *if f is differentiable in an interval I, with derivative f', and if f' is **negative** at every interior point of I, then f is **decreasing** in I.* Give the details.

For example, the function $f(x) = (x - 1)^3$ shown in Figure 6A is increasing in both intervals $(-\infty, 1]$ and $[1, \infty)$, since $f'(x) = 3(x - 1)^2 > 0$ if $x < 1$ or $x > 1$. Therefore $f(x)$ is increasing in the whole interval $(-\infty, \infty)$. Similarly, the function $g(x) = 1 - x^4$ shown in Figure 6B is increasing in $(-\infty, 0]$, since $g'(x) = -4x^3 > 0$ if $x < 0$, and decreasing in $[0, \infty)$, since $g'(x) = -4x^3 < 0$ if $x > 0$.

PROBLEMS

1. The function

$$f(x) = \begin{cases} 1 & \text{if } x > 0, \\ -1 & \text{if } x < 0 \end{cases}$$

is not a constant, but the derivative f' vanishes at every point of the domain of f. Why doesn't this contradict Theorem 3.51?

2. Find all antiderivatives of the function

(a) x^3;   (b) $\dfrac{1}{x^2}$ $(x \neq 0)$;   (c) $\sqrt{x}$ $(x \geqslant 0)$.

3. Verify that

$$\int dx = \int 1 \cdot dx = x + C.$$

4. Show that if

$$\int f(x)\, dx = F(x) + C,$$

then

$$\int f(ax + b)\, dx = \frac{1}{a} F(ax + b) + C$$

for arbitrary constants $a \neq 0$ and b.

5. Evaluate

(a) $\displaystyle\int (x^4 - 3x^2 + x - 4)\, dx$;   (b) $\displaystyle\int (1 - x)(1 - 2x)(1 - 3x)\, dx$;

(c) $\displaystyle\int \frac{x + 1}{\sqrt{x}}\, dx$;   (d) $\displaystyle\int x^2 (1 - \sqrt{x})^3\, dx$.

6. "The indefinite integral of a polynomial of degree n is a polynomial of degree $n + 1$." True or false?

7. Is the derivative of an increasing function necessarily increasing?

8. In which intervals are the following functions increasing? Decreasing?

(a) $2 + x - x^2$;   (b) $3x - x^3$;   (c) $\dfrac{2x}{1 + x^2}$.

*9. What can be said about the function f if $f^{(m)}(x) \equiv 0$?

3.6 LOCAL EXTREMA

3.61. a. Figure 7 shows the graph of a function f continuous in a closed interval $[a, b]$. Just as in Figure 3, p. 110, f takes its maximum in $[a, b]$, equal to M, at the "highest point" of the graph, namely $P = (p, M)$, and its minimum in $[a, b]$, equal to m, at the "lowest point" of the graph, namely $A = (a, m)$, which this time happens to be an end point of the graph. But now there also seems to be something special about the behavior of the graph at certain other points, namely Q, R and S. In fact, Q is "higher" than all "nearby points" of the graph, although not as high as P, while each of the points R and S is "lower" than all "nearby points" of the graph, although not as low as A.

b. We now make these qualitative notions precise. Let f be a function defined in an interval I, and suppose there is a point $p \in I$ such that $f(x) \leqslant f(p)$ for all x "sufficiently near" p, that is, for all x in some neighborhood of p. Then f is said to have a *local maximum*, equal to $f(p)$, at the point p. Similarly, suppose there is a point $q \in I$ such that $f(x) \geqslant f(q)$ for all x in some neighborhood of q. Then f is said to have a *local minimum*, equal to $f(q)$, at the point q. The term *local extremum* refers to a *local maximum* or a *local minimum*.

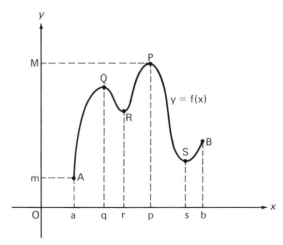

Figure 7.

c. Make sure you understand the crucial distinction between *local extrema*, as just defined, and the kind of extrema defined in Sec. 3.32a. The latter, which will henceforth be called *global extrema* whenever there is any possibility of confusion, involve comparison of a proposed extremum $f(p)$ with the value of f at *every point* of the interval I in which f is defined, while the former only require comparison of $f(p)$ with the values of f at points "sufficiently near" p, or, for that matter, "arbitrarily near" p. The adjective "global," which suggests the "overall" behavior of f in the whole interval I, and the adjective "local," which suggests the behavior of f in the "immediate vicinity" of the point p, are well-suited to emphasize this distinction. In other books you will often encounter the terms *absolute extremum* and *relative extremum*, as synonyms for global extremum and local extremum.

3.62. a. Unlike the case of a global maximum or minimum, a function can have several distinct (that is, different) local maxima or minima. On the other hand, as noted in Sec. 3.32a, a function can take its global extremum at more than one point. For example, the function f shown in Figure 7 has distinct local maxima at the points p and q, and distinct local minima at the points r and s. The global minimum of f at the point a is not a local minimum, since f is not defined in a neighborhood of a (it is not enough to be defined on only one side of a). For the same reason, a function defined in an interval I can have local extrema only at *interior* points of I, that is, at points of I other than the end points of I. The function f in Figure 7 has neither a global extremum nor a local extremum at the point b, but b is still special in the sense that it is an end point of the interval $[a, b]$. (In this regard, see Prob. 1.) Note that f has both a global maximum and a local maximum at the point p. In fact, if a function f is defined in an interval I, then any global extremum of f at an interior point of I is automatically a local extremum of f. (Why is this so?)

b. A local maximum $f(p)$ is said to be *strict* if $f(x) < f(p)$, with $<$ instead of $\leqslant$, for all x "sufficiently near" p but not equal to p, that is, for all x in some *deleted* neighborhood of p (Sec. 1.63a). Similarly, a local minimum $f(q)$ is said to be *strict* if $f(x) > f(q)$, with $>$ instead of $\geqslant$, for all x in some deleted neighborhood of q. For example, all four local extrema of the function in Figure 7 are strict. On the other hand, a constant function has both a local maximum and a local minimum at every point, but none of these extrema is strict.

3.63. a. In practical problems involving local extrema, we will always be concerned with a function f defined in some interval I, where f is differentiable at every point of I with the possible exception of certain special points. We say that f fails to have a derivative at a point p, or that the derivative $f'(p)$ fails to exist, if the limit defining $f'(p)$ either does not exist or is infinite.

THEOREM. *If f has a local extremum at a point p, then either f fails to have a derivative at p, or $f'(p)$ exists and equals zero.*

Proof. Either $f'(p)$ exists or it does not. Suppose $f'(p)$ exists and is nonzero. Then $f'(p)$ is either positive or negative. In the first case, f is increasing in some neighborhood of p, by Sec. 3.41a, while in the second case, f is decreasing in some neighborhood of p, by Sec. 3.41b. In either case, this neighborhood, *or any smaller neighborhood*, contains values of f larger than $f(p)$ and values of f smaller than $f(p)$, so that $f(p)$ cannot be either a local maximum or a local minimum of f. It follows that $f'(p) = 0$. $\square$

This argument, of course, closely resembles that used in the proof of Rolle's theorem.

b. Interpreted geometrically, the theorem says that if a function f has a local extremum at a point p, then either the graph of f has no tangent at the point $P = (p, f(p))$, if $f'(p)$ fails to exist, or the graph of f has a *horizontal* tangent at P, if $f'(p) = 0$. These two possibilities are illustrated in Figure 8A, where each of the functions has a (strict) local maximum at p.

c. By a *critical point* of a function f we mean either a point where f has no derivative or a point where the derivative of f vanishes, and by a *stationary point* of f we mean a point where the derivative of f vanishes. Thus a critical point of f is either a point where f has no derivative or a stationary point of f. According to the theorem, if f has a local extremum at p, then p is a critical point of f. On the other hand, if p is a critical point of f, there is no necessity for f to have a local extremum at p. This is illustrated by Figure 8B, which shows two functions, each with a critical point at p, but neither with a local extremum at p.

Thus what we really want are conditions on a function f which *compel* f to have a local extremum at a given critical point p. (We will always assume that f

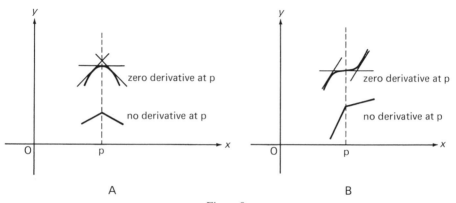

A   B

Figure 8.

is continuous at p.) Such conditions will now be presented in the form of two tests for a local extremum, one called the *first derivative test*, the other called the *second derivative test*.

3.64. The first derivative test

a.   Suppose f is differentiable in a deleted neighborhood of a point p, and suppose the derivative f' has one sign to the left of p and the opposite sign to the right of p. Then f' is said to *change sign in going through p*, from the sign on the left of p to the sign on the right of p. Note that we do not require f to be differentiable at the point p itself and in fact $f''(p)$ may fail to exist.

THEOREM **(First derivative test for a local extremum).** *Let p be a critical point of f, and suppose f' changes sign in going through p. Then f has a strict local extremum at p. The extremum is a maximum if f' changes sign from plus to minus, and a minimum if f' changes sign from minus to plus.*

Proof.   Suppose f' changes sign from plus to minus in going through p, in a deleted δ-neighborhood of p, that is, in the union of intervals $(p - \delta, p) \cup (p, p + \delta)$ (Sec. 1.63b). By the mean value theorem in increment form,

$$f(p + \Delta x) - f(p) = f'(p + \alpha \Delta x)\,\Delta x \qquad (0 < \alpha < 1).$$

Therefore

$$f(p + \Delta x) - f(p) < 0 \tag{1}$$

if $-\delta < \Delta x < 0$, since then $f'(p + \alpha \Delta x) > 0$, $\Delta x < 0$, and similarly

$$f(p + \Delta x) - f(p) < 0 \tag{2}$$

if $0 < \Delta x < \delta$, since then $f'(p + \alpha \Delta x) < 0$, $\Delta x > 0$. In other words,

$$f(p + \Delta x) < f(p) \tag{3}$$

if $0 < |\Delta x| < \delta$, so that f has a strict local maximum at p. On the other hand, if f' changes sign from minus to plus in going through p, then we get $>$ instead of $<$ in (1), (2) and (3), so that f has a strict local minimum at p (check the details). $\square$

b.   Interpreted geometrically, the first derivative test says that if the slope of the tangent to the graph of f at a variable point $P_x = (x, f(x))$ changes from plus to minus as P_x goes through the point $P = (p, f(p))$, then f has a strict local maximum at p even if there is no tangent to the graph of f at P. (What is the analogous statement for the case where the slope of the tangent changes sign from minus to plus?) That this is actually so is apparent from Figure 8A. On the other hand, it is easy to see that the slope of the tangent to the graph of both functions in Figure 8B does not change sign in going through P, and is in fact positive on both sides of P. This is because both functions are increasing in a neighborhood of p, and hence cannot have a local extremum at p.

c. Example.   Find the local extrema of the function

$$f(x) = (x - 1)x^{2/3}. \tag{4}$$

SOLUTION.   Differentiating (4), we get

$$f'(x) = x^{2/3} + \frac{2}{3}(x - 1)x^{-1/3} = \frac{5x - 2}{3x^{1/3}}, \tag{5}$$

with the help of formulas (14) and (15), p. 78, and the formula $x^r x^s = x^{r+s}$, to be proved in Sec. 4.45a. Therefore f has two critical points, the point $x = 0$ at which the derivative f' fails to exist (check this directly), and the point $x = \frac{2}{5}$ at which f'

vanishes. It follows from (5) that

$$f'(x) > 0 \quad \text{if} \quad x < 0,$$

$$f'(x) < 0 \quad \text{if} \quad 0 < x < \frac{2}{5},$$

$$f'(x) > 0 \quad \text{if} \quad x > \frac{2}{5}$$

(the cube root of a negative number is negative). Thus f' changes sign from plus to minus in going through $x = 0$ and from minus to plus in going through $x = \frac{2}{5}$. Therefore, by the first derivative test, f has a strict relative maximum, equal to 0, at $x = 0$, and a strict relative minimum, equal to

$$-\frac{3}{5}\left(\frac{2}{5}\right)^{2/3},$$

at $x = \frac{2}{5}$, as confirmed by Figure 9.

3.65. a. The next test is applicable only when the first derivative $f'(p)$ and second derivative $f''(p)$ both exist, but this is the most common situation.

THEOREM (**Second derivative test for a local extremum**). *Let p be a stationary point of f, and suppose $f''(p)$ exists and is nonzero. Then f has a strict local extremum at p. The extremum is a maximum if $f''(p) < 0$ and a minimum if $f''(p) > 0$.*

Proof. Since the second derivative $f''(p)$ exists, f is differentiable in a neighborhood of p, that is, f' exists in a neighborhood of p. Moreover, $f'(p) = 0$, since p is a stationary point of f. Applying the argument in Sec. 3.41 to the derivative f' instead of to the function f itself, we see that f' is decreasing in a neighborhood of p if $f''(p) < 0$ and increasing in a neighborhood of p if $f''(p) > 0$. Since $f'(p) = 0$, it follows that f' changes sign from plus to minus in going through p if $f''(p) < 0$ and from minus to plus if $f''(p) > 0$. The second derivative test is now an immediate consequence of the first derivative test. $\square$

b. The function f may or may not have a local extremum at p if $f''(p) = 0$. For example, $f''(0) = 0$ if $f(x) = x^3$ or if $f(x) = \pm x^4$, but in the first case f has no local extremum at $x = 0$, being increasing in $(-\infty, \infty)$, by Sec. 3.54, while in the second case f clearly has a strict local extremum at $x = 0$, in fact a minimum if $f(x) = x^4$ and a maximum if $f(x) = -x^4$.

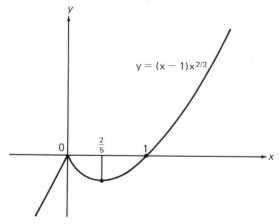

$$y = (x - 1)x^{2/3}$$

Figure 9.

c. Example. Find the local extrema of the function

$$f(x) = 3x^5 - 5x^3.$$

SOLUTION. Here f is differentiable for all x, and the only critical points of f are stationary points. These are the roots of the equation

$$f'(x) = 15x^4 - 15x^2 = 15x^2(x - 1)(x + 1) = 0,$$

namely the points $x = -1, 0, 1$. Calculating the second derivative, we get

$$f''(x) = 60x^3 - 30x = 30x(2x^2 - 1),$$

so that

$$f''(-1) = -30 < 0, \qquad f''(0) = 0, \qquad f''(1) = 30 > 0.$$

Therefore, by the second derivative test, f has a strict local maximum, equal to 2, at $x = -1$, and a strict local minimum, equal to -2, at $x = 1$. Although the second derivative test does not work at the point $x = 0$, it is easy to see that f has no extremum at $x = 0$. In fact,

$$f'(x) = 15x^2(x^2 - 1) < 0$$

if $-1 < x < 1$. Therefore f is decreasing in $[-1, 1]$, by Sec. 3.54, and hence can have no extremum at $x = 0$. You should confirm all this by sketching a graph of $f(x)$.

PROBLEMS

1. Verify the following rule for finding the global extrema of a function f continuous in a closed interval $[a, b]$: Let $x_1, x_2, \ldots, x_n$ be all the points of the open interval (a, b) at which f has local extrema. Then the largest of the numbers

$$f(a), f(x_1), f(x_2), \ldots, f(x_n), f(b)$$

is the global maximum of f in $[a, b]$, while the smallest of these numbers is the global minimum of f in $[a, b]$.

2. By investigating all critical points, find the local extrema, if any, of
   (a) $y = |x|$;   (b) $y = 2 + x - x^2$;   (c) $y = x^3 - 2x^2 + 3x - 1$.

3. Do the same for
   (a) $y = 2x^2 - x^4$;   (b) $y = x + \dfrac{1}{x}$;   (c) $y = x^{1/3}(1 - x)^{2/3}$.

4. Find the global extrema of
   (a) $y = x^2 - 4x + 6$ in $[-3, 10]$;   (b) $y = x + \dfrac{1}{x}$ in $[0.01, 100]$;

   (c) $y = \sqrt{5 - 4x}$ in $[-1, 1]$;   (d) $y = |x^2 - 3x + 2|$ in $[-10, 10]$.

5. Show that $|3x - x^3| \leqslant 2$ if $|x| \leqslant 2$.

6. Show that the function

$$y = \frac{ax + b}{cx + d} \qquad (c^2 + d^2 \neq 0)$$

has no strict local extrema, regardless of the values of a, b, c, d.

*7. Find the local extrema of the function

$$y = x^m(1 - x)^n,$$

where m and n are positive integers.

*8. What value of c minimizes the maximum of the function $f(x) = |x^2 + c|$ in the interval $[-1, 1]$?

*9. Suppose the function

$$y = \frac{ax + b}{(x - 1)(x - 4)}$$

has a local extremum, equal to -1, at the point $x = 2$. Find a and b, and show that the extremum is a maximum.

*10. Which term of the sequence

$$y_n = \frac{\sqrt{n}}{n + 10{,}000} \qquad (n = 1, 2, \ldots)$$

is the largest?

3.7 CONCAVITY AND INFLECTION POINTS

3.71. a. Let f be continuous in an interval I and differentiable at a point $p \in I$, and let $y = T(x)$ be the equation of the tangent to the curve $y = f(x)$ at the point with abscissa p. (For brevity, we will henceforth say "at the point p" or simply "at p," instead of "at the point with abscissa p.") Then, according to Sec. 2.52d,

$$y = T(x) = f'(p)(x - p) + f(p).$$

Suppose that $f(x) > T(x)$ in some deleted neighborhood of p, so that the curve $y = f(x)$ lies *above* its tangent at p in this neighborhood, as shown in Figure 10A. Then f is said to be *concave upward at* p. Similarly, suppose that $f(x) < T(x)$ in some deleted neighborhood of p, so that the curve $y = f(x)$ lies *below* its tangent at p in this neighborhood, as shown in Figure 10B. Then f is said to be *concave downward at* p. If f is concave upward (or downward) at every point of the interval I, we say that f is *concave upward* (or *downward*) *in* I.

b. A point p is said to be an *inflection point* of the function f if the curve $y = f(x)$ lies on one side of its tangent (at p) if $x < p$ and on the other side of its tangent if $x > p$. The two ways in which this can happen are illustrated in Figures 11A and 11B. If p is an inflection point of f, we also say that f has an inflection point at p.

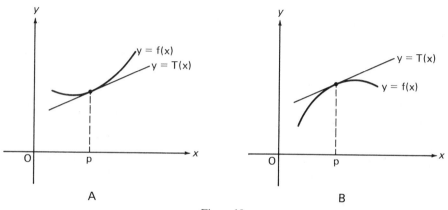

Figure 10.

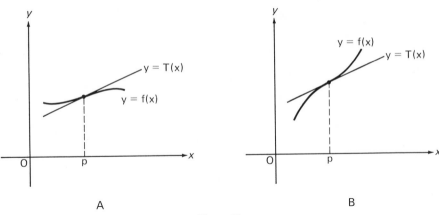

A   B

Figure 11.

3.72. With the help of the mean value theorem, it is not hard to show (see Prob. 11) that if f' exists and is *increasing* in some neighborhood of p, then f is concave *upward* at p, while if f' exists and is *decreasing* in some neighborhood of p, then f is concave *downward* at p. It can also be shown (see Prob. 12) that if f' exists in some neighborhood of p and has a strict local extremum at p, then p is an inflection point of f. Using these facts, we can develop a complete parallelism between the theory of increasing (or decreasing) functions and critical points, on the one hand, and the theory of upward (or downward) concavity and inflection points, on the other hand, with the first derivative f' now playing the role of the function f, and the second derivative f'' now playing the role of the first derivative f'. Thus you can easily convince yourself of the validity of the following propositions. In every case, the proof is the exact analogue of a proof that has already been given.

(1) *If $f''(p)$ exists and is positive, then f' is increasing in some neighborhood of p, so that f is concave upward at p.* This is the analogue of the result in Sec. 3.41a.

(2) *If $f''(p)$ exists and is negative, then f' is decreasing in some neighborhood of p, so that f is concave downward at p.* This is the analogue of the result in Sec. 3.41b.

(3) *If f has an inflection point at p, then either $f''(p)$ fails to exist or $f''(p)$ exists and equals zero.* This is the analogue of Theorem 3.63a.

(4) *Given that f'' exists in a deleted neighborhood of p, suppose $f''(p)$ either fails to exist or equals zero, and suppose f'' changes sign in going through p. Then f has an inflection point at p.* This second derivative test for an inflection point is the analogue of the first derivative test for a local extremum (Theorem 3.64a).

(5) *If $f''(p) = 0$ and if the third derivative $f'''(p)$ exists and is nonzero, then f has an inflection point at p.* This third derivative test for an inflection point is the analogue of the second derivative test for a local extremum (Theorem 3.65a).

3.73. Examples

a. Find the inflection points and investigate the concavity of the function

$$f(x) = x^4 - 2x^3 + 3x - 4.$$

SOLUTION. Here f'' exists for all x. Therefore, by Proposition (3), the only candidates for inflection points of f are the roots of the equation

$$f''(x) = 12x^2 - 12x = 12x(x - 1) = 0,$$

namely the points $x = 0$ and $x = 1$. By Propositions (1) and (2), f is concave upward in the interval $(-\infty, 0)$, since $f''(x) > 0$ if $x < 0$, concave downward in the interval $(0, 1)$, since $f''(x) < 0$ if $0 < x < 1$, and concave upward in the interval $(1, \infty)$, since $f''(x) > 0$ if $x > 1$. Therefore $x = 0$ and $x = 1$ are both inflection points of f, by Proposition (4). This also follows from Proposition (5), since $f'''(x) = 24x - 12$, and hence $f'''(0) = -12 \neq 0$, $f'''(1) = 12 \neq 0$.

b. Graph the function

$$f(x) = \frac{1}{24} x^4 - x^2 + 6.$$

SOLUTION. Since f is even, the graph of f is symmetric in the y-axis (Example 2.32d), and we need only study the behavior of f in the interval $[0, \infty)$. To find the extrema of f, we solve the question

$$f'(x) = \frac{1}{6} x^3 - 2x = \frac{1}{6} x(x^2 - 12) = 0,$$

obtaining two nonnegative stationary points $x = 0$ and $x = \sqrt{12}$. Since

$$f''(x) = \frac{1}{2} x^2 - 2,$$

we have

$$f''(0) = -2 < 0, \qquad f''(\sqrt{12}) = 4 > 0.$$

It follows from Theorem 3.65a that f has a strict local maximum, equal to $f(0) = 6$, at the point $x = 0$, and a strict local minimum, equal to $f(\sqrt{12}) = 0$, at the point $x = \sqrt{12}$. The only point in $[0, \infty)$ which can be an inflection point of f is the

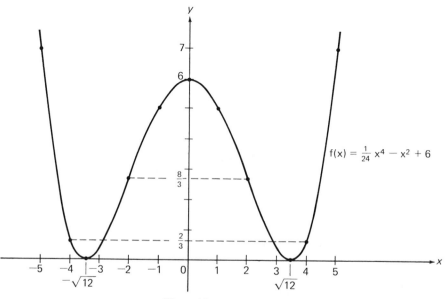

$$f(x) = \frac{1}{24} x^4 - x^2 + 6$$

Figure 12.

nonnegative solution of the equation

$$f''(x) = \frac{1}{2}x^2 - 2 = 0,$$

namely the point $x = 2$. This point is actually an inflection point, by Proposition (4), since $f''(x) < 0$ in $[0, 2)$ and $f''(x) > 0$ in $(2, \infty)$. At the same time, we note that f is concave downward in the interval $[0, 2)$ and concave upward in the interval $(2, \infty)$. Moreover $f'''(x) = x$, and hence $f'''(2) = 2 \neq 0$. Therefore the fact that $x = 2$ is an inflection point of f also follows from Proposition (5).

The function f has no asymptotes (Sec. 2.93), since it does not become infinite at any finite points and does not approach a finite limit as $x \to \pm\infty$. In fact, $f(x) \to \infty$ as $x \to \pm\infty$. To draw an accurate graph of f, we need a few more values of f besides $f(0) = 6$ and $f(\sqrt{12}) = 0$. The following three will suffice:

$$f(1) = \frac{121}{24} \approx 5, \qquad f(2) = \frac{8}{3}, \qquad f(4) = \frac{2}{3}, \qquad f(5) = \frac{169}{24} \approx 7.$$

Plotting the corresponding points and connecting them by a "smooth curve," we get the graph shown in Figure 12, after using the symmetry of the graph in the y-axis.

c. It is apparent from the previous example that we can form no clear idea of the behavior of a function without first locating all its extrema and inflection points, as well as examining it for possible asymptotes and testing it for parity (evenness or oddness). Figure 13 shows what can go wrong if we try to graph a function without doing this first. The solid curve is the "true graph" and the dashed curve is the quite misleading result of connecting five points of the graph by a "smooth curve."

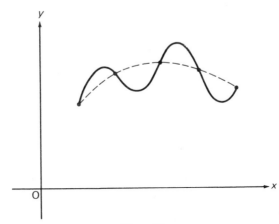

Figure 13.

PROBLEMS

1. "The function f is concave upward at p if it has a strict local minimum at p and concave downward at p if it has a strict local maximum at p." True or false?

2. What does the condition $f''(p) = 0$ by itself tell us about concavity at p or the presence of an inflection point at p?

3. Must a function have a local extremum between two consecutive inflection points?

4. Find the inflection points and investigate the concavity of the function $y = 2x^4 - 3x^2 + 2x + 2$.

5. Do the same for the function

$$y = \frac{x^3}{x^2 + 3a^2}.$$

6. For what value of c does the function $y = x^3 + cx^2 + 1$ have an inflection point at $x = 1$?

7. A point $P = (p, f(p))$ is said to be an inflection point of the *curve* $y = f(x)$ if p is an inflection point of the *function f*. For what values of a and b is the point $(1, 3)$ an inflection point of the curve $y = ax^3 + bx^2$?

8. Graph the function $y = (x + 1)(x - 1)^2$, after first investigating extrema, concavity, inflection points, asymptotes, etc.

*9. Do the same for the function

$$y = \frac{x + 1}{x^2 + 1}. \tag{1}$$

*10. Show that the three inflection points of the function (1) are collinear, that is, lie on the same straight line.

*11. Show that if f' exists and is increasing in some neighborhood of p, then f is concave upward at p, while if f' exists and is decreasing in some neighborhood of p, then f is concave downward at p.

*12. Show that if f' exists in some neighborhood of p and has a strict local extremum at p, then p is an inflection point of f.

*13. Suppose f and its first and second derivatives f' and f'' are continuous in an interval I. Justify the following statement: f vanishes at a point p if the sign of f changes in passing through p, f has a local extremum at p if the sign of f' changes in passing through p, f has an inflection point at p if the sign of f'', and hence the concavity of f, changes in passing through p.

3.8 OPTIMIZATION PROBLEMS

A host of practical problems involve the determination of *largest* size, *least* cost, *shortest* time, *greatest* revenue, and so on. Problems of this type ask for the "best value" of some variable, and hence are called *optimization problems*. Many of them can be solved with the help of the powerful tools developed in the last few sections. There is no universal rule that works in all cases, and as in all "word problems," there is no substitute for using a little common sense early in the game before trying to turn some computational crank. The following examples will give you a good idea of how to go about solving optimization problems.

3.81. Example. A square box with no top is made by cutting little squares out of the four corners of a square sheet of metal c inches on a side, and then folding up the resulting flaps, as shown in Figure 14. What size squares should be cut out to make the box of largest volume?

SOLUTION. Let x be the side length of each little square. Then the volume of the box in cubic inches is just

$$V = V(x) = x(c - 2x)^2. \tag{1}$$

Moreover,

$$0 \leqslant x \leqslant \frac{c}{2}, \tag{2}$$

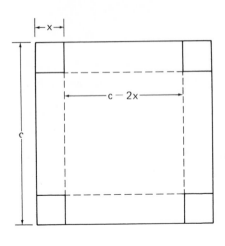
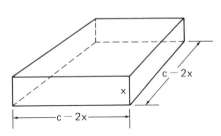

Figure 14.

since it is impossible to cut away either overlapping squares or squares of negative side length. Our problem is thus to determine the value of x at which the function (1) takes its global maximum in the interval (2). Since V is differentiable for all x, the only critical points of V, and hence, by Sec. 3.63c, the only points at which V can have a local extremum in the whole interval $(-\infty, \infty)$ are the solutions of the equation

$$\frac{dV}{dx} = c^2 - 8cx + 12x^2 = (c - 6x)(c - 2x) = 0,$$

namely $x = c/6$ and $x = c/2$. Moreover, since V is positive in the open interval

$$0 < x < \frac{c}{2}$$

and vanishes at the end points $x = 0$ and $x = c/2$, the global maximum of V in the closed interval (2), guaranteed by Theorem 3.32c and by the "physical meaning" of the problem, must be at an interior point of (2), and hence must be a local maximum of V. But $x = c/6$ is the only interior point of (2) at which V can have a local extremum, and therefore it is apparent without any further tests that V takes its maximum in (2) at the point $x = c/6$. This can be confirmed by noting that

$$\frac{d^2V}{dx^2}\bigg|_{x=c/6} = (-8c + 24x)\big|_{x=c/6} = -4c < 0,$$

and then applying the second derivative test (Theorem 3.65a).

Thus, finally, the largest box is obtained by cutting squares of side length $c/6$ out of the corners of the original sheet of metal. The volume of the resulting box equals

$$V\big|_{x=c/6} = \frac{c}{6}\left(\frac{2c}{3}\right)^2 = \frac{2}{27}c^3.$$

3.82. Example.   An island lies l miles offshore from a straight beach. Down the beach h miles from the point nearest the island, there is a group of vacationers who plan to get to the island by using a beach buggy going α mi/hr, trailing a motor-boat which can do β mi/hr. At what point of the beach should the vacationers transfer from the buggy to the boat in order to get to the island in the shortest time?

SOLUTION. The geometry of the problem is shown in Figure 15, where the vacationers start at A, the island is at C, and x is the distance between the point P at which they launch the boat and the point B of the beach nearest the island. The time it takes to get to the island is given by the formula

$$T = T(x) = \frac{|AP|}{\alpha} + \frac{|PC|}{\beta}$$

$$= \frac{1}{\alpha}(h - x) + \frac{1}{\beta}\sqrt{x^2 + l^2} \qquad (0 \leqslant x \leqslant h) \qquad (3)$$

(the time taken equals the distance travelled divided by the speed), where the boat leaves from the starting point A if $x = h$ and from the point B nearest the island if $x = 0$. Differentiating (3) with respect to x, we get

$$\frac{dT}{dx} = \frac{1}{\beta}\left(\frac{x}{\sqrt{x^2 + l^2}} - k\right),$$

where $k = \beta/\alpha$. If $k \geqslant 1$, that is, if $\beta \geqslant \alpha$, then dT/dx is negative for all $x \in (0, h)$, and hence T is decreasing in $[0, h]$, by Sec. 3.54. In this case, T takes its global minimum in $[0, h]$ at $x = h$, so that the vacationers should forget about the buggy and go straight to the island by boat.

The same is true if $k < 1$, provided that

$$x_0 = \frac{kl}{\sqrt{1 - k^2}} \geqslant h. \qquad (4)$$

In fact, if $k < 1$, the equation

$$\frac{x}{\sqrt{x^2 + l^2}} - k = 0$$

has the unique solution $x = x_0$, where x_0 lies outside the interval $(0, h)$ if (4) holds. Therefore dT/dx is again negative at every point of $(0, h)$, since dT/dx cannot change sign in $(0, h)$ and dT/dx is clearly negative for small enough x. But then T is again decreasing in $[0, h]$. Thus, in this case too, the vacationers should go straight to the island by boat.

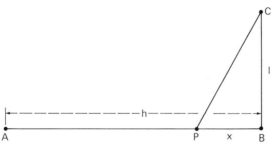

Figure 15.

However, if

$$x_0 = \frac{kl}{\sqrt{1 - k^2}} < h,$$

then x_0 lies in the interval $(0, h)$. Moreover, in this case, dT/dx is negative at every point of $(0, x_0)$ and positive at every point of (x_0, h). Therefore T must have a local minimum at x_0, by the first derivative test (Theorem 3.64a). But then T takes its global minimum in $[0, h]$ at x_0 (why?). This means that the vacationers should now stop the buggy and launch the boat at the point with coordinate x_0, as measured from B.

3.83. Example. A monopolistic firm has a total revenue function

$$R(Q) = -AQ^2 + BQ$$

(Sec. 3.2, Prob. 7) and a total cost function

$$C(Q) = aQ^2 + bQ + c$$

(Sec. 3.22a), where the coefficients A, B, a, b, c are all positive constants and $B > b$. The government wishes to levy an excise tax on the commodity produced by the firm. What tax rate should the government impose on the firm's output to maximize the tax revenue $T = rQ$, knowing that the firm will add the tax to its costs and adjust its output to maximize the profit after taxes?

SOLUTION.   The cost and profit after taxes are

$$C_T(Q) = C(Q) + rQ = aQ^2 + (b + r)Q + c$$

and

$$\Pi_T(Q) = R(Q) - C_T(Q) = -(A + a)Q^2 + (B - b - r)Q - c \qquad (5)$$

(Sec. 3.2, Prob. 10). Differentiating (5) with respect to Q, with r regarded as a constant, and setting the result equal to zero, we get the equation

$$\frac{d\Pi_T(Q)}{dQ} = -2(A + a)Q + (B - b - r) = 0,$$

whose only solution is

$$Q_0 = \frac{B - b - r}{2(A + a)}. \qquad (6)$$

Since

$$\frac{d^2\Pi_T(Q)}{dQ^2} = -2(A + a) < 0,$$

it follows from the second derivative test that the output level (6) actually maximizes the firm's profit after taxes, at the tax rate r.

Knowing that the firm will maximize its profit after taxes, the government chooses its tax rate r to maximize the revenue

$$T = rQ_0 = \frac{(B - b - r)r}{2(A + a)}, \qquad (7)$$

calculated at the output level (6). To maximize T as a function of the tax rate r, which is now regarded as variable, we differentiate (7) with respect to r, obtaining

$$\frac{dT}{dr} = \frac{B - b - 2r}{2(A + a)}.$$

The optimum tax rate r_0 is the solution of the equation $dT/dr = 0$, namely

$$r_0 = \frac{B - b}{2}.$$

By the second derivative test, r_0 actually maximizes the government's revenue, since

$$\frac{d^2 T}{dr^2} = -\frac{1}{A + a} < 0.$$

PROBLEMS

1. Among all rectangles of a given area A, find the one with the smallest perimeter.
2. Find the right triangle of largest area, given that the sum of one leg of the triangle and the hypotenuse is a constant c.
3. What is the largest volume of a right circular cone of slant height l?
4. What is the largest volume of a right circular cylinder inscribed in a sphere of radius R?
5. Given two points $A = (0, 3)$ and $B = (4, 5)$, find the point P on the x-axis for which the distance $|AP| + |PB|$ is the smallest.
6. Find the least amount of sheet metal needed to make a cylindrical cup of a given volume V.
7. In Example 3.82, should the vacationers ever take the buggy all the way to the point B nearest the island?
8. The results of n measurements of an unknown quantity x are $x_1, x_2, \ldots, x_n$. What value of x minimizes the expression $(x - x_1)^2 + (x - x_2)^2 + \cdots + (x - x_n)^2$?
9. Two ships, originally at distances a and b from a point P, sail toward P with speeds α and β along straight line routes making an angle of $90°$ with each other. At what time t is the distance between the two ships the smallest? What is the distance d of closest approach?
10. Given a point $P = (a, b)$ in the first quadrant, find the line through P which cuts off the triangle of least area from the quadrant.
11. Let $R(Q)$ be the total revenue received by a monopolistic firm from the sale of a quantity Q of some commodity, and let $C(Q)$ be the firm's total cost function. The firm wants to adjust its output to maximize its profit

$$\Pi(Q) = R(Q) - C(Q). \tag{8}$$

   Show that the profit function (8) is maximized at any output level such that
   (a) Marginal revenue (MR) equals marginal cost (MC);
   (b) Marginal revenue is increasing more slowly than marginal cost.
12. Suppose a monopolistic firm has a total revenue function $R(Q) = 1200Q - 10Q^2$ and a total cost function $C(Q) = Q^3 - 60Q^2 + 1500Q + 1000$. What output level maximizes the firm's profit?

*13.   "Normal cost conditions" are characterized by three properties:

   (a)   There are certain fixed costs (the overhead);
   (b)   Total cost increases with output;
   (c)   Marginal cost is always positive; as the output increases, the marginal cost first decreases and then increases.

Suppose a firm has a cubic total cost function

$$C(Q) = aQ^3 + bQ^2 + cQ + d.$$

Show that

$$a > 0, \quad b < 0, \quad c > 0, \quad d > 0, \quad b^2 < 3ac$$

under normal cost conditions.

*14.   For which chord BC parallel to the tangent to a circle at a point A is the area of the triangle ABC largest?

*15.   What is the largest surface area (including the top and bottom) of a right circular cylinder inscribed in a sphere of radius R?

*16.   Given a point P inside an acute angle, let L be the line segment through P cutting off the triangle of least area from the angle. Show that P bisects the part of L inside the angle. Show that this property also characterizes the point P in Problem 10.

*17.   According to *Fermat's principle*, the path taken by a ray of light which leaves a point A and passes through a point B after being reflected by a plane mirror is such as to minimize the time taken to traverse the whole path from A to the mirror to B. According to the *law of reflection*, the *angle of incidence* (between the incident ray and the perpendicular to the mirror) equals the *angle of reflection* (between the reflected ray and the perpendicular to the mirror). Deduce the law of reflection from Fermat's principle.

Chapter 4

INTEGRAL CALCULUS

4.1 THE DEFINITE INTEGRAL

The study of calculus is closely associated with the study of limits of various kinds. So far we have encountered the limit of a function at a point, the limit of a sequence, and the sum of an infinite series, as well as one-sided limits, infinite limits, limits at infinity, and asymptotes. We now consider still another kind of limit, leading to the concept of the *definite integral*. This kind of limit comes up time and again in problems involving the "summation of a very large number of individually small terms." The prototype of all such problems is the problem of finding the "area under a curve."

4.11. The area under a curve

a. Let

$$y = f(x) \qquad (a \leqslant x \leqslant b)$$

be a function which is continuous and nonnegative in a closed interval $[a, b]$. Then, by the *area under the curve* $y = f(x)$, from $x = a$ to $x = b$, we mean the area A of the plane region bounded by the curve $y = f(x)$, the x-axis, and the lines $x = a$ and $x = b$. This can also be described as the *area between the curve* $y = f(x)$ *and the x-axis*, from $x = a$ to $x = b$.

We can think of the region as a kind of trapezoid with three straight sides and one curved side, unless $f(a) = 0$ or $f(b) = 0$, in which case one or both of the vertical sides may shrink to a point. Such regions are not considered in elementary geometry. Thus, in the process of calculating A, we must decide what is meant by A in the first place!

b. With this in mind, we divide the interval $[a, b]$ into a large number n of small subintervals $[x_{i-1}, x_i]$, by introducing points of subdivision $x_1, x_2, \ldots, x_{n-1}$ such that

$$a = x_0 < x_1 < x_2 < \cdots < x_{n-1} < x_n = b,$$

where, in the interest of a uniform notation, the end points a and b of the original interval $[a, b]$ are assigned alternative symbols x_0 and x_n, as if they were points of subdivision too. Let

$$\Delta x_i = x_i - x_{i-1} \qquad (i = 1, 2, \ldots, n)$$

be the length of the ith subinterval, and let λ be the maximum length of all the subintervals, that is, the largest of the numbers $\Delta x_1, \Delta x_2, \ldots, \Delta x_n$. We denote this by

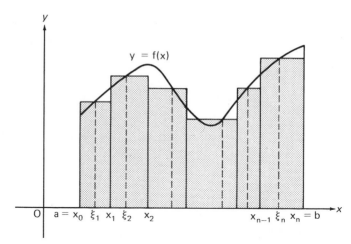

Figure 1.

writing

$$\lambda = \max \{x_1 - x_0, x_2 - x_1, \ldots, x_n - x_{n-1}\} = \max \{\Delta x_1, \Delta x_2, \ldots, \Delta x_n\}$$

(see Prob. 1). The lines $x = x_0$, $x = x_1$, $x = x_2$, ..., $x = x_{n-1}$, $x = x_n$ divide the region into n narrow strips, as shown in Figure 1. Being continuous, $f(x)$ does not change "much" in the interval $[x_{i-1}, x_i]$, and hence it seems like a good approximation to regard $f(x)$ as having the constant value $f(\xi_i)$ in $[x_{i-1}, x_i]$, where ξ_i is an *arbitrary* point of $[x_{i-1}, x_i]$. This is equivalent to replacing the strips, with curved tops, by the shaded rectangles shown in the figure. The sum of the areas of these rectangles is given by

$$\sum_{i=1}^{n} f(\xi_i)(x_i - x_{i-1}) = \sum_{i=1}^{n} f(\xi_i)\,\Delta x_i, \tag{1}$$

where we use the summation notation introduced in Sec. 2.96a. It seems reasonable to regard (1) as a good approximation to the area A of the region, where the approximation gets "better and better" as the bases of the rectangles all get "smaller and smaller," that is, as the number λ, the largest of these bases, gets "smaller and smaller." This suggests that we *define* A as the limit

$$A = \lim_{\lambda \to 0} \sum_{i=1}^{n} f(\xi_i)(x_i - x_{i-1}) = \lim_{\lambda \to 0} \sum_{i=1}^{n} f(\xi_i)\,\Delta x_i, \tag{2}$$

and this is exactly what we will do.

4.12. a. These considerations lead naturally to the following definition: Given a function $f(x)$ defined in a closed interval $[a, b]$, let $x_1, x_2, \ldots, x_{n-1}$ be points of subdivision of $[a, b]$ such that

$$a = x_0 < x_1 < x_2 < \cdots < x_{n-1} < x_n = b,$$

and let ξ_i be an arbitrary point of the subinterval $[x_{i-1}, x_i]$, of length $\Delta x_i = x_i - x_{i-1}$. Suppose the sum

$$\sigma = \sum_{i=1}^{n} f(\xi_i)(x_i - x_{i-1}) = \sum_{i=1}^{n} f(\xi_i)\,\Delta x_i \tag{3}$$

approaches a finite limit as

$$\lambda = \max\{x_1 - x_0, x_2 - x_1, \ldots, x_n - x_{n-1}\} = \max\{\Delta x_1, \Delta x_2, \ldots, \Delta x_n\}$$

approaches zero. Then the limit is called the *definite integral* of $f(x)$ from a to b, denoted by

$$\int_a^b f(x)\,dx, \tag{4}$$

and the function $f(x)$ is said to be *integrable* in $[a, b]$, or over $[a, b]$.

It should be noted that here we do not require the function $f(x)$ to be continuous in $[a, b]$, and in fact, the integral (4) exists even in cases where $f(x)$ is discontinuous. This matter will be discussed further in Sec. 4.14.

b. The quantities σ and λ depend, of course, on the choice of the points of subdivision

$$x_1, x_2, \ldots, x_{n-1}. \tag{5}$$

To emphasize this, we can write $\sigma = \sigma(X)$ and $\lambda = \lambda(X)$, where X is the set of points (5). Loosely speaking, X is a "partition" of the interval $[a, b]$, and the quantity $\lambda = \lambda(X)$ is a measure of the "fineness" of the partition. What does it mean to say that σ approaches the limit (4) as $\lambda \to 0$? Just this: The quantity

$$\sigma(X) - \int_a^b f(x)\,dx$$

is "arbitrarily near" zero for all X such that $\lambda(X)$ is "sufficiently small," regardless of the choice of the points

$$\xi_i \in [x_{i-1}, x_i] \qquad (i = 1, 2, \ldots, n).$$

Or in "ε, δ language," which is particularly appropriate here, given any $\varepsilon > 0$, we can find a number $\delta > 0$ such that

$$\left| \sigma(X) - \int_a^b f(x)\,dx \right| < \varepsilon$$

whenever $0 < \lambda(X) < \delta$.

c. This is a different kind of limit than those considered so far, but it is handled in the same way. For example, let σ be the sum (3) and let c be an arbitrary constant. Then

$$c\sigma = \sum_{i=1}^n cf(\xi_i)\,\Delta x_i,$$

by elementary algebra, and

$$\lim_{\lambda \to 0} c\sigma = c \lim_{\lambda \to 0} \sigma,$$

by the usual rule for the limit of a product, provided that the limit on the right exists. Therefore

$$\int_a^b cf(x)\,dx = c \int_a^b f(x)\,dx, \tag{6}$$

provided that $f(x)$ is integrable in $[a, b]$.

Similarly, let

$$\tau = \sum_{i=1}^n g(\xi_i)(x_i - x_{i-1}) = \sum_{i=1}^n g(\xi_i)\,\Delta x_i,$$

where $g(x)$ is another function defined in $[a, b]$. Then

$$\sigma + \tau = \sum_{i=1}^{n} [f(\xi_i) + g(\xi_i)] \, \Delta x_i,$$

by elementary algebra, and

$$\lim_{\lambda \to 0} (\sigma + \tau) = \lim_{\lambda \to 0} \sigma + \lim_{\lambda \to 0} \tau,$$

by the usual rule for the limit of a sum, provided that both limits on the right exist. Therefore

$$\int_a^b [f(x) + g(x)] \, dx = \int_a^b f(x) \, dx + \int_a^b g(x) \, dx, \tag{7}$$

provided that both functions $f(x)$ and $g(x)$ are integrable in $[a, b]$.

By repeated application of (6) and (7), we arrive at the more general formula

$$\int_a^b [c_1 f_1(x) + c_2 f_2(x) + \cdots + c_n f_n(x)] \, dx$$
$$= c_1 \int_a^b f_1(x) \, dx + c_2 \int_a^b f_2(x) \, dx + \cdots + c_n \int_a^b f_n(x) \, dx, \tag{8}$$

where $f_1(x), f_2(x), \ldots, f_n(x)$ are functions which are all integrable in $[a, b]$, and $c_1, c_2, \ldots, c_n$ are arbitrary constants. For example,

$$\int_a^b [c_1 f_1(x) + c_2 f_2(x) + c_3 f_3(x)] \, dx = \int_a^b [c_1 f_1(x) + c_2 f_2(x)] \, dx + \int_a^b c_3 f_3(x) \, dx$$
$$= \int_a^b c_1 f_1(x) \, dx + \int_a^b c_2 f_2(x) \, dx + \int_a^b c_3 f_3(x) \, dx$$
$$= c_1 \int_a^b f_1(x) \, dx + c_2 \int_a^b f_2(x) \, dx + c_3 \int_a^b f_3(x) \, dx,$$

and similarly for more than three terms. Formula (8) is, of course, the exact analogue for definite integrals of formula (7), p. 119, for indefinite integrals.

d. Note the distinction between the *definite* integral (4), which is a *number*, and the *indefinite* integral

$$\int f(x) \, dx,$$

which is a *function*. A definite integral always has two numbers attached to the integral sign, like the numbers a and b in (4), called the *lower limit of integration* and the *upper limit of integration*, respectively, where in these expressions, the word "limit" is used in the loose, colloquial sense, meaning "boundary" or "extent," and not in the precise technical sense in which it is used elsewhere in this book. The numbers a and b are, of course, the end points of the interval $[a, b]$, called the *interval of integration*. Otherwise, the terminology is the same for both definite and indefinite integrals. Thus the function $f(x)$ in (4) is again called the *integrand*, as in Sec. 3.53a, the operation leading from $f(x)$ to the number (4) is called *(definite) integration*, with respect to x, and the argument x is called the *variable of integration*. Since definite integration is an operation producing a number from a given function $f(x)$, the symbol x is a "dummy variable," in the sense that it can be replaced by any other symbol without changing the meaning of (4). For example,

$$\int_a^b f(x) \, dx = \int_a^b f(t) \, dt = \int_a^b f(\omega) \, d\omega.$$

The situation is exactly the same as for a dummy index of summation (Sec. 2.96a).

Things are different for indefinite integration, since our convention is to give the indefinite integral, which is an antiderivative, the same argument as the integrand. Thus

$$\int x \, dx = \frac{1}{2} x^2 + C, \qquad \int t \, dt = \frac{1}{2} t^2 + C.$$

and in this sense

$$\int x \, dx \neq \int t \, dt.$$

4.13. Examples

a. Comparing Secs. 4.11b and 4.12a, we find that the area under the curve $y = f(x)$ from a to b is given by the formula

$$A = \int_a^b f(x) \, dx. \tag{9}$$

More generally, let $f(x)$ and $g(x)$ be two functions defined and continuous in the same interval $[a, b]$, and suppose $f(x) \geq g(x)$ for every $x \in [a, b]$, as shown in Figure 2. Then how do we define the area A of the plane region $DCEF$ bounded by the lines $x = a$, $x = b$ and the curves $y = f(x)$, $y = g(x)$? Clearly, if the precise definition of area is to be compatible with the everyday meaning of area, we must insist that area be "additive" in the following sense: The area of a figure Φ made up of two other figures Φ_1 and Φ_2, which have no points in common except possibly parts of their boundaries, must equal the sum of the separate areas of Φ_1 and Φ_2. As applied to Figure 2, this means that

$$\text{(Area of } abCD) + \text{(Area of } DCEF) = \text{Area of } abEF,$$

and therefore

$$A = \text{Area of } DCEF = \text{(Area of } abEF) - \text{(Area of } abCD).$$

Since

$$\text{Area of } abEF = \int_a^b f(x) \, dx, \qquad \text{Area of } abCD = \int_a^b g(x) \, dx,$$

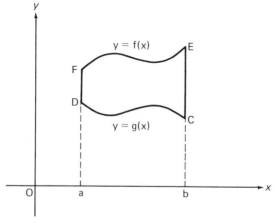

Figure 2.

it follows that

$$A = \int_a^b f(x)\,dx - \int_a^b g(x)\,dx = \int_a^b [f(x) - g(x)]\,dx. \tag{10}$$

Note that if $g(x) \equiv 0$, then the "lower curve" is just the x-axis, and (10) reduces to formula (9) for the area under the curve $y = f(x)$.

     b.   Evaluate

$$\int_a^b dx.$$

SOLUTION.   Here we are integrating the constant function $f(x) \equiv 1$. Thus

$$\int_a^b dx = \int_a^b 1 \cdot dx = \lim_{\lambda \to 0} \sum_{i=1}^n f(\xi_i)(x_i - x_{i-1}) = \lim_{\lambda \to 0} \sum_{i=1}^n (x_i - x_{i-1}),$$

since $f(\xi_i) = 1$ for each ξ_i. But

$$\sum_{i=1}^n (x_i - x_{i-1}) = (x_1 - x_0) + (x_2 - x_1) + (x_3 - x_2) + \cdots + (x_{n-1} - x_{n-2}) + (x_n - x_{n-1})$$

$$= -x_0 + (x_1 - x_1) + (x_2 - x_2) + \cdots + (x_{n-1} - x_{n-1}) + x_n,$$

where all the terms in the sum on the right vanish except the first and the last, leaving

$$\sum_{i=1}^n (x_i - x_{i-1}) = x_n - x_0 = b - a.$$

Therefore

$$\int_a^b dx = \lim_{\lambda \to 0} \sum_{i=1}^n (x_i - x_{i-1}) = \lim_{\lambda \to 0} (b - a),$$

so that, finally,

$$\int_a^b dx = b - a. \tag{11}$$

     c.   Evaluate

$$\int_a^b x\,dx.$$

SOLUTION.   Here the integrand is $f(x) = x$. Therefore

$$\int_a^b x\,dx = \lim_{\lambda \to 0} \sum_{i=1}^n f(\xi_i)(x_i - x_{i-1}) = \lim_{\lambda \to 0} \sum_{i=1}^n \xi_i(x_i - x_{i-1}).$$

Suppose we choose ξ_i to be the midpoint of the interval $[x_{i-1}, x_i]$, so that

$$\xi_i = \frac{1}{2}(x_i + x_{i-1})$$

(Sec. 1.5, Prob. 9). Then

$$\int_a^b x\,dx = \lim_{\lambda \to 0} \frac{1}{2} \sum_{i=1}^n (x_i + x_{i-1})(x_i - x_{i-1})$$

$$= \frac{1}{2} \lim_{\lambda \to 0} \sum_{i=1}^n (x_i^2 - x_{i-1}^2) = \frac{1}{2} \lim_{\lambda \to 0} (x_n^2 - x_0^2) = \frac{1}{2} \lim_{\lambda \to 0} (b^2 - a^2),$$

since the sum again "telescopes," reducing to simply $x_n^2 - x_0^2$. Thus, finally,

$$\int_a^b x \, dx = \frac{1}{2}(b^2 - a^2). \tag{12}$$

In Sec. 4.24 we will establish a general technique for evaluating definite integrals, which will allow us to completely bypass "brute force calculations" like those just made in deriving formulas (11) and (12).

4.14.   We now come to a crucial question: Which functions defined in a closed interval $[a, b]$ are integrable in $[a, b]$? In other words, if we take a function $f(x)$ defined in $[a, b]$ and form the sum (3), when does the sum approach a finite limit as $\lambda \to 0$? The answer is well beyond the scope of this book, and cannot even be expressed in the language of elementary calculus. However, the fact that we can't describe the *largest* set of integrable functions shouldn't bother us very much, since the following key proposition presents us with a *huge* set of integrable functions: *If $f(x)$ is continuous in $[a, b]$, then $f(x)$ is integrable in $[a, b]$.* The proof of this proposition is not particularly difficult, but it does require a deeper study of continuous functions than it would be profitable to pursue here.

The importance of continuity in calculus again stands revealed. Recall some of the other nice properties of continuous functions, presented in Sec. 3.3. It should be noted that there are integrable functions which are not continuous. An example of such a function is given in Problem 10. With a little ingenuity, we can also construct a function which fails to be integrable (see Prob. 11).

PROBLEMS

1.   Given a set A, all of whose elements are numbers, suppose A contains a largest element, that is, an element M such that $x \leqslant M$ for all $x \in A$. Then M is called the *maximum* of A, denoted by max A. Find max A if
(a) $A = \{0, 1, 2, \frac{4}{3}, (\sqrt{2})^2\}$;   (b) $A = \{x : x^3 - 2x^2 + x = 0\}$;
(c) $A = \{x : 0 < x < 1\}$.

2.   Let f be continuous in $[a, b]$. What is the number max $\{f(x) : a \leqslant x \leqslant b\}$?

3.   As in Sec. 4.12a, let $\lambda = $ max $\{\Delta x_1, \Delta x_2, \ldots, \Delta x_n\}$. Does $n \to \infty$ imply $\lambda \to 0$? Does $\lambda \to 0$ imply $n \to \infty$?

4.   What is the smallest value of $\lambda = $ max $\{\Delta x_1, \Delta x_2, \ldots, \Delta x_n\}$ for all choices of the points of subdivision $x_1, x_2, \ldots, x_{n-1}$? Does λ have a largest value?

5.   Show that formula (9) leads to the correct expressions for the area of a rectangle and for the area of a right triangle.

6.   Test formula (10) by using it to calculate the area of the trapezoid bounded by the lines $x = 2$, $x = 4$, $y = 1$ and $y = x$.

7.   Can negative area be defined in a meaningful way?

8.   How can we tell at once that the function $f(x) = 1/x$ is integrable in every closed interval that does not contain the point $x = 0$?

*9.   Find max A if $A = \{a, a^2, a^3, \ldots\}$ and $0 \leqslant a \leqslant 1$. What happens if $a > 1$? If a is negative?

*10.   The function

$$f(x) = \begin{cases} 1 & \text{if } x = 0, \\ 0 & \text{if } x \neq 0, \end{cases}$$

is discontinuous at $x = 0$. Verify that $f(x)$ is integrable in $[-1, 1]$.

*11.   Let

$$f(x) = \begin{cases} 1 & \text{if } x \text{ is rational,} \\ -1 & \text{if } x \text{ is irrational.} \end{cases}$$

Show that
(a) $f(x)$ is discontinuous at every point c;
(b) $f(x)$ fails to be integrable in every interval $[a, b]$.

4.2   PROPERTIES OF DEFINITE INTEGRALS

4.21.   First we consider what happens when the interval of integration is "split up."

a. THEOREM.   *If f is continuous in $[a, b]$ and if c is an interior point of $[a, b]$, then*

$$\int_a^b f(x)\, dx = \int_a^c f(x)\, dx + \int_c^b f(x)\, dx \qquad (a < c < b). \tag{1}$$

Proof.   As before, we divide the interval $[a, b]$ into a large number of small subintervals, by introducing points of subdivision, but this time we insist that one of the points of subdivision be the fixed point c. In other words, we now choose points of subdivision x_i $(i = 1, \dots, n - 1)$ such that

$$a = x_0 < x_1 < \cdots < x_{m-1} < x_m = c < x_{m+1} < \cdots < x_{n-1} < x_n = b,$$

where the subscript m depends, of course, on the number of points x_i which are less than c. Every such "partition" of $[a, b]$ automatically gives rise to a partition of the interval $[a, c]$, made up of the points $x_1, \dots, x_{m-1}$, and a partition of the interval $[c, b]$, made up of the points $x_{m+1}, \dots, x_{n-1}$. Correspondingly, the sum

$$\sigma = \sum_{i=1}^n f(\xi_i)\, \Delta x_i,$$

used to define the integral of f from a to b, can be written as

$$\sigma = \sigma' + \sigma'',$$

in terms of the sums

$$\sigma' = \sum_{i=1}^m f(\xi_i)\, \Delta x_i, \qquad \sigma'' = \sum_{i=m+1}^n f(\xi_i)\, \Delta x_i,$$

needed to define the integral of f from a to c and the integral of f from c to b. (Here Δx_i and ξ_i have the same meaning as in Sec. 4.12a.)
   Now let

$$\lambda = \max \{\Delta x_1, \dots, \Delta x_n\},$$
$$\lambda' = \max \{\Delta x_1, \dots, \Delta x_m\},$$
$$\lambda'' = \max \{\Delta x_{m+1}, \dots, \Delta x_n\}.$$

Then clearly $\lambda \to 0$ implies $\lambda' \to 0$ and $\lambda'' \to 0$, so that

$$\int_a^b f(x)\, dx = \lim_{\lambda \to 0} \sigma = \lim_{\lambda \to 0} (\sigma' + \sigma'') = \lim_{\lambda \to 0} \sigma' + \lim_{\lambda \to 0} \sigma''$$
$$= \lim_{\lambda' \to 0} \sigma' + \lim_{\lambda'' \to 0} \sigma'' = \int_a^c f(x)\, dx + \int_c^b f(x)\, dx,$$

where the existence of all three integrals follows from the assumption that f is continuous in $[a, b]$, and hence in $[a, c]$ and $[c, b]$ as well. □

b. So far, in writing the integral

$$\int_a^b f(x)\,dx,$$

it has been assumed that $a < b$. We now allow the case $a \geq b$, setting

$$\int_a^b f(x)\,dx = -\int_b^a f(x)\,dx, \tag{2}$$

by definition. Suppose $b = a$ in (2). Then

$$\int_a^a f(x)\,dx = -\int_a^a f(x)\,dx,$$

which implies

$$\int_a^a f(x)\,dx = 0. \tag{3}$$

The merit of the definition (2) is shown by the following extension of the preceding result:

c. THEOREM.   *If f is continuous in an interval containing the points a, b and c, then*

$$\int_a^b f(x)\,dx = \int_a^c f(x)\,dx + \int_c^b f(x)\,dx \qquad (a, b, c \ arbitrary). \tag{4}$$

Proof.   Formula (4) is an immediate consequence of (2) and (3) if two or three of the points a, b and c coincide. Moreover, (4) reduces to (1) if $a < c < b$. The other cases can be dealt with by using (2) together with (1). For example, if $c < b < a$, then, by (1),

$$\int_c^a f(x)\,dx = \int_c^b f(x)\,dx + \int_b^a f(x)\,dx,$$

and hence, by (2),

$$-\int_a^c f(x)\,dx = \int_c^b f(x)\,dx - \int_a^b f(x)\,dx,$$

which implies

$$\int_a^b f(x)\,dx = \int_a^c f(x)\,dx + \int_c^b f(x)\,dx.$$

The remaining cases $a < b < c$, $b < a < c$, $b < c < a$ and $c < a < b$ are treated similarly (give the details). □

4.22. The mean value theorem for integrals

The mean value theorem of Sec. 3.43a expresses the difference between the values of a differentiable function at two points a and b in terms of the derivative of the function at some point of $[a, b]$, in fact, at an interior point of $[a, b]$. There is a similar proposition expressing the definite integral from a to b of a continuous function in terms of the value of the function at some point of $[a, b]$:

a. THEOREM (**Mean value theorem for integrals**).   *If f is continuous in $[a, b]$, then there is a point $c \in [a, b]$ such that*

$$\int_a^b f(x)\,dx = f(c)(b - a). \tag{5}$$

Proof.   As in Sec. 3.32, let M be the maximum and m the minimum of f in $[a, b]$, taken at points p and q, respectively, and let

$$\sigma = \sum_{i=1}^{n} f(\xi_i) \Delta x_i$$

be the sum involved in the definition of the integral in (5). Clearly

$$m \Delta x_i \leqslant f(\xi_i) \Delta x_i \leqslant M \Delta x_i,$$

and therefore

$$\sum_{i=1}^{n} m \Delta x_i \leqslant \sigma \leqslant \sum_{i=1}^{n} M \Delta x_i,$$

or

$$m(b - a) \leqslant \sigma \leqslant M(b - a), \tag{6}$$

since

$$\sum_{i=1}^{n} \Delta x_i = (x_1 - x_0) + (x_2 - x_1) + \cdots + (x_n - x_{n-1}) = x_n - x_0 = b - a.$$

Taking the limit of (6) as $\lambda \to 0$, we get

$$m(b - a) \leqslant \int_a^b f(x)\, dx \leqslant M(b - a), \tag{7}$$

with the help of Problem 13, or equivalently

$$m \leqslant \frac{1}{b - a} \int_a^b f(x)\, dx \leqslant M.$$

Thus the quantity

$$h = \frac{1}{b - a} \int_a^b f(x)\, dx \tag{8}$$

is a number belonging to the interval $[m, M]$. It follows from the intermediate value theorem (Sec. 3.33b) that there is a point c, either equal to p or q if $h = M$ or $h = m$, or lying between p and q if $m < h < M$, but in any event certainly in the interval $[a, b]$, such that

$$f(c) = h. \tag{9}$$

Comparing (8) and (9), we immediately obtain (5). $\square$

Note that formula (5) remains true for $b < a$, provided that f is continuous in $[b, a]$. In fact, we then have

$$\int_a^b f(x)\, dx = -\int_b^a f(x)\, dx = -f(c)(a - b) = f(c)(b - a).$$

b.   The mean value theorem for integrals has a simple geometrical interpretation. Suppose $f(x) \geqslant 0$, as in Figure 3. Then, by Example 4.13a, the area of the region $abCD$ bounded by the curve $y = f(x)$, the x-axis and the lines $x = a$ and $x = b$ is given by the integral

$$\int_a^b f(x)\, dx.$$

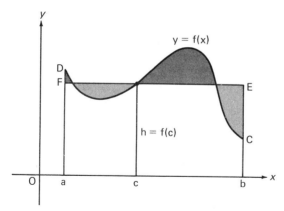

Figure 3.

According to (5), there is a point $c \in [a, b]$ such that the rectangle with base $b - a$ and altitude $h = f(c)$ has the same area as $abCD$. How this comes about is shown in the figure, where the dark parts of $abCD$ are "compensated" by the shaded parts of the rectangle $abEF$. The number h, equal to

$$\frac{1}{b - a} \int_a^b f(x) \, dx,$$

is called the *mean value* or *average* of the function $f(x)$ over the interval $[a, b]$.

4.23.   According to Sec. 4.14, every continuous function has a definite integral. We now use the mean value theorem for integrals to prove that every continuous function has an antiderivative and hence an indefinite integral.

a. THEOREM.   *Let f be continuous in an interval I, and let*

$$\Phi(x) = \int_{x_0}^x f(t) \, dt, \tag{10}$$

where x_0 is a fixed point of I and x is a variable point of I. Then Φ is an antiderivative of f in I.

Proof.   There is a slight technicality here, namely, if I has end points a and b, then $\Phi'(a)$ is defined by the right-hand limit

$$\Phi'(a) = \lim_{\Delta x \to 0+} \frac{\Phi(a + \Delta x) - \Phi(a)}{\Delta x}$$

if $a \in I$, while $\Phi'(b)$ is defined by the left-hand limit

$$\Phi'(b) = \lim_{\Delta x \to 0-} \frac{\Phi(b + \Delta x) - \Phi(b)}{\Delta x}$$

if $b \in I$. This is simply because x must belong to I, and hence can approach a only from the right and b only from the left.

To get on with the proof, we first note that the existence of the integral (10) follows from the continuity of f, by Sec. 4.14. Suppose x and $x + \Delta x$ both belong to I. Then

$$\Phi(x + \Delta x) = \int_{x_0}^{x + \Delta x} f(t) \, dt = \int_{x_0}^x f(t) \, dt + \int_x^{x + \Delta x} f(t) \, dt,$$

by Theorem 4.21c, and hence

$$\Phi(x + \Delta x) - \Phi(x) = \int_x^{x+\Delta x} f(t) \, dt. \tag{11}$$

Applying the mean value theorem for integrals to the right side of (11), which is independent of the fixed point x_0, we get

$$\Phi(x + \Delta x) - \Phi(x) = f(c)(x + \Delta x - x) = f(c) \, \Delta x, \tag{12}$$

where $x \leqslant c \leqslant x + \Delta x$ or $x + \Delta x \leqslant c \leqslant x$, depending on whether Δx is positive or negative. But then $c \to x$ as $\Delta x \to 0$, and therefore $f(c) \to f(x)$ as $\Delta x \to 0$, by the continuity of f. It follows that

$$\Phi'(x) = \lim_{\Delta x \to 0} \frac{\Phi(x + \Delta x) - \Phi(x)}{\Delta x} = \lim_{\Delta x \to 0} \frac{f(c) \, \Delta x}{\Delta x} = \lim_{\Delta x \to 0} f(c) = f(x),$$

that is, Φ is an antiderivative of f in I. $\square$

   b.   The content of this key theorem can be written concisely as

$$\frac{d}{dx} \int_{x_0}^x f(t) \, dt = f(x).$$

Notice how the presence of the letter x in the upper limit of integration forces us to use another letter for the variable of integration. Here we use t, but any letter other than x would do just as well. The theorem has a simple geometrical interpretation: Suppose $I = [a, b]$ and $f(t) \geqslant 0$, as in Figure 4, and let $x_0 = a$. Then $\Phi(x)$ is the shaded area under the curve $y = f(t)$ from $t = a$ to $t = x$, which varies from

$$\int_a^a f(t) \, dt = 0$$

to

$$\int_a^b f(t) \, dt$$

as x varies from a to b, and the rate of change of this area with respect to x at a given point of $[a, b]$ equals the "height" of the curve at the given point.

   c.   COROLLARY.   *If f is continuous in an interval I, then F has an indefinite integral in I.*

   Proof.   Since the function (10) is an antiderivative of f in I, we have

$$\int f(x) \, dx = \int_{x_0}^x f(t) \, dt + C,$$

where C is an arbitrary constant. $\square$

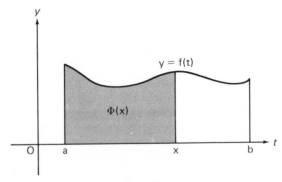

Figure 4.

4.24. The fundamental theorem of calculus

The following key theorem reveals the connection between differential and integral calculus. At the same time, it gives us a powerful tool for evaluating definite integrals.

a. THEOREM (**Fundamental theorem of calculus**). *If f is continuous in* $[a, b]$ *and if F is any antiderivative of f in* $[a, b]$, *then*

$$\int_a^b f(x)\,dx = F(b) - F(a). \tag{13}$$

Proof. By the preceding theorem, the function

$$\Phi(x) = \int_a^x f(t)\,dt$$

is an antiderivative of f in $[a, b]$. Let F be any other antiderivative of f in $[a, b]$. Then, by Theorem 3.52b,

$$\Phi(x) = F(x) + C,$$

where C is a constant. To determine C, we note that

$$F(a) + C = \Phi(a) = \int_a^a f(t)\,dt = 0,$$

which implies

$$C = -F(a).$$

Therefore

$$\Phi(x) = F(x) + C = F(x) - F(a).$$

Changing the dummy variable of integration from t to x, we then get

$$\int_a^b f(x)\,dx = \Phi(b) = F(b) - F(a). \quad \square$$

Note that formula (13) remains true for $b < a$, provided that f is continuous in $[b, a]$. In fact, we then have

$$\int_a^b f(x)\,dx = -\int_b^a f(x)\,dx = -[F(a) - F(b)] = F(b) - F(a).$$

b. A little extra notation comes in handy here. Given any function $\varphi(x)$ defined for $x = a$ and $x = b$, let

$$\varphi(x)\Big|_a^b \quad \text{or} \quad \left[\varphi(x)\right]_a^b$$

denote the *difference* $\varphi(b) - \varphi(a)$. With this notation, we can write (13) compactly as

$$\int_a^b f(x)\,dx = F(x)\Big|_a^b. \tag{14}$$

Moreover, since

$$F(x)\Big|_a^b = \left[F(x) + C\right]_a^b = \left[\int f(x)\,dx\right]_a^b,$$

we can also write (14) as

$$\int_a^b f(x)\,dx = \left[\int f(x)\,dx\right]_a^b$$

This formula shows the connection between the definite and indefinite integrals of $f(x)$ very explicitly.

4.25. Examples

a.   Evaluate

$$\int_a^b x^r \, dx.$$

SOLUTION.   It follows from (14) and formula (5), p. 119, that

$$\int_a^b x^r \, dx = \frac{x^{r+1}}{r+1}\Big|_a^b = \frac{b^{r+1} - a^{r+1}}{r+1} \tag{15}$$

if $r \neq -1$. Choosing $r = 0$ and $r = 1$ in (15), we immediately get the results of Examples 4.13b and 4.13c. The interval $[a, b]$ (or $[b, a]$ if $b < a$) must not contain the point $x = 0$ if r is negative, since otherwise x^r will fail to be continuous in $[a, b]$.

b.   Suppose

$$MC(Q) = 3Q^2 - 100Q + 1200$$

is the marginal cost of producing a commodity at output level Q. Find the total cost function $C(Q)$ if the overhead is 1000 money units. Express the cost (exclusive of overhead) of producing the second 10 units of the commodity as an integral, and evaluate it. Find the average cost (inclusive of overhead) of producing 20 units of the commodity.

SOLUTION.   According to Sec. 3.22a,

$$MC(Q) = \frac{dC(Q)}{dQ}.$$

Therefore

$$C(Q) = \int MC(Q) \, dQ = \int (3Q^2 - 100Q + 1200) \, dQ = Q^3 - 50Q^2 + 1200Q + k,$$

where k is a constant of integration. But $k = C(0) = 1000$, and hence

$$C(Q) = Q^3 - 50Q^2 + 1200Q + 1000.$$

The cost of producing the second 10 units is

$$\int_{10}^{20} MC(Q) \, dQ = C(Q)\Big|_{10}^{20} = C(20) - C(10) = 4000,$$

while the average cost of producing 20 units is

$$AC(20) = \frac{C(20)}{20} = \frac{13000}{20} = 650.$$

PROBLEMS

1.   Verify that

$$\int_{-1}^{1} x^3 \, dx = \int_{-1}^{0} x^3 \, dx + \int_0^1 x^3 \, dx$$

by direct calculation.

2. Evaluate

   (a) $\int_1^4 \sqrt{x}\, dx;$ (b) $\int_1^{27} x^{-2/3}\, dx;$ (c) $\int_1^2 \left(x^2 + \dfrac{1}{x^2}\right) dx;$

   (d) $\int_0^2 |1 - x|\, dx.$

3. Verify that

$$\frac{n+1}{2} \int_{-1}^1 x^n\, dx = \begin{cases} 1 & \text{if } n \text{ is even,} \\ 0 & \text{if } n \text{ is odd.} \end{cases}$$

4. Find the definite integral from 0 to 2 of the function

$$f(x) = \begin{cases} x^3 & \text{if } 0 \leqslant x \leqslant 1, \\ 2 - x & \text{if } 1 < x \leqslant 2. \end{cases}$$

5. Find the area A between the curves $y = \sqrt{x}$ and $y = x^2$ (see Figure 5). Why is the part of the curve $y = x^2$ lying in the first quadrant the reflection of the curve $y = \sqrt{x}$ in the line $y = x$?

6. Find the area between the line $x + y = 2$ and the curve $y = x^2$.

7. According to formula (5),

$$\int_1^7 f(x)\, dx = 6f(c),$$

   where $c \in [1, 7]$. Find c if $f(x) = x$.

8. Verify that the average of the function $f(x) = x$ over the interval $[a, b]$ is just the midpoint of the interval.

9. Consider a particle with equation of motion $s = s(t)$. We now have two definitions of the average velocity of the particle over the interval $[a, b]$, namely

$$v_{av} = \frac{s(b) - s(a)}{b - a}$$

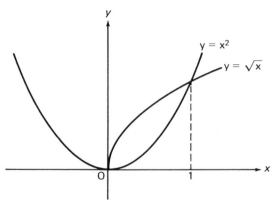

Figure 5.

(Sec. 3.12a) and

$$v_{av} = \frac{1}{b - a} \int_a^b v(t)\, dt$$

(Sec. 4.22b), where $v = v(t)$ is the particle's instantaneous velocity. Show that the two definitions are equivalent.

10. Is it true that

$$\int \frac{dx}{x} = \int_1^x \frac{dt}{t} + C?$$

11. Why is the function $\Phi(x)$ defined by (10) continuous in the interval I?

12. Show that if $\sigma(\lambda) \geq 0$ and $\sigma(\lambda) \to \sigma_0$ as $\lambda \to 0$, then $\sigma_0 \geq 0$.

13. Show that if $A \leq \sigma(\lambda) \leq B$ and $\sigma(\lambda) \to \sigma_0$ as $\lambda \to 0$, then $A \leq \sigma_0 \leq B$.

14. Evaluate

(a) $\dfrac{d}{dx} \displaystyle\int_a^b f(x)\, dx;$ (b) $\dfrac{d}{da} \displaystyle\int_a^b f(x)\, dx;$ (c) $\dfrac{d}{db} \displaystyle\int_a^b f(x)\, dx.$

15. Show that if f is continuous in $[a, b]$ and if $A \leq f(x) \leq B$ for every $x \in [a, b]$, then

$$A(b - a) \leq \int_a^b f(x)\, dx \leq B(b - a).$$

16. Show that if f is continuous and nonnegative in $[a, b]$, that is, nonnegative at every point $x \in [a, b]$, then

$$\int_a^b f(x)\, dx \geq 0.$$

17. Show that if f_1 and f_2 are continuous in $[a, b]$ and if $f_1(x) \leq f_2(x)$ for every $x \in [a, b]$, then

$$\int_a^b f_1(x)\, dx \leq \int_a^b f_2(x)\, dx.$$

*18. Show that if f is continuous and nonnegative in $[a, b]$, and if f is nonzero at some point $c \in [a, b]$, then

$$\int_a^b f(x)\, dx > 0.$$

*19. Let f be continuous and nonnegative in $[a, b]$, and suppose that

$$\int_a^b f(x)\, dx = 0.$$

Show that f has the constant value 0 in $[a, b]$.

*20. Let f_1 and f_2 be the same as in Problem 17, and suppose that in addition $f_1 \neq f_2$ in $[a, b]$, that is, suppose that $f_1(c) \neq f_2(c)$ for at least one point $c \in [a, b]$. Show that

$$\int_a^b f_1(x)\, dx < \int_a^b f_2(x)\, dx.$$

*21. Verify that

$$\frac{1}{6} < \int_0^2 \frac{dx}{10 + x} < \frac{1}{5}.$$

*22. Show that if f is continuous in $[a, b]$, then

$$\left| \int_a^b f(x)\, dx \right| \leqslant \int_a^b |f(x)|\, dx.$$

*23. Show that we can always choose the point c in formula (5) to be an *interior* point of $[a, b]$, that is, a point of (a, b).

4.3 THE LOGARITHM

4.31. One of the most important functions in mathematics is the *natural logarithm*, or simply the *logarithm*, denoted by ln x and defined for all positive x by the formula

$$\ln x = \int_1^x \frac{dt}{t}. \tag{1}$$

The expression on the right is just the definite integral from the fixed point $t = 1$ to the variable point $t = x$ of the function $1/t$. In geometrical terms, if $x > 1$, then ln x is the area under the curve $y = 1/t$ from $t = 1$ to $t = x$, as in Figure 6A. If $0 < x < 1$, then, since

$$\int_1^x \frac{dt}{t} = -\int_x^1 \frac{dt}{t},$$

ln x is the *negative* of the area under the curve $y = 1/t$ from $t = x$ to $t = 1$, as in Figure 6B. Thus ln $x > 0$ if $x > 1$, while ln $x < 0$ if $0 < x < 1$. Moreover,

$$\ln 1 = 0, \tag{2}$$

since

$$\int_1^1 \frac{dt}{t} = 0.$$

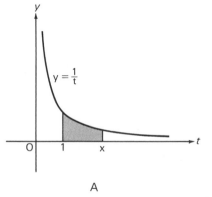
A

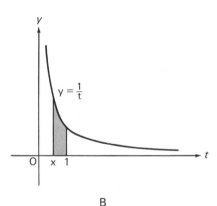
B

Figure 6.

4.32. It follows from Theorem 4.23a that the function $\ln x$ is *differentiable* in the interval $(0, \infty)$, with derivative

$$\frac{d}{dx} \ln x = \frac{1}{x}. \tag{3}$$

In particular, this shows that $\ln x$ is *continuous* in $(0, \infty)$, for the reason given in Sec. 2.66. Moreover, the function $\ln x$ is *increasing* in $(0, \infty)$. In fact, suppose that $0 < x_1 < x_2 < \infty$. Then

$$\ln x_2 = \int_1^{x_2} \frac{dt}{t} = \int_1^{x_1} \frac{dt}{t} + \int_{x_1}^{x_2} \frac{dt}{t} = \ln x_1 + \int_{x_1}^{x_2} \frac{dt}{t}.$$

Applying the mean value theorem for integrals to the last integral on the right, we get

$$\ln x_2 = \ln x_1 + \frac{x_2 - x_1}{c},$$

where $x_1 \leqslant c \leqslant x_2$, so that

$$\ln x_2 - \ln x_1 = \frac{x_2 - x_1}{c} > 0.$$

In other words, $x_1 < x_2$ implies $\ln x_1 < \ln x_2$, so that $\ln x$ is increasing, as claimed.

4.33. The next property of $\ln x$ is so important that it deserves a theorem of its own:

a. THEOREM. *Let a and b be arbitrary positive numbers. Then*

$$\ln (ab) = \ln a + \ln b. \tag{4}$$

Proof. Using the chain rule to differentiate the composite function $\ln (ax)$, we find that

$$\frac{d}{dx} \ln (ax) = \frac{1}{ax} \frac{d}{dx} (ax) = \frac{a}{ax} = \frac{1}{x}.$$

Therefore both functions $\ln (ax)$ and $\ln x$ have the same derivative $1/x$. In other words, $\ln (ax)$ and $\ln x$ are both antiderivatives of $1/x$. It follows from Theorem 3.52b that

$$\ln (ax) = \ln x + C, \tag{5}$$

where C is a constant. To determine C, we set $x = 1$ in (5), obtaining

$$\ln a = \ln 1 + C = C,$$

with the help of (2). Therefore (5) becomes

$$\ln (ax) = \ln x + \ln a.$$

Setting $x = b$ in this formula, we immediately get (4). $\square$

b. In particular, (4) implies

$$\ln (a^2) = \ln a + \ln a = 2 \ln a,$$
$$\ln (a^3) = \ln (a^2) + \ln a = 2 \ln a + \ln a = 3 \ln a,$$

and, more generally,

$$\ln (a^n) = n \ln a, \tag{6}$$

where, as always,

$$a^n = \underbrace{a \cdot a \cdots a.}_{n \text{ factors}}$$

It also follows from (2) and (4) that

$$\ln a + \ln \frac{1}{a} = \ln \left(a \cdot \frac{1}{a} \right) = \ln 1 = 0,$$

so that

$$\ln \frac{1}{a} = -\ln a. \tag{7}$$

Therefore

$$\ln \left(\frac{a}{b} \right) = \ln \left(a \cdot \frac{1}{b} \right) = \ln a + \ln \frac{1}{b} = \ln a - \ln b.$$

Note that (6) holds for every integer, positive, negative or zero, if we make the usual definitions

$$a^0 = 1, \qquad a^{-n} = \frac{1}{a^n}.$$

In fact,

$$\ln (a^{-n}) = \ln \frac{1}{a^n} = -\ln (a^n) = -n \ln a,$$

by (6) and (7), while

$$\ln a^0 = \ln 1 = 0 = 0 \cdot \ln a,$$

by (2).

    c. Choosing $a = 2$, say, in formula (6), we get

$$\ln (2^n) = n \ln 2,$$

where $\ln 2 > 0$. Given any positive number M, no matter how large, let n be any integer greater than $M/\ln 2$. Then

$$\ln x > \ln (2^n) = n \ln 2 > M$$

whenever $x > 2^n$, since $\ln x$ is increasing. This means that

$$\lim_{x \to \infty} \ln x = \infty \tag{8}$$

(Sec. 2.91b). Moreover,

$$\lim_{x \to 0+} \ln x = -\infty, \tag{9}$$

since, by Sec. 2.91c,

$$\lim_{x \to 0+} \ln x = \lim_{t \to \infty} \ln \frac{1}{t} = -\lim_{t \to \infty} \ln t = -\infty,$$

where we use (7) and (8).

According to (8) and (9), ln x takes "arbitrarily large" values of both signs. Since ln x is continuous, it follows from the intermediate value theorem (Sec. 3.33b) that ln x takes *every* value. In other words, the range of the function ln x is the whole interval $(-\infty, \infty)$.

4.34. Let e be the number such that

$$\ln e = 1, \tag{10}$$

or equivalently

$$\int_1^e \frac{dt}{t} = 1,$$

so that the area under the curve $y = 1/t$ from $t = 1$ to $t = e$ is precisely 1. The number e, called the *base of the natural logarithms*, is a constant of great importance in calculus and its applications. It turns out that e is irrational and equals

$$e = 2.7182818284\ldots$$

As we will see in Sec. 4.51a,

$$e = \lim_{n \to \infty} \left(1 + \frac{1}{n}\right)^n, \tag{11}$$

that is, e is the limit of the sequence

$$2, \left(\frac{3}{2}\right)^2, \left(\frac{4}{3}\right)^3, \ldots, \left(\frac{n+1}{n}\right)^n, \ldots$$

4.35. Figure 7 shows the graph of the function $y = \ln x$. It is apparent from the figure that ln x is increasing in $(0, \infty)$, has the range $(-\infty, \infty)$, and satisfies formulas (2) and (10). Note also that ln x is a one-to-one function, like every increasing function (Sec. 2.3, Prob. 15), and has the y-axis as its only asymptote. Moreover, ln x is concave downward in the whole interval $(0, \infty)$, by Sec. 3.72, Proposition (2),

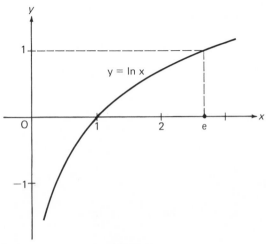

Figure 7.

since

$$\frac{d^2}{dx^2} \ln x = \frac{d}{dx}\frac{1}{x} = -\frac{1}{x^2} < 0$$

for every $x \in (0, \infty)$.

4.36. The function $\log_a x$

a. We now introduce the function $\log_a x$, where x is any positive number and a is positive but different from 1. This function, called the *logarithm to the base a*, is defined by the formula

$$\log_a x = \frac{\ln x}{\ln a}, \tag{12}$$

and has properties very similar to those of the function $\ln x$, to which it reduces for $a = e$. For example, it follows at once from (12) that

$$\log_a 1 = \frac{\ln 1}{\ln a} = 0, \qquad \log_a a = \frac{\ln a}{\ln a} = 1,$$

$$\log_a \frac{1}{x} = \frac{1}{\ln a} \ln \frac{1}{x} = -\frac{\ln x}{\ln a} = -\log_a x,$$

$$\log_a (xy) = \frac{\ln (xy)}{\ln a} = \frac{\ln x + \ln y}{\ln a} = \log_a x + \log_a y,$$

with the help of (2), (4) and (7). For $a = 10$, we get the *common logarithm* $\log_{10} x$ of elementary mathematics, usually denoted by $\log x$ without the subscript 10.

b. If a and b are two positive numbers different from 1, then

$$\log_a x = \frac{\ln x}{\ln a} = \frac{\ln b \ln x}{\ln a \ln b},$$

so that

$$\log_a x = \log_a b \cdot \log_b x. \tag{13}$$

Setting $x = a$ in (13), we find that

$$1 = \log_a b \cdot \log_b a,$$

or equivalently

$$\log_a b = \frac{1}{\log_b a}. \tag{14}$$

In particular, choosing $b = e$ in (14), we get

$$\log_a e = \frac{1}{\log_e a} = \frac{1}{\ln a}. \tag{15}$$

c. The derivative of the function $\log_a x$ is easily calculated. In fact,

$$\frac{d}{dx} \log_a x = \frac{d}{dx}\frac{\ln x}{\ln a} = \frac{1}{x \ln a} = \frac{1}{x} \log_a e,$$

with the help of (15).

PROBLEMS

1. Why can't the integral defining $\ln x$ be evaluated by using formula (15), p. 150?
2. Are the functions $\ln (x^2)$ and $2 \ln x$ identical?
3. Find the domain of
    (a) $\ln (\sqrt{x - 4} + \sqrt{6 - x})$;   (b) $\ln (\ln x)$;   (c) $\ln (\ln (\ln x)))$.
4. Differentiate
    (a) $\ln (x^3 - 2x + 5)$;   (b) $x \ln x$;   (c) $(\ln x)^3$;   (d) $\ln (\ln x)$.
5. Differentiate

    (a) $\ln \dfrac{1 + x}{1 - x}$;   (b) $\ln \dfrac{1 + x^2}{1 - x^2}$;   (c) $\ln \sqrt{\dfrac{1 + x}{1 - x}}$;

    (d) $\ln (x + \sqrt{1 + x^2}))$.

6. According to the mean value theorem (Sec. 3.43a),

    $$f(2) - f(1) = f'(c),$$

    where $1 < c < 2$. Find c if $f(x) = \ln x$.
7. Show that the tangent to the curve $y = \ln x$ at the point e goes through the origin.
8. What is the fourth derivative of the function $y = x^2 \ln x$?
9. Verify that

    $$\int \ln x \, dx = x \ln x - x + C.$$

10. Find the area of the region bounded by the x-axis, the line $x = e$ and the curve $y = \ln x$.
11. In which intervals is the function $x^2 - \ln (x^2)$ increasing? Decreasing?
12. Where does the function $x - \ln x$ have its global minimum in $(0, \infty)$? Does it have a maximum in $(0, \infty)$?
13. Find the inflection points and investigate the concavity of the function $\ln (1 + x^2)$.
*14. Find the domain of
    (a) $\sqrt{\log_a x}$;   (b) $\log_{10} (1 - \log_{10} (x^2 - 5x + 16))$.
*15. Show that

    $$\lim_{x \to \infty} \log_a x = \infty, \qquad \lim_{x \to 0+} \log_a x = -\infty$$

    if $a > 1$, while

    $$\lim_{x \to \infty} \log_a x = -\infty, \qquad \lim_{x \to 0+} \log_a x = \infty$$

    if $0 < a < 1$.
*16. Verify that the function $\ln (x + \sqrt{1 + x^2})$ is odd.
*17. Use the mean value theorem to verify that

    $$\frac{b - a}{b} < \ln \frac{b}{a} < \frac{b - a}{a}$$

    if $0 < a < b$.
*18. Show that

    $$\frac{d}{dx} \log_x a = -\frac{\ln a}{x(\ln x)^2}.$$

4.4 THE EXPONENTIAL

4.41. a. The logarithm function

$$y = \ln x = \int_1^x \frac{dt}{t},$$

defined in the preceding section, has domain $(0, \infty)$ and range $(-\infty, \infty)$. Moreover, it is increasing, one-to-one and continuous in the whole interval $(0, \infty)$, and hence in every closed subinterval $[a, b] \subset (0, \infty)$. Therefore, by the proposition cited in Sec. 2.81c, the inverse function

$$x = \ln^{-1} y \tag{1}$$

is increasing and continuous in the interval $[\ln a, \ln b]$. But $\alpha = \ln a \to -\infty$ as $a \to 0+$, while $\beta = \ln b \to \infty$ as $b \to \infty$, by Sec. 4.33c. Therefore the function (1) is increasing and continuous in every closed subinterval $[\alpha, \beta] \subset (-\infty, \infty)$, and hence in the whole interval $(-\infty, \infty)$.

b. In studying the function (1), it is natural to preserve the custom of denoting the independent variable by x and the dependent variable by y. Thus we now write

$$y = \ln^{-1} x,$$

instead of (1). This function, which is one of the most important in mathematics, deserves a name and notation of its own. It is called the *exponential to the base e*, or simply the *exponential*, and is denoted by

$$y = \exp x.$$

Another, even more common notation for the exponential will be introduced in a moment. The function $\exp x$ is defined and *positive* for all x. This follows from the fact that the range of $\exp x$ is just the domain of $\ln x$, namely the interval $(0, \infty)$.

c. Being the inverse of the function $\ln x$, the exponential satisfies the formulas

$$\exp(\ln x) \equiv x, \qquad \ln(\exp x) \equiv x. \tag{2}$$

To see this, we use the formulas (3), p. 43, changing y to x in the second formula, and writing $\ln$ for f and $\exp$ for f^{-1}. It follows from the formulas

$$\ln 1 = 0, \qquad \ln e = 1$$

that

$$\exp 0 = 1 \tag{3}$$

and

$$\exp 1 = e, \tag{4}$$

where e is the number introduced in Sec. 4.34.

4.42. The following theorem expresses one of the key properties of the exponential:

a. THEOREM. *Let x and y be arbitrary real numbers. Then*

$$\exp(x + y) = (\exp x)(\exp y). \tag{5}$$

Proof. Let $a = \exp x$, $b = \exp y$, so that $x = \ln a$, $y = \ln b$. Then

$$x + y = \ln a + \ln b = \ln(ab),$$

by Theorem 4.33a, and hence

$$\exp(x + y) = ab = (\exp x)(\exp y). \quad \square$$

b.   In particular, (4) and (5) together imply

$$\exp(2) = (\exp 1)(\exp 1) = e \cdot e = e^2,$$
$$\exp(3) = (\exp 2)(\exp 1) = e \cdot e \cdot e = e^3,$$

and, more generally,

$$\exp(n) = \underbrace{e \cdot e \cdots e}_{n \text{ factors}} = e^n. \tag{6}$$

Since $\exp x$ coincides for $x = n$ with e^n, the nth power of the number e, it is natural to write

$$e^x = \exp x, \tag{7}$$

even when x is a real number rather than a positive integer. Thus (7) is to be regarded as the *definition* of the function e^x, but one which is particularly appropriate, because of (6). In terms of this notation, formulas (3) and (4) become

$$e^0 = 1, \qquad e^1 = e, \tag{8}$$

while (5) takes the form

$$e^{x+y} = e^x e^y. \tag{9}$$

Choosing $y = -x$ in (9), we get

$$e^x e^{-x} = e^{x-x} = e^0 = 1,$$

so that

$$e^{-x} = \frac{1}{e^x}. \tag{10}$$

This is in keeping with the usual definition of negative powers for the case where x is a positive integer.

4.43. a.   Given any positive number M, no matter how large, we have

$$e^x > e^{\ln M} = M$$

whenever $x > \ln M$, since e^x is increasing. It follows that

$$\lim_{x \to \infty} e^x = \infty. \tag{11}$$

Moreover,

$$\lim_{x \to -\infty} e^x = \lim_{x \to \infty} e^{-x} = \lim_{x \to \infty} \frac{1}{e^x},$$

with the help of (10). Therefore, by (11),

$$\lim_{x \to -\infty} e^x = 0, \tag{12}$$

for the reason given in Sec. 2.91c.

b. To differentiate the function e^x, we use Theorem 2.81a, noting that all the conditions of the theorem are satisfied (check this). Thus, writing $y = e^x$, $x = \ln y$, we have

$$\frac{d}{dx} e^x = \frac{dy}{dx} = \frac{1}{\dfrac{dx}{dy}} = \frac{1}{\dfrac{d \ln y}{dy}} = \frac{1}{\dfrac{1}{y}} = y,$$

so that

$$\frac{d}{dx} e^x = e^x. \tag{13}$$

As this formula shows, the function e^x has the remarkable property of being equal to its own derivative, and therefore of being unaffected by any number of differentiations. Thus, for example,

$$\frac{d^{100}}{dx^{100}} e^x = e^x.$$

c. Figure 8 shows the graph of the function e^x. It is apparent from the figure that e^x is increasing in $(-\infty, \infty)$, has the range $(0, \infty)$, and satisfies the formulas (8). Note also that e^x has the x-axis as its only asymptote, and is concave upward in the whole interval $(-\infty, \infty)$, by Sec. 3.72, Proposition (1), since

$$\frac{d^2}{dx^2} e^x = e^x > 0$$

for every $x \in (-\infty, \infty)$.

4.44. The function a^x

a. We now introduce the function a^x, where a is positive and x is arbitrary. This function, called the *exponential to the base a*, is defined by the formula

$$a^x = e^{x \ln a}, \tag{14}$$

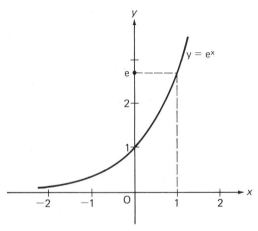

Figure 8.

and has properties very similar to those of the function e^x, to which it reduces for $a = e$. For example, it follows at once from (14) that

$$a^0 = e^0 = 1, \qquad a^1 = e^{\ln a} = a,$$

$$a^{-x} = e^{-x \ln a} = \frac{1}{e^{x \ln a}} = \frac{1}{a^x},$$

$$a^{x+y} = e^{(x+y) \ln a} = e^{x \ln a + y \ln a} = e^{x \ln a} e^{y \ln a} = a^x a^y,$$

with the help of (8), (9) and (10). In particular, if n is a positive integer,

$$a^n = e^{n \ln a} = \underbrace{e^{\ln a} e^{\ln a} \cdots e^{\ln a}}_{n \text{ factors}} = \underbrace{a \cdot a \cdots a}_{n \text{ factors}},$$

so that a^x coincides for $x = n$ with the nth power of the number a, as we would expect from the notation.

Taking the logarithm of both sides of (14), we find that

$$\ln (a^x) = \ln (e^{x \ln a}).$$

Therefore

$$\ln (a^x) = x \ln a, \tag{15}$$

which generalizes formula (6), p. 154, from the case $x = n$, where n is an integer, to the case of arbitrary real x.

b. The following proposition gives another key property of a^x:

THEOREM. *Let a be positive, and let x and y be arbitrary real numbers. Then*

$$(a^x)^y = a^{xy}. \tag{16}$$

Proof. By (14),

$$(e^x)^y = e^{y \ln e^x} = e^{yx} = e^{xy},$$

so that (16) holds for $a = e$. Therefore

$$(a^x)^y = (e^{x \ln a})^y = e^{xy \ln a} = a^{xy}. \quad \square$$

c. The derivative of the function a^x is easily calculated. In fact,

$$\frac{d}{dx} a^x = \frac{d}{dx} e^{x \ln a} = e^{x \ln a} \frac{d}{dx} (x \ln a) = a^x \ln a,$$

with the help of (13) and the chain rule.

4.45. The function x^r

a. Let x be positive, and let r be an arbitrary real number. Then, changing a to x and x to r in (14), we get the formula

$$x^r = e^{r \ln x}, \tag{17}$$

defining the rth power of x. The function x^r has the domain $(0, \infty)$ and is continuous, being a continuous function of a continuous function (Sec. 2.82d). It follows from (17) that

$$x^{-r} = e^{-r \ln x} = \frac{1}{e^{r \ln x}} = \frac{1}{x^r}, \tag{18}$$

with the help of (10). Similarly, if r and s are arbitrary real numbers, then

$$(x^r)^s = (e^{r \ln x})^s = e^{rs \ln x},$$

because of (16), with $a = e$, so that

$$(x^r)^s = x^{rs}, \tag{19}$$

while

$$x^{r+s} = e^{(r+s) \ln x} = e^{r \ln x + s \ln x} = e^{r \ln x} e^{s \ln x},$$

because of (9), so that

$$x^{r+s} = x^r x^s. \tag{20}$$

b. Suppose r is a rational number m/n, possibly an integer. Then the function (17), which is defined only for $x > 0$, coincides in $(0, \infty)$ with the function $x^{m/n}$ introduced in Sec. 2.7, Problems 5–7, which may well be defined for all $x \geqslant 0$ or even for all x. This follows at once from formulas (18) and (19), and the way the function $x^{m/n}$ was previously defined. The merit of (17) is, of course, that we are now able to define x^r for *irrational* r. Another virtue of (17) is that we can now *construct* the nth root $\sqrt[n]{x}$, a number whose existence has so far been tacitly *assumed*. In fact, if $x > 0$, then $\sqrt[n]{x}$ is simply the perfectly well-defined number.

$$x^{1/n} = e^{(1/n) \ln x}.$$

If n is odd, we then set $\sqrt[n]{-x} = -\sqrt[n]{x}$, and the definition of $\sqrt[n]{x}$ is completed by noting that $0^n = 0$ for every positive integer n, whether odd or even, so that $\sqrt[n]{0} = 0$.

There is nothing to prevent us from *defining* $0^r = 0$ for a positive irrational number r. In fact, this is the only sensible definition, since if $r > 0$, then $r \ln x \to -\infty$ as $x \to 0+$, by formula (9), p. 155, so that $x^r = e^{r \ln x} \to 0$ as $x \to 0+$, by formula (12).

c. We are now ready to prove the key formula (14), p. 78.

THEOREM. *Let r be an arbitrary real number. Then*

$$\frac{d}{dx} x^r = rx^{r-1} \qquad (x > 0).$$

Proof. By (13) and the chain rule,

$$\frac{d}{dx} x^r = \frac{d}{dx} e^{r \ln x} = e^{r \ln x} \frac{d}{dx} (r \ln x) = x^r \frac{r}{x} = rx^{r-1},$$

with the help of (18) and (20). $\square$

Thus we have at last proved all the formulas given in Sec. 2.74e.

PROBLEMS

1. Verify that the graph of the function ke^x ($k > 0$) can be obtained by shifting the graph of e^x along the x-axis.
2. Differentiate
   (a) e^{4x+5}; (b) e^{-3x}; (c) xe^x; (d) $e^x(1 - x^2)$.
3. Differentiate

   (a) e^{x^2}; (b) $\dfrac{e^x - 1}{e^x + 1}$; (c) $e^{\sqrt{x+1}}$; (d) $\sqrt{1 + e^x}$.

4.   According to the mean value theorem (Sec. 3.43b),

$$f(1 + \Delta x) - f(1) = f'(1 + \alpha \Delta x)\, \Delta x,$$

where $0 < \alpha < 1$. Find α if $f(x) = e^x$, $\Delta x = 1$.

5.   Verify that

$$\int e^{ax}\, dx = \frac{e^{ax}}{a} + C.$$

6.   Find the area between the curves $y = e^x$ and $y = e^{-x}$ from $x = \ln 2$ to $x = \ln 3$.

7.   Find the local extrema of the function $y = (x + 1)^{10} e^{-x}$.

8.   Find the inflection points and investigate the concavity of the function $y = e^{-x^2/2}$. Graph this function.

9.   How is the function a^x defined in Sec. 4.44a related to the function $\log_a x$ defined in Sec. 4.36a?

10.   Solve the equation $2^x - 2x = 0$.

11.   An advisor to a certain king was asked what he would like as a reward for interpreting one of the king's dreams. He asked for a chessboard with one grain of rice on the first square, twice as much rice on the second square as on the first, twice as much on the third square as on the second, and so on. Why did the king have his advisor executed for insolence?

12.   Differentiate

(a) $x10^x$;   (b) a^{x^2};   (c) $a^x x^a$;   (d) $\dfrac{1 - 10^x}{1 + 10^x}$.

*13.   Show that

$$\lim_{x \to \infty} a^x = \infty, \qquad \lim_{x \to -\infty} a^x = 0$$

if $a > 1$, while

$$\lim_{x \to \infty} a^x = 0, \qquad \lim_{x \to -\infty} a^x = \infty$$

if $0 < a < 1$.

*14.   Show that

$$\lim_{x \to \infty} x^r = \infty, \qquad \lim_{x \to 0+} x^r = 0$$

if $r > 0$, while

$$\lim_{x \to \infty} x^r = 0, \qquad \lim_{x \to 0+} x^r = \infty$$

if $r < 0$.

*15.   Find the inverse of the function

$$y = \log_2 (x + \sqrt{1 + x^2}).$$

*16.   For what values of c does the function $e^x + cx^3$ have an inflection point?

17.   Use differentiation to confirm that the fourteenth term of the sequence

$$y_n = \frac{n^{10}}{2^n} \qquad (n = 1, 2, \ldots)$$

is the largest.

4.5   MORE ABOUT THE LOGARITHM AND EXPONENTIAL

4.51. The number e

a.   To derive formula (11), p. 156, we start from the fact that

$$\frac{d \ln x}{dx}\bigg|_{x=1} = \frac{1}{x}\bigg|_{x=1} = 1.$$

Expressing this derivative as a limit, with h instead of Δx for brevity, we have

$$\frac{d \ln x}{dx}\bigg|_{x=1} \quad \lim_{h \to 0} \frac{\ln(1+h) - \ln 1}{h} = \lim_{h \to 0} \frac{\ln(1+h)}{h},$$

so that

$$\lim_{h \to 0} \frac{1}{h} \ln(1+h) = 1.$$

Therefore

$$\lim_{h \to 0} \ln(1+h)^{1/h} = 1,$$

with the help of formula (15), p. 162. Thus the function

$$f(h) = \begin{cases} \ln(1+h)^{1/h} & \text{if } h \neq 0, \\ 1 & \text{if } h = 0 \end{cases} \tag{1}$$

is continuous at $h = 0$. But then, by Sec. 2.82d, the composite function $e^{f(h)}$ is also continuous at $h = 0$, since e^x is continuous at $x = 1$. It follows that

$$\lim_{h \to 0} e^{f(h)} = e^{f(0)} = e^1 = e,$$

or equivalently

$$\lim_{h \to 0} (1+h)^{1/h} = e. \tag{2}$$

Suppose $h = 1/n$, where n is a positive integer. Then $h \to 0$ implies $n \to \infty$, and (2) takes the form

$$\lim_{n \to \infty} \left(1 + \frac{1}{n}\right)^n = e, \tag{3}$$

in agreement with formula (11), p. 156.

b.   There is a more general formula

$$\lim_{n \to \infty} \left(1 + \frac{r}{n}\right)^n = e^r, \tag{4}$$

valid for an arbitrary real number r. (Note that (4) reduces to (3) for $r = 1$.) The formula holds for $r = 0$, since it then reduces to the trivial equality

$$\lim_{n \to \infty} 1^n = e^0 = 1.$$

Let $h = r/n$, where $r \neq 0$. Then $h \to 0+$ as $n \to \infty$ if $r > 0$, while $h \to 0-$ as $n \to \infty$ if $r < 0$. Therefore

$$\lim_{n \to \infty} \left(1 + \frac{r}{n}\right)^n = \lim_{h \to 0\pm} (1 + h)^{r/h} = \lim_{h \to 0\pm} \left[e^{f(h)}\right]^r,$$

where $f(h)$ is the function (1). As already noted, the function $e^{f(h)}$ is continuous at $h = 0$, where it has the value e. Therefore the composite function $\left[e^{f(h)}\right]^r$ is also continuous at $h = 0$, since x^r is continuous at $x = e$. It follows that

$$\lim_{h \to 0\pm} \left[e^{f(h)}\right]^r = \left[e^{f(0)}\right]^r = e^r,$$

which is equivalent to (4).

4.52. Compound interest

The following three examples explore this very practical topic, and show how the exponential function gets into the act:

a. Example. Suppose money invested at an annual interest rate r, or equivalently at $100r$ percent, is compounded N times a year. Show that a principal of P dollars will grow to

$$A = P\left(1 + \frac{r}{N}\right)^{Nt} \tag{5}$$

dollars at the end of t years. Show that

$$P = A\left(1 + \frac{r}{N}\right)^{-Nt} \tag{6}$$

dollars must be invested now to become worth A dollars in t years.

SOLUTION. Let A_m be the amount in the bank at the end of the mth interest period, assuming that no money is withdrawn after the initial deposit. Then

$$A_{m+1} = A_m + A_m \frac{r}{N} = A_m\left(1 + \frac{r}{N}\right),$$

since the interest is computed on the accrued amount at a rate equal to r, the nominal annual interest rate, divided by N, the number of compoundings per annum. The initial amount A_0 is, of course, just the principal P. Therefore the amount A in the bank after t years, that is, after Nt interest periods, is equal to

$$A = A_{Nt} = A_{Nt-1}\left(1 + \frac{r}{N}\right) = A_{Nt-2}\left(1 + \frac{r}{N}\right)^2 = \cdots$$

$$= A_2\left(1 + \frac{r}{N}\right)^{Nt-2} = A_1\left(1 + \frac{r}{N}\right)^{Nt-1} = A_0\left(1 + \frac{r}{N}\right)^{Nt},$$

which proves (5), since $A_0 = P$. To get (6), we solve (5) for P, obtaining

$$P = \frac{A}{\left(1 + \dfrac{r}{N}\right)^{Nt}} = A\left(1 + \frac{r}{N}\right)^{-Nt}.$$

b. Example. Suppose interest is "compounded continuously," that is, suppose the number of compoundings N per annum becomes "arbitrarily large." Show that formula (5) for the "compound amount" A becomes

$$A = Pe^{rt}, \tag{7}$$

while formula (6) for the "present value" P becomes

$$P = Ae^{-rt}. \tag{8}$$

SOLUTION.   Here we have

$$A = P \lim_{N \to \infty} \left(1 + \frac{r}{N}\right)^{Nt}$$

Let $n = Nt$, so that $N \to \infty$ implies $n \to \infty$. Then

$$A = P \lim_{n \to \infty} \left(1 + \frac{rt}{n}\right)^n = e^{rt},$$

with the help of (4). This proves (7). To get (8), we solve (7) for P, obtaining

$$P = \frac{A}{e^{rt}} = Ae^{-rt}.$$

c. **Example.**   Let $P = \$1,000$, $r = 6\%$, $t = 1$ year. Test the accuracy of formula (7), as compared with the exact formula (5).

SOLUTION.   The results are given in the following table for annual, semi-annual, quarterly, monthly, daily, and continuous compounding (the last indicated by ∞):

N	1	2	4	12	365	∞
A	\$1,060.00	\$1,060.90	\$1,061.36	\$1,061.68	\$1,061.83	\$1,061.83

4.53. Logarithmic differentiation

In calculating derivatives, it is often helpful to first take logarithms and then differentiate the result. The following two examples show the power of this technique, called *logarithmic differentiation*.

a. **Example.**   Differentiate

$$y = \frac{(x - 1)^2 \sqrt{2x + 1}}{(x^2 + 3)^4 e^x}.$$

SOLUTION.   Taking logarithms, we get

$$\ln y = 2 \ln (x - 1) + \frac{1}{2} \ln (2x + 1) - 4 \ln (x^2 + 3) - x. \tag{9}$$

Differentiation of (9) then gives

$$\frac{y'}{y} = \frac{2}{x - 1} + \frac{2}{2(2x + 1)} - \frac{4 \cdot 2x}{x^2 + 3} - 1,$$

where the prime denotes differentiation with respect to x and we repeatedly use the chain rule. Multiplying by y, we get the desired derivative

$$y' = \frac{(x - 1)^2 \sqrt{2x + 1}}{(x^2 + 3)^4 e^x} \left(\frac{2}{x - 1} + \frac{1}{2x + 1} - \frac{8x}{x^2 + 3} - 1 \right).$$

The quantity

$$\frac{d}{dx} \ln y = \frac{y'}{y} = \frac{1}{y} \frac{dy}{dx}$$

is called the *logarithmic derivative* of *y*.

   b. Example.   Differentiate

$$y = u^v,$$

where *u* and *v* are both differentiable functions of *x*.

   SOLUTION.   By logarithmic differentiation, we have

$$\ln y = v \ln u,$$

$$\frac{y'}{y} = v' \ln u + v \frac{u'}{u}.$$

Therefore

$$y' = y \left(v' \ln u + v \frac{u'}{u} \right) = u^v \left(v' \ln u + v \frac{u'}{u} \right) = u^v v' \ln u + v u^{v-1} u'. \qquad (10)$$

For example, if $y = x^x$, so that $u = v = x$, then (10) reduces to

$$y' = x^x x' \ln x + x x^{x-1} x' = x^x (\ln x + 1),$$

since $x' = (d/dx)x = 1$.

   4.54. Elasticity

   a.   Next we introduce a concept of considerable interest in business and economics. Given a function $y = f(x)$, by the derivative of *y* with respect to *x* we mean, of course, the limit as $\Delta x \to 0$ of the difference quotient

$$\frac{\Delta y}{\Delta x} = \frac{f(x + \Delta x) - f(x)}{\Delta x} = \frac{\text{Change in } y}{\text{Change in } x}.$$

Let the *proportional change in y* be defined by

$$\frac{\Delta y}{y}. \qquad (11)$$

Then the logarithmic derivative of *y* with respect to *x* is the limit as $\Delta x \to 0$ of the ratio

$$\frac{\dfrac{\Delta y}{y}}{\Delta x} = \frac{\text{Proportional change in } y}{\text{Change in } x},$$

since

$$\lim_{\Delta x \to 0} \frac{\dfrac{\Delta y}{y}}{\Delta x} = \frac{1}{y} \lim_{\Delta x \to 0} \frac{\Delta y}{\Delta x} = \frac{1}{y} \frac{dy}{dx} = \frac{d}{dx} \ln y.$$

Similarly, defining the *proportional change in x* by

$$\frac{\Delta x}{x}, \qquad (12)$$

we can go a step further and introduce the limit as $\Delta x \to 0$ of the ratio

$$\frac{\dfrac{\Delta y}{y}}{\dfrac{\Delta x}{x}} = \frac{\text{Proportional change in } y}{\text{Proportional change in } x}.$$

We then get a quantity

$$\varepsilon_{yx} = \lim_{\Delta x \to 0} \frac{\dfrac{\Delta y}{y}}{\dfrac{\Delta x}{x}} = \frac{x}{y} \lim_{\Delta x \to 0} \frac{\Delta y}{\Delta x} = \frac{x}{y} \frac{dy}{dx}, \tag{13}$$

called the *elasticity* of the function $y = f(x)$, at the point x. As a measure of the "change in y due to a change in x," the elasticity has the merit of being independent of the units of x and y, which are divided out in forming the ratios (11) and (12). This is a great convenience in certain business problems, where, for example, y might be the quantity of a commodity demanded at the price x. Changing the units of x from dollars to pesos, say, or the units of y from bushels to carloads, would then have no effect on the elasticity of the demand curve.

Note that we can also regard the elasticity ε_{yx} as the limit

$$\varepsilon_{yx} = \lim_{\Delta x \to 0} \frac{\text{Percentage change in } y}{\text{Percentage change in } x}.$$

This follows at once from the fact that

$$\frac{\text{Percentage change in } y}{\text{Percentage change in } x} = \frac{100 \dfrac{\Delta y}{y}}{100 \dfrac{\Delta x}{x}} = \frac{\dfrac{\Delta y}{y}}{\dfrac{\Delta x}{x}} = \frac{\text{Proportional change in } y}{\text{Proportional change in } x}.$$

b. There is another way of writing (13) as a kind of "double logarithmic derivative" of the form

$$\varepsilon_{yx} = \frac{d (\ln y)}{d (\ln x)}. \tag{14}$$

where the expression on the right is the ratio of the differential of $\ln y$ to the differential of $\ln x$. To verify (14), we recall from Sec. 2.55a that if $y = f(x)$, then dy, the differential of $f(x)$, is given by the formula

$$dy = f'(x) \, dx.$$

Therefore

$$d(\ln x) = (\ln x)' \, dx = \frac{1}{x} \, dx,$$

while

$$d(\ln y) = (\ln y)' \, dx = \frac{y'}{y} \, dx = \frac{1}{y} \frac{dy}{dx} \, dx,$$

by the chain rule, so that

$$\frac{d(\ln y)}{d(\ln x)} = \frac{\dfrac{1}{y}\dfrac{dy}{dx} dx}{\dfrac{1}{x} dx} = \frac{x}{y}\frac{dy}{dx} = \varepsilon_{yx},$$

as claimed.

c. Example. Suppose the demand for a commodity produced by a monopolistic firm is described by the function $Q = Q(P)$, where Q is the quantity demanded at the price P. Then

$$\varepsilon_D = -\frac{d(\ln Q)}{d(\ln P)} \qquad (15)$$

is called the *elasticity of demand*, at the price P. Do not be disconcerted by the extra minus sign in (15). Its sole purpose is to make the elasticity come out positive, in keeping with economic convention and in anticipation of the fact that a demand curve typically has negative slope (Sec. 3.2, Prob. 9). The demand is said to be *elastic* if $\varepsilon_D > 1$ and *inelastic* if $\varepsilon_D < 1$.

Suppose the demand function is

$$Q = Q(P) = 60 - 3P.$$

Find ε_D. When is the demand elastic? Inelastic? Express the firm's marginal revenue (Sec. 3.2, Prob. 7) in terms of the price and the elasticity of demand. Why should the firm adjust its price to keep the demand elastic?

SOLUTION. Here

$$\varepsilon_D = -\frac{P}{Q}\frac{dQ}{dP} = -\frac{-3P}{60 - 3P} = \frac{P}{20 - P},$$

so that the demand is elastic ($\varepsilon_D > 1$) if $10 < P < 20$ and inelastic ($\varepsilon_D < 1$) if $0 < P < 10$. The firm's total revenue is

$$R(Q) = PQ(P) = QP(Q),$$

as in Sec. 3.2, Problem 10, where $P(Q)$ is the inverse function of $Q(P)$. Correspondingly, the firm's marginal revenue is

$$MR(Q) = \frac{d}{dQ} QP(Q) = P + Q\frac{dP}{dQ} = P\left(1 + \frac{Q}{P}\frac{dP}{dQ}\right) = P\left(1 - \frac{1}{\varepsilon_D}\right), \qquad (16)$$

where we use the fact that

$$\frac{dP}{dQ} = \frac{1}{\dfrac{dQ}{dP}},$$

by the rule for differentiating an inverse function (Sec. 2.81b). The firm certainly wants to operate at an output level where marginal revenue is positive, so that more revenue would be received if the output were increased slightly. It should therefore make $\varepsilon_D > 1$ in (16). In other words, other things being equal, it should keep the demand elastic, by choosing a price in the range $10 < P < 20$.

PROBLEMS

1. Evaluate

   (a) $\displaystyle\lim_{n\to\infty}\left(\frac{n}{n+1}\right)^n$;   (b) $\displaystyle\lim_{n\to\infty}\left(1+\frac{1}{n}\right)^{n+1}$;   (c) $\displaystyle\lim_{n\to\infty}\left(1-\frac{1}{n}\right)^n$.

2. Evaluate

   (a) $\displaystyle\lim_{x\to\infty}\left(1+\frac{1}{x}\right)^x$;   (b) $\displaystyle\lim_{x\to0}(1+2x)^{1/x}$;   (c) $\displaystyle\lim_{x\to\infty}\left(\frac{x+1}{x-1}\right)^x$.

3. It was shown in Sec. 4.51a that

$$\lim_{x\to0}\frac{\ln(1+x)}{x}=1.$$

   Use this to show that

$$\lim_{x\to0}\frac{\log_a(1+x)}{x}=\log_a e. \tag{17}$$

4. Use (17) to show that

$$\lim_{x\to0}\frac{a^x-1}{x}=\ln a. \tag{18}$$

5. Use (18) to show that

$$\lim_{x\to1}\frac{x^r-1}{x-1}=r. \tag{19}$$

6. Use (17)–(19) to evaluate

   (a) $\displaystyle\lim_{x\to1}\frac{\sqrt[5]{x}-1}{\sqrt[3]{x}-1}$;   (b) $\displaystyle\lim_{x\to0}\frac{2^x-1}{\sqrt{1+x}-1}$;   (c) $\displaystyle\lim_{x\to0}\frac{\log_{10}(1+x)}{10^x-1}$.

7. How much is a principal of $1,000 worth in 5 years if it is compounded quarterly at an annual interest rate of 8%?

8. A principal which is being compounded continuously doubles in 10 years. What is the annual interest rate?

9. How long does it take $10,000 compounded continuously at an annual interest rate of 7% to grow to $25,000?

10. What amount of money must be invested now to be worth $10,000 in 5 years if compounded continuously at an annual interest rate of 6%?

11. Justify calling

$$r_E=\left(1+\frac{r}{N}\right)^N-1$$

   the *effective* annual interest rate, as opposed to the *nominal* annual rate r.

12. Interpret the number e in the language of finance.

13. Use logarithmic differentiation to find the derivative of

   (a) $\displaystyle\frac{(x+1)^2}{(x+2)^3(x+3)^4}$;   (b) $\displaystyle\frac{e^{x^2+2x}}{x^{4/3}\ln x}$.

14. Differentiate
     (a) x^{x^2};   (b) $x^{1/x}$;   (c) $(\ln x)^x$;   (d) e^{x^x}.
15. Show that the function $y = x^x$ has no inflection points.
16. Verify the following "chain rule for elasticities":

$$\varepsilon_{zx} = \varepsilon_{zy}\varepsilon_{yx}.$$

17. What is the elasticity of the function e^{ax}?
18. Given an example of a function with constant nonzero elasticity.
19. Show that if $f(x)$ has elasticity ε_{yx}, then $xf(x)$ has elasticity $1 + \varepsilon_{yx}$.
*20. If $C = C(Q)$ is a firm's total cost function (Sec. 3.22), then

$$\varepsilon_C = \frac{d(\ln C)}{d(\ln Q)}$$

(without a minus sign) is called the *elasticity of cost*, at the output Q. Verify that
(a) If $\varepsilon_C < 1$ at a given output, then average cost (AC) is greater than marginal cost (MC), and average cost decreases as output increases;
(b) If $\varepsilon_C > 1$ at a given output, then average cost is less than marginal cost, and average cost increases as output increases.
*21. Consider the functions

$$\cosh x = \frac{1}{2}(e^x + e^{-x})$$

and

$$\sinh x = \frac{1}{2}(e^x - e^{-x}),$$

called the *hyperbolic cosine* of x and the *hyperbolic sine* of x, respectively. Graph the functions $\frac{1}{2}e^x$, $\frac{1}{2}e^{-x}$, $\cosh x$ and $\sinh x$ in the same system of rectangular coordinates. Show that
(a) $\cosh x \geqslant 1$ for all x, $\cosh 0 = 1$;
(b) $\sinh x > 0$ if $x > 0$, $\sinh x < 0$ if $x < 0$, $\sinh 0 = 0$;
(c) $\lim\limits_{x \to \pm\infty} \cosh x = \infty$, $\lim\limits_{x \to \infty} \sinh x = \infty$, $\lim\limits_{x \to -\infty} \sinh x = -\infty$;
(d) $\dfrac{d}{dx}\cosh x = \sinh x$, $\dfrac{d}{dx}\sinh x = \cosh x$.

*22. It can be shown that the number e is the sum of the "rapidly convergent" series

$$\sum_{n=0}^{\infty} \frac{1}{n!} = 1 + 1 + \frac{1}{2} + \frac{1}{6} + \frac{1}{24} + \cdots + \frac{1}{n!} + \cdots.$$

By referring to the value of e given in Sec. 4.34, show that the sum of just 8 terms of this series gives a value of e which is accurate to 4 decimal places.

4.6 INTEGRATION TECHNIQUE

The technique of integration is inherently more difficult than that of differentiation. Thus, while it is no trick at all to become a minor expert on differentiation, it is the easiest thing in the world to write down integrals that would stump even a professional mathematician. However, two powerful methods of integration are available at the level of this course. We now discuss these methods and use them to evaluate a number of integrals which may appear quite intractable at first glance.

4.61. Integration by substitution

a. We begin by observing that if

$$\int g(t)\, dt = G(t) + C,$$

then

$$\int g(t(x))t'(x)\, dx = G(t(x)) + C \tag{1}$$

for every differentiable function $t = t(x)$. Here the common practice of denoting the dependent variable and the function by the same letter, t in this case, is particularly appropriate. To verify (1), we merely note (as in Sec. 3.53c) that both sides of (1) have the same derivative. In fact,

$$\frac{d}{dx} \int g(t(x))t'(x)\, dx = g(t(x))t'(x),$$

by the very definition of the indefinite integral as an antiderivative of its integrand, while

$$\frac{d}{dx} G(t(x)) = G'(t(x))t'(x) = g(t(x))t'(x),$$

by the chain rule and the fact that $G(t)$ is an antiderivative of $g(t)$.

Now suppose we want to evaluate an integral

$$\int f(x)\, dx, \tag{2}$$

which does not look like anything familiar, but which can be recognized as being of the form

$$\int g(t(x))t'(x)\, dx, \tag{3}$$

in terms of some function $g(t)$ of a new variable $t = t(x)$, where $g(t)$ is a function which is more easily integrated than $f(x)$ itself. Then it follows from (1) that

$$\int f(x)\, dx = G(t(x)) + C. \tag{4}$$

Integration by substitution is also known as integration by *change of variables*, for a self-evident reason.

b. Recalling from Sec. 2.55a that the differential of t is given by

$$dt = t'(x)\, dx, \tag{5}$$

we can write (3) simply as

$$\int g(t)\, dt,$$

where it is understood that the substitution $t = t(x)$ will eventually be made. The fact that (2) and (3) are equivalent then takes the concise form

$$\int f(x)\, dx = \int g(t)\, dt.$$

The advantage of writing the expression behind an integral sign as a product of a function (the integrand) and a *differential* is now apparent for the first time. In fact, if we change variables, then formula (5), or its analogue

$$dx = x'(t)\, dt \qquad (6)$$

for the case of a substitution $x = x(t)$, *automatically* multiplies the old integrand by the appropriate "correction factor," without any need for a separate calculation involving the chain rule.

   c.   For example, to evaluate

$$\int \frac{x\, dx}{1 + x^2}, \qquad (7)$$

let $t = 1 + x^2$, so that $dt = 2x\, dx$, or equivalently $x\, dx = \frac{1}{2}dt$. Then

$$\int \frac{x\, dx}{1 + x^2} = \frac{1}{2}\int \frac{dt}{t},$$

where the integral on the right can be recognized at once as being equal to $\ln t$. Therefore, by (4),

$$\int \frac{x\, dx}{1 + x^2} = \frac{1}{2}\ln t = \frac{1}{2}\ln (1 + x^2) = \ln \sqrt{1 + x^2} + C,$$

after going back to the original variable x and introducing a constant of integration C.

   Once you have got the idea of how the technique of integration by substitution works, you can omit some of the intermediate steps, even leaving out explicit introduction of the auxiliary variable t. Thus a more concise way of evaluating (7) is

$$\int \frac{x\, dx}{1 + x^2} = \frac{1}{2}\int \frac{d(1 + x^2)}{1 + x^2} = \frac{1}{2}\ln (1 + x^2) = \ln \sqrt{1 + x^2} + C,$$

where the whole expression $1 + x^2$ is treated as a variable of integration.

   d.   To evaluate a *definite* integral by substitution, we first evaluate the corresponding indefinite integral, and then use the fundamental theorem of calculus (Sec. 4.24a). Thus, for example,

$$\int_0^1 \frac{x\, dx}{1 + x^2} = \ln \sqrt{1 + x^2}\,\Big|_0^1 = \ln \sqrt{1 + 1} - \ln \sqrt{1 + 0} = \ln \sqrt{2}.$$

There is also a more direct method of evaluating a definite integral by substitution (see Examples 4.62c and 4.62d).

   e.   Instead of recognizing (2) as being of the form (3), involving a differentiable substitution $t = t(x)$, we can also try making a differentiable substitution $x = x(t)$ directly in the integral (2), thereby "transforming" it into

$$\int f(x(t))x'(t)\, dt, \qquad (8)$$

with the aid of (6). Again, this will help only if the new integral (8) is easier to evaluate than the original integral (2), which means that the substitution $x = x(t)$ must be chosen intelligently.

4.62. Examples

a. Evaluate

$$\int \frac{dx}{\sqrt{x}(1 + \sqrt{x})}.$$

SOLUTION. The substitution $x = t^2$ seems a good choice, since it gets rid of the radical. With this substitution, $\sqrt{x} = t$,

$$d(\sqrt{x}) = (\sqrt{x})' \, dx = \frac{1}{2\sqrt{x}} \, dx = \frac{1}{2t} \, dx = dt,$$

so that $dx = 2t \, dt$. Therefore

$$\int \frac{dx}{\sqrt{x}(1 + \sqrt{x})} = 2 \int \frac{t \, dt}{t(1 + t)} = 2 \int \frac{dt}{1 + t} = 2 \ln (1 + t),$$

or

$$\int \frac{dx}{\sqrt{x}(1 + \sqrt{x})} = 2 \ln (1 + \sqrt{x}) + C,$$

after going back to the original variable x and introducing a constant of integration C. The expression $2 \ln (1 + \sqrt{x})$ can be replaced by $\ln (1 + \sqrt{x})^2$, if you prefer.
Alternatively, you might recognize that

$$\frac{1}{2} \int \frac{dx}{\sqrt{x}(1 + \sqrt{x})}$$

is of the form

$$\int \frac{dt}{1 + t}$$

if $t = \sqrt{x}$, but this requires a good eye. The fact that each of the functions $x = t^2$ and $t = \sqrt{x}$ is the inverse of the other is, of course, no accident (see Prob. 18).

b. Evaluate

$$\int \frac{\ln x}{x} \, dx.$$

SOLUTION. If $t = \ln x$, then $dt = dx/x$ and

$$\int \frac{\ln x}{x} \, dx = \int t \, dt = \frac{1}{2} t^2 + C = \frac{1}{2} (\ln x)^2 + C. \tag{9}$$

On the other hand, if we choose the "inverse substitution" $x = e^t$, then $dx = e^t \, dt$ and

$$\int \frac{\ln x}{x} \, dx = \int \frac{\ln e^t}{e^t} e^t \, dt = \int t \, dt,$$

and we get the same answer again.

c. Evaluate

$$\int_1^e \frac{\ln x}{x} \, dx. \tag{10}$$

SOLUTION. Using (9), we have

$$\int_1^e \frac{\ln x}{x}\, dx = \frac{1}{2}(\ln x)^2 \Big|_1^e = \frac{1}{2}(\ln e)^2 - \frac{1}{2}(\ln 1)^2 = \frac{1}{2}.$$

On the other hand, suppose that instead of first evaluating the indefinite integral (9), we try to calculate (10) from scratch. Then, just as before, we observe that the substitution $t = \ln x$ reduces the expression behind the integral sign in (10) to $t\, dt$. This suggests writing

$$\int_1^e \frac{\ln x}{x}\, dx = \int_\alpha^\beta t\, dt. \tag{11}$$

But what are the appropriate limits of integration α and β? The answer is simple enough: As the variable x varies from 1 to e in the left side of (11), the variable $t = \ln x$ varies from $\ln 1 = 0$ to $\ln e = 1$ in the right side, and hence $\alpha = 0$, $\beta = 1$. This is not only plausible, but perfectly correct, as shown in Problem 19. Therefore we can write

$$\int_1^e \frac{\ln x}{x}\, dx = \int_0^1 t\, dt = \frac{1}{2}t^2 \Big|_0^1 = \frac{1}{2}, \tag{12}$$

without bothering to calculate the indefinite integral (9). Note that there is now no need to return from t to the original variable x. In fact, once the second integral in (12) has been evaluated, the first integral is automatically known, since both are *definite* integrals and therefore *numbers*.

d. Starting from the definition

$$\ln x = \int_1^x \frac{dt}{t},$$

give another proof of the formula

$$\ln (ab) = \ln a + \ln b \qquad (a, b \text{ positive}),$$

already established in Theorem 4.33a.

SOLUTION. We have

$$\ln b = \int_1^b \frac{dt}{t} = \int_1^b \frac{d(at)}{at} = \int_a^{ab} \frac{du}{u},$$

where in the last step we go over to a new variable $u = at$ and make the corresponding change in the limits of integration. Therefore

$$\ln b = \int_a^{ab} \frac{dt}{t},$$

after returning to the original dummy variable t. It follows that

$$\ln (ab) = \int_1^{ab} \frac{dt}{t} = \int_1^a \frac{dt}{t} + \int_a^{ab} \frac{dt}{t} = \ln a + \ln b,$$

with the help of Theorem 4.21c.

4.63. Integration by parts

a. We now consider another important integration technique. Let $u(x)$ and $v(x)$ be two differentiable functions such that $u'(x)v(x)$ and $u(x)v'(x)$ both have antiderivatives. Differentiating the product $u(x)v(x)$ with respect to x, and omitting arguments for simplicity, we have

$$(uv)' = u'v + uv',$$

so that

$$uv' = (uv)' - vu'. \tag{13}$$

Multiplying (13) by dx and integrating with respect to x, we get

$$\int uv'\, dx = \int (uv)'\, dx - \int vu'\, dx.$$

But

$$\int (uv)'\, dx = uv + C,$$

and hence

$$\int uv'\, dx = uv - \int vu'\, dx,$$

where C is absorbed into the other constants of integration. In terms of the differentials

$$du = u'\, dx, \qquad dv = v'\, dx,$$

(13) takes the even simpler form

$$\int u\, dv = uv - \int v\, du. \tag{14}$$

Equation (14), called the formula for *integration by parts*, is well worth memorizing. It is one of the most valuable tricks of the trade, often allowing us to express difficult integrals in terms of easy ones.

b. To find a corresponding formula for definite integrals, we merely observe that

$$\int_a^b u\, dv = \left[\int u\, dv \right]_a^b = \left[uv - \int v\, du \right]_a^b = uv \Big|_a^b - \left[\int v\, du \right]_a^b$$

(justify the last step). Therefore

$$\int_a^b u\, dv = uv \Big|_a^b - \int_a^b v\, du. \tag{15}$$

4.64. Examples

a. Evaluate

$$\int \ln x\, dx.$$

SOLUTION. This integral is of the form $\int u\, dv$ if we choose $u = \ln x$, $dv = dx$. We then have $du = dx/x$, $v = x$, and hence, by (14),

$$\int \ln x\, dx = x \ln x - \int dx = x \ln x - x + C,$$

where a constant of integration is supplied in the last step. It would be pointless to include a constant of integration C in going from dv to v, since C would be cancelled out automatically in the expression $uv - \int v\, du$.

b. Evaluate

$$\int x \ln x \, dx.$$

SOLUTION.   There are various possibilities here. We can choose $u = x$, $dv =$ $\ln x \, dx$, or $u = \ln x$, $dv = x \, dx$, or even $u = x \ln x$, $dv = dx$. The only good choice is $u = \ln x$, $dv = x \, dx$, since only this choice makes $\int v \, du$ simpler than $\int u \, dv$, which is the whole point of integration by parts. We then have $du = dx/x$, $v = \frac{1}{2}x^2$, and hence, by (14),

$$\int x \ln x \, dx = \frac{1}{2} x^2 \ln x - \frac{1}{2} \int x \, dx = \frac{1}{2} x^2 \ln x - \frac{1}{4} x^2 + C. \tag{16}$$

c. Evaluate

$$\int_1^e x \ln x \, dx.$$

SOLUTION.   We can start from (16), writing

$$\int_1^e x \ln x \, dx = \left[\frac{1}{2} x^2 \ln x - \frac{1}{4} x^2 \right]_1^e = \frac{1}{2} e^2 \ln e - \frac{1}{4} e^2 - \frac{1}{2} \ln 1 + \frac{1}{4}$$

$$= \frac{1}{2} e^2 - \frac{1}{4} e^2 + \frac{1}{4} = \frac{1}{4}(e^2 + 1).$$

Alternatively, we can start from (15), writing

$$\int_1^e x \ln x \, dx = \frac{1}{2} x^2 \ln x \Big|_1^e - \frac{1}{2} \int_1^e x \, dx = \frac{1}{2} e^2 \ln e - \frac{1}{2} \ln 1 - \frac{1}{4} x^2 \Big|_1^e$$

$$= \frac{1}{2} e^2 - \frac{1}{4} e^2 + \frac{1}{4} = \frac{1}{4}(e^2 + 1).$$

d. Evaluate

$$\int x^2 e^x \, dx.$$

SOLUTION.   Let $u = x^2$, $dv = e^x \, dx$. Then $du = 2x \, dx$, $v = e^x$, and therefore

$$\int x^2 e^x \, dx = x^2 e^x - 2 \int x e^x \, dx. \tag{17}$$

To evaluate the integral on the right, we integrate by parts *again*, this time choosing $u = x$, $dv = e^x \, dx$, $du = dx$, $v = e^x$:

$$\int x e^x \, dx = x e^x - \int e^x \, dx = x e^x - e^x. \tag{18}$$

Substituting (18) into (17) and supplying a constant of integration, we find that

$$\int x^2 e^x \, dx = x^2 e^x - 2 x e^x + 2 e^x + C. \tag{19}$$

You should get into the habit of checking formulas like this by differentiating the expression on the right. In this case,

$$\frac{d}{dx}(x^2 e^x - 2x e^x + 2 e^x + C) = 2 x e^x + x^2 e^x - 2 e^x - 2 x e^x + 2 e^x = x^2 e^x,$$

which confirms (19).

PROBLEMS

1. Let $f(x)$ be continuous and even in $[-a, a]$. Show that

$$\int_{-a}^{0} f(x)\, dx = \int_{0}^{a} f(x)\, dx = \frac{1}{2} \int_{-a}^{a} f(x)\, dx.$$

2. Let $f(x)$ be continuous and odd in $[-a, a]$. Show that

$$\int_{-a}^{0} f(x)\, dx = -\int_{0}^{a} f(x)\, dx, \qquad \int_{-a}^{a} f(x)\, dx = 0.$$

3. Verify that

$$\int_{0}^{1} x^{m}(1 - x)^{n}\, dx = \int_{0}^{1} x^{n}(1 - x)^{m}\, dx, \tag{20}$$

where m and n are positive integers.

4. Use integration by substitution to evaluate

   (a) $\displaystyle \int \frac{x^{2}}{\sqrt{x^{3} + 1}}\, dx;$ (b) $\displaystyle \int e^{x^{2}} x\, dx;$ (c) $\displaystyle \int \frac{dx}{x \ln^{2} x};$

   (d) $\displaystyle \int \frac{\sqrt{1 + \ln x}}{x}\, dx.$

5. Show that

$$\int \frac{f'(x)}{f(x)}\, dx = \ln |f(x)| + C. \tag{21}$$

6. Use (21) to evaluate

   (a) $\displaystyle \int \frac{2x}{1 + x^{2}}\, dx;$ (b) $\displaystyle \int \frac{x + 1}{x^{2} + 2x - 3}\, dx;$ (c) $\displaystyle \int \frac{dx}{x \ln x};$

   (d) $\displaystyle \int \frac{e^{2x}}{e^{2x} + 1}\, dx.$

7. Find the average of the function $y = \ln x$ over the interval $[1, e]$.
8. Use integration by parts to evaluate

   (a) $\displaystyle \int xa^{x}\, dx;$ (b) $\displaystyle \int x^{3}e^{x}\, dx;$ (c) $\displaystyle \int x^{3} \ln x\, dx;$ (d) $\displaystyle \int \sqrt{x} \ln x\, dx.$

9. Use integration by parts to verify that

$$\int \ln (x + \sqrt{1 + x^{2}})\, dx = x \ln (x + \sqrt{1 + x^{2}}) - \sqrt{1 + x^{2}} + C.$$

10. Evaluate

   (a) $\displaystyle \int_{0}^{1} xe^{-x}\, dx;$ (b) $\displaystyle \int_{0}^{e-1} \ln (x + 1)\, dx;$ (c) $\displaystyle \int_{1}^{2} x \log_{2} x\, dx;$

   (d) $\displaystyle \int_{1}^{e} \ln^{2} x\, dx.$

11. Evaluate

$$\int (2^{x} + 3^{x})^{2}\, dx,$$

   after first showing that $a^{x}b^{x} = (ab)^{x}$.

12. Verify that

$$\int \frac{dx}{(x + a)(x + b)} = \frac{1}{a - b} \ln \left| \frac{x + b}{x + a} \right| + C \qquad (a \neq b). \qquad (22)$$

13. Use (22) to evaluate

(a) $\int \dfrac{dx}{x^2 - 5x + 6}$;   (b) $\int \dfrac{dx}{4x^2 + 4x - 3}$.

14. Find the area between the curves $y = \ln x$ and $y = \ln^2 x$.

*15. Show that

$$\int \cosh x \, dx = \sinh x + C, \qquad \int \sinh x \, dx = \cosh x + C.$$

*16. Use the substitution $\sqrt{1 + x^2} = t - x$ to show that

$$\int \frac{dx}{\sqrt{1 + x^2}} = \ln (x + \sqrt{1 + x^2}) + C.$$

*17. Evaluate

(a) $\displaystyle\int \frac{e^{3x} + 1}{e^x + 1} \, dx$;   (b) $\displaystyle\int \frac{1 + x}{1 - x} \, dx$;   (c) $\displaystyle\int \frac{x^2}{1 - x^2} \, dx$.

*18. Given two functions $f(x)$ and $g(t)$, with antiderivatives $F(x)$ and $G(t)$, let $t = t(x)$ be a differentiable one-to-one function, with a differentiable inverse $x = x(t)$, such that one of the formulas

$$\int f(x) \, dx = \int g(t(x)) t'(x) \, dx, \qquad \int f(x(t)) x'(t) \, dt = \int g(t) \, dt \qquad (23)$$

holds. Show that the other formula also holds.

*19. Let

$$\int f(x) \, dx = \int g(t) \, dt$$

be shorthand for both of the formulas (23), depending on whether we replace dt by $t'(x) \, dx$ or dx by $x'(t) \, dt$. Show that

$$\int_a^b f(x) \, dx = \int_{t(a)}^{t(b)} g(t) \, dt, \qquad \int_{x(\alpha)}^{x(\beta)} f(x) \, dx = \int_\alpha^\beta g(t) \, dt.$$

*20. Evaluate the integral (20).

*21. Let $P(x)$ be a polynomial of degree n. Verify that

$$\int P(x) e^x \, dx = [P(x) - P'(x) + P''(x) - \cdots + (-1)^n P^{(n)}(x)] e^x + C.$$

*22. Let the quantity of a commodity demanded by the market at price P be $Q = Q(P)$, where $Q(P)$ is a decreasing function, and let P_1 be the actual market price. Then the total revenue received from the sale of the commodity is

$$P_1 Q(P_1) = Q_1 P(Q_1), \qquad (24)$$

where $Q_1 = Q(P_1)$ and $P = P(Q)$ is the inverse of the function $Q = Q(P)$. Since some consumers are willing to pay more than P_1 for the commodity, the total revenue from the sale of a quantity Q_1 of the commodity would be greater than (24) by some amount S, known as the *consumer's surplus*, if the price of the commodity were gradually lowered from P_0, the price at which demand just begins, to the actual market price P_1. Show that

$$S = \int_0^{Q_1} P(Q)\, dQ - P_1 Q_1,$$ (25)

or equivalently

$$S = \int_{P_1}^{P_0} Q(P)\, dP.$$ (26)

*23. Use both (25) and (26) to calculate the consumer's surplus S for the demand function $Q = 100 \ln (P_0/P)$ and market price $P_1 = P_0/e$.

4.7 IMPROPER INTEGRALS

4.71. a. In introducing the concept of the definite integral

$$\int_a^b f(x)\, dx,$$

it was assumed from the outset that $f(x)$ is defined at every point of the closed interval $[a, b]$, where, of course, a and b are finite numbers (Sec. 1.64a). Thus, at this stage of the game, neither of the integrals

$$\int_1^\infty \frac{dx}{x^2}$$ (1)

and

$$\int_0^1 \frac{dx}{\sqrt{x}}$$ (2)

makes sense, the first because the upper limit of integration is infinite, the second because the integrand is not defined at $x = 0$ and in fact approaches infinity as $x \to 0+$. However, there is a simple way of ascribing meaning to both of these integrals, which are called "improper" to distinguish them from the ordinary or "proper" integrals considered up to now. As we will see in a moment, the device is essentially the same in both cases: First we calculate the integral over a *finite* interval in which the integrand is well-defined, and then we take the limit of the resulting *proper* integral as the interval of integration is suitably enlarged. For simplicity, we will consider only continuous integrands.

b. Suppose $f(x)$ is continuous, and hence integrable (Sec. 4.14) in every interval $[a, X]$, where a is fixed and $X > a$ is variable, and suppose the limit

$$\lim_{X \to \infty} \int_a^X f(x)\, dx$$ (3)

exists and is finite. Then the improper integral

$$\int_a^\infty f(x)\, dx$$ (4)

is said to be *convergent* and is assigned the value (3). On the other hand, if the limit (3) is infinite or fails to exist, we call the integral (4) *divergent* and assign it no value at all.

Similarly, suppose $f(x)$ is continuous in every interval $[X, a]$, where a is fixed and $X < a$ is variable, and suppose the limit

$$\lim_{X \to -\infty} \int_X^a f(x)\, dx \tag{5}$$

exists and is finite. Then the improper integral

$$\int_{-\infty}^a f(x)\, dx \tag{6}$$

is said to be *convergent* and is assigned the value (5). Again we call the integral (6) *divergent* if the limit (5) is infinite or fails to exist. We can also consider the case where *both* limits of integration are infinite. Thus, suppose $f(x)$ is continuous in every finite interval, and suppose both improper integrals

$$\int_{-\infty}^c f(x)\, dx, \qquad \int_c^\infty f(x)\, dx \tag{7}$$

are convergent for an arbitrary finite point c. Then the improper integral

$$\int_{-\infty}^\infty f(x)\, dx \tag{8}$$

is said to be *convergent*, and is assigned the value

$$\int_{-\infty}^c f(x)\, dx + \int_c^\infty f(x)\, dx. \tag{9}$$

Of course, this definition depends on the fact that both the sum (9), and the convergence or divergence of the integrals (7), are independent of the choice of the point c (see Prob. 5). On the other hand, the integral (8) is said to be *divergent* if either of the integrals (7) is divergent.

   c. We now turn to improper integrals of the type (2), where the integrand becomes infinite at one or more points of the interval of integration. Suppose $f(x)$ approaches infinity as $x \to a+$, at the same time that $f(x)$ is continuous, and hence integrable, in every interval $[a + \varepsilon, b]$, where a and $b > a$ are fixed and $\varepsilon > 0$ is variable (but less than $b - a$). Suppose further that the limit

$$\lim_{\varepsilon \to 0+} \int_{a+\varepsilon}^b f(x)\, dx \tag{10}$$

exists and is finite. Then the improper integral

$$\int_a^b f(x)\, dx \tag{11}$$

is said to be *convergent* and is assigned the value (10). As before, we call the integral (11) *divergent* if the limit (10) is infinite or fails to exist. Similarly, if $f(x)$ becomes infinite at the other end point b, we set

$$\int_a^b f(x)\, dx = \lim_{\varepsilon \to 0+} \int_a^{b-\varepsilon} f(x)\, dx,$$

by definition, while if $f(x)$ becomes infinite at an *interior* point $c \in (a,b)$, we write

$$\int_a^b f(x)\, dx = \int_a^c f(x)\, dx + \int_c^b f(x)\, dx, \tag{12}$$

provided that both integrals on the right are convergent.

4.72. Examples

a. The integral (1) is convergent. In fact

$$\int_1^\infty \frac{dx}{x^2} = \lim_{X \to \infty} \int_1^X \frac{dx}{x^2} = \lim_{X \to \infty} \left[-\frac{1}{x} \right]_1^X = \lim_{X \to \infty} \left(1 - \frac{1}{X} \right) = 1,$$

and hence

$$\int_1^\infty \frac{dx}{x^2} = 1.$$

This number can also be regarded as the area of the "infinite region" under the curve $y = 1/x^2$ from 1 to ∞, by a natural extension of the definition of the area under a curve for the finite case (Sec. 4.11).

b. The integral

$$\int_1^\infty \frac{dx}{x}$$

is divergent, since

$$\lim_{X \to \infty} \int_1^X \frac{dx}{x} = \lim_{X \to \infty} \ln x \Big|_1^X = \lim_{X \to \infty} \ln X = \infty.$$

c. The integral

$$\int_{-\infty}^\infty e^{-|x|}\, dx$$

is convergent. In fact,

$$\int_0^\infty e^{-|x|}\, dx = \int_0^\infty e^{-x}\, dx = \lim_{X \to \infty} \int_0^X e^{-x}\, dx$$

$$= \lim_{X \to \infty} (-e^{-x}) \Big|_0^X = \lim_{X \to \infty} (1 - e^{-X}) = 1,$$

since $e^{-X} \to 0$ as $X \to \infty$, and similarly

$$\int_{-\infty}^0 e^{-|x|}\, dx = \int_{-\infty}^0 e^x\, dx = \lim_{X \to -\infty} \int_X^0 e^x\, dx$$

$$= \lim_{X \to -\infty} e^x \Big|_X^0 = \lim_{X \to -\infty} (1 - e^X) = 1.$$

Therefore, as in (9),

$$\int_{-\infty}^\infty e^{-|x|}\, dx = \int_{-\infty}^0 e^{-|x|}\, dx + \int_0^\infty e^{-|x|}\, dx = 2.$$

d. The integral (2) is convergent. In fact,

$$\int_0^1 \frac{dx}{\sqrt{x}} = \lim_{\varepsilon \to 0+} \int_\varepsilon^1 \frac{dx}{\sqrt{x}} = \lim_{\varepsilon \to 0+} \left[2\sqrt{x} \right]_\varepsilon^1 = \lim_{\varepsilon \to 0+} (2 - 2\sqrt{\varepsilon}),$$

and hence

$$\int_0^1 \frac{dx}{\sqrt{x}} = 2. \tag{13}$$

   e.   The integral

$$\int_0^1 \frac{dx}{x^2} \tag{14}$$

is divergent, since

$$\lim_{\varepsilon \to 0+} \int_\varepsilon^1 \frac{dx}{x^2} = \lim_{\varepsilon \to 0+} \left[-\frac{1}{x} \right]_\varepsilon^1 = \lim_{\varepsilon \to 0+} \left(\frac{1}{\varepsilon} - 1 \right) = \infty.$$

   f.   The integral

$$\int_0^2 \frac{dx}{\sqrt{|x - 1|}}$$

is convergent. In fact, using (12) with $c = 1$, we get

$$\int_0^2 \frac{dx}{\sqrt{|x - 1|}} = \int_0^1 \frac{dx}{\sqrt{1 - x}} + \int_1^2 \frac{dx}{\sqrt{x - 1}}$$

$$= -\int_1^0 \frac{dt}{\sqrt{t}} + \int_0^1 \frac{du}{\sqrt{u}} = \int_0^1 \frac{dt}{\sqrt{t}} + \int_0^1 \frac{du}{\sqrt{u}}$$

(let $t = 1 - x$, $u = x - 1$). But both integrals on the right equal 2, by formula (13), and hence

$$\int_0^2 \frac{dx}{\sqrt{|x - 1|}} = 4.$$

   g.   Since the integral (14) is divergent, so is the integral

$$\int_{-1}^1 \frac{dx}{x^2}$$

(why?). Suppose we make the mistake of calculating this integral formally, ignoring the fact that the integrand becomes infinite at the origin. Then we get the absurd result

$$\int_{-1}^1 \frac{dx}{x^2} = \left[-\frac{1}{x} \right]_{-1}^1 = -2,$$

seemingly an example of a positive function with a negative integral!

PROBLEMS

1. Investigate the improper integral

$$\int_a^\infty \frac{dx}{x^r} \qquad (a > 0)$$

for arbitrary r.

2. Evaluate

    (a) $\int_0^\infty xe^{-x^2}\,dx$;   (b) $\int_2^\infty \dfrac{dx}{x\ln x}$;   (c) $\int_2^\infty \dfrac{dx}{x\ln^2 x}$.

3. Investigate the improper integral

$$\int_0^a \frac{dx}{x^r} \qquad (a > 0)$$

    for arbitrary r.

4. Evaluate

    (a) $\int_1^2 \dfrac{dx}{x\ln x}$;   (b) $\int_{-1}^8 \dfrac{dx}{\sqrt[3]{x}}$;   (c) $\int_3^5 \dfrac{dx}{\sqrt[3]{(4-x)^2}}$.

5. Let $f(x)$ be continuous in every finite interval. Show that the two integrals

$$\int_c^\infty f(x)\,dx, \qquad \int_{c'}^\infty f(x)\,dx \qquad (c \neq c')$$

    are either both convergent or both divergent, and similarly for

$$\int_{-\infty}^c f(x)\,dx, \qquad \int_{-\infty}^{c'} f(x)\,dx \qquad (c \neq c').$$

    Show that

$$\int_{-\infty}^c f(x)\,dx + \int_c^\infty f(x)\,dx = \int_{-\infty}^{c'} f(x)\,dx + \int_{c'}^\infty f(x)\,dx.$$

    Show that the sum (12) is also independent of c.

6. How does the theory of improper integrals resemble the theory of infinite series?
7. First define and then find the area A between the curves $y = x^{-1/2}$ and $y = x^{-1/3}$ from $x = 0$ to $x = 1$.
*8. First define and then find the area A between the curves $y = \cosh x$ and $y = \sinh x$ in the first quadrant.

Chapter 5

INTEGRATION AS A TOOL

5.1 ELEMENTARY DIFFERENTIAL EQUATIONS

5.11. The equations

$$xy' - 2y = 0, \qquad y'' + x^2 = 0, \qquad y^{(4)} = y^2 \tag{1}$$

are all called *differential equations*, because each contains at least one derivative

$$y' = \frac{dy}{dx} = f'(x), \qquad y'' = \frac{d^2y}{dx^2} = f''(x), \ldots$$

of an "unknown" function $y = f(x)$. Note that the second equation does not contain the function y itself, while the third equation does not contain the independent variable x, although x is, of course, present implicitly as the argument of the function $y = f(x)$ and its fourth derivative $y^{(4)} = f^{(4)}(x)$. A differential equation is said to be *of order n* if it contains the nth derivative $y^{(n)} = f^{(n)}(x)$, but no derivatives of higher order. Thus the equations (1) are of orders 1, 2 and 4, respectively.

In this book we have no intention of doing more than scratch the surface of the vast subject of differential equations and their applications. In fact, we will consider only the most elementary differential equations, of either the first or the second order.

5.12. a. Consider the *first-order differential equation*

$$\frac{dy}{dx} = F(x, y), \tag{2}$$

where $F(x, y)$ is a function of two variables, which may reduce to a function of x alone, to a function of y alone, or even to a constant. By a *solution* of (2) we mean any function $y = \varphi(x)$ such that

$$\frac{d\varphi(x)}{dx} = F(x, \varphi(x))$$

holds for all values of x in some interval. We write $\varphi(x)$ instead of $f(x)$ here, because the solution is regarded as a "known" function.

For example, $y = e^{x^2}$ is a solution of the differential equation

$$\frac{dy}{dx} = 2xy, \tag{3}$$

since

$$\frac{de^{x^2}}{dx} = 2xe^{x^2} \equiv 2xy$$

in every interval. Moreover, if C is an arbitrary constant, then $y = Ce^{x^2}$ is also a solution of (3), since

$$\frac{d(Ce^{x^2})}{dx} = 2Cxe^{x^2} \equiv 2xy.$$

We call

$$y = Ce^{x^2} \tag{4}$$

the *general solution* of (3), because every solution of (4) can be obtained from (4) by making a suitable choice of C (see Example 5.13d).

More generally, let

$$y = \varphi(x, C) \tag{5}$$

be a solution of the differential equation (2), involving an "arbitrary constant" C (which is temporarily variable!), and suppose every solution of (2) can be obtained from (5) by making a suitable choice of C. Then (5) is called the *general solution* of (2), and each solution of (2), corresponding to a particular choice of C in (5), is called a *particular solution* of (2). For example, giving C the values 0 and $\sqrt{3}$ in (5), we get two particular solutions

$$y \equiv 0, \qquad y = \sqrt{3}e^{x^2}$$

of equation (3).

b. In Sec. 1.12 we posed two key problems of calculus. The meaning of the first problem was clarified in Sec. 2.42d. The second problem was originally stated in the following unsophisticated language:

(2) Given the rate of change of one quantity with respect to another, what is the relationship between the two quantities?

We are now in a position to restate this problem elegantly in more precise language:

(2′) Solve the differential equation $dy/dx = F(x, y)$.

Here, of course, dy/dx is the rate of change, and to give this rate of change we will in general have to know the values of *both* variables x and y.

5.13. Examples

a. The simplest first-order differential equation is

$$\frac{dy}{dx} = f(x), \tag{6}$$

and its general solution is just

$$y = \int f(x)\, dx.$$

This follows at once from the meaning of the indefinite integral as the "general anti-derivative" of $f(x)$. Here it is better to write the general solution of (6) in the form

$$y = \int f(x)\, dx + C, \tag{7}$$

which makes the arbitrary constant of integration explicit. We then think of $\int f(x)\, dx$ as any *fixed* antiderivative of $f(x)$. We will follow this convention in any problem involving differential equations.

   b.   Find the particular solution of the differential equation

$$\frac{dy}{dx} = x \tag{8}$$

satisfying the condition

$$y\big|_{x=1} = 1. \tag{9}$$

SOLUTION.   First we use (7) to find the general solution of (8), obtaining

$$y = \int x\, dx + C = \frac{1}{2}x^2 + C. \tag{10}$$

Then we use the condition (9) to determine the constant C in (10). Thus

$$y\big|_{x=1} = \left(\frac{1}{2}x^2 + C\right)\bigg|_{x=1} = \frac{1}{2} + C = 1,$$

so that $C = \frac{1}{2}$. Substituting this value of C into (10), we get the desired particular solution

$$y = \frac{1}{2}(x^2 + 1). \tag{11}$$

The fact that (11) satisfies both (8) and (9) is easily verified by direct calculation.

   More generally, by an *initial condition* for the differential equation (2), we mean a condition of the form

$$y\big|_{x=x_0} = y_0, \tag{12}$$

where x_0 and y_0 are given numbers (the word "initial" stems, by analogy, from the common situation where the independent variable is the *time*). If (2) has the general solution $\varphi(x, C)$, the particular solution of (2) satisfying (12) has the value of C obtained by solving

$$\varphi(x_0, C) = y_0.$$

   c.   Solve the differential equation

$$\frac{dy}{dx} = \frac{f(x)}{g(y)}, \tag{13}$$

subject to the initial condition (12).

   SOLUTION.   This equation, or the equivalent equation

$$\frac{dy}{dx} = f(x)g^*(y),$$

with

$$g^*(y) = \frac{1}{g(y)},$$

is said to have *separated variables*. Multiplying (13) by $g(y)\,dx$, we get

$$g(y)\,dy = f(x)\,dx, \tag{14}$$

where the left side involves only the variable y and the right side involves only the variable x; it is in this sense that the variables are "separated." To solve (14), we merely integrate both sides, obtaining

$$\int g(y)\,dy = \int f(x)\,dx + C$$

(one constant of integration is enough), or

$$G(y) = F(x) + C, \tag{15}$$

where $G(y)$ is any antiderivative of $g(y)$ and $F(x)$ is any antiderivative of $f(x)$. To determine C, we impose the initial condition (12), which says that $y = y_0$ when $x = x_0$. Thus

$$G(y_0) = F(x_0) + C,$$

so that

$$C = G(y_0) - F(x_0).$$

Substituting this expression for C into (15), we get

$$G(y) = F(x) + G(y_0) - F(x_0). \tag{16}$$

The unique solution of the differential equation (13) satisfying the initial condition (12) is then obtained by solving (16) for y as a function of x, call it $y = \varphi(x)$. In the cases to be considered here, we will always be able to do this without difficulty.

The function $y = \varphi(x)$ determined by (16) clearly satisfies the initial condition (12), which is "built into" equation (16). To verify that $y = \varphi(x)$ actually satisfies the differential equation (13), we need only use the chain rule to differentiate (16) with respect to x. This gives

$$\frac{dG(y)}{dy}\frac{dy}{dx} = \frac{dF(x)}{dx},$$

so that

$$g(y)\frac{dy}{dx} = f(x),$$

which is equivalent to (13).

    d.   Find the general solution of the differential equation (3).

    SOLUTION.   To separate variables, we divide (3) by y, obtaining

$$\frac{1}{y}\frac{dy}{dx} = 2x. \tag{17}$$

Multiplying (17) by dx and integrating, we then get

$$\int \frac{dy}{y} = \int 2x\,dx + k,$$

or

$$\ln|y| = x^2 + k, \tag{18}$$

where we denote the arbitrary constant of integration by k, saving the symbol C for later. In writing $\ln |y|$, we use the fact that if y is negative, then $\ln |y| = \ln(-y)$ has the same derivative $1/y$ as $\ln y$ (check this). Taking the exponential of both sides of (18), we find that

$$|y| = e^{x^2 + k} = e^k e^{x^2} = Ce^{x^2},$$

where $C = e^k$ is now an arbitrary *positive* constant (why?). But $|y|$ can never vanish, since $Ce^{x^2} > 0$. Therefore y is either positive or negative for all x. Thus we can take the vertical bars off $|y|$, obtaining formula (4), by simply allowing C to take arbitrary negative values, as well as arbitrary positive values. In dividing (3) by y, we have tacitly assumed that y is nonvanishing. Thus the solution $y \equiv 0$ may have been lost in solving (17) instead of (3), and indeed it has, as we see at once by substituting $y \equiv 0$ into (3). Therefore the general solution of (3) is obtained by allowing C to take any value in (4), *including zero*.

  e. The simplest *second-order* differential equation is

$$\frac{d^2 y}{dx^2} = f(x). \tag{19}$$

Integrating (19), we get

$$\frac{dy}{dx} = \int \frac{d^2 y}{dx^2}\, dx = \int f(x)\, dx + C_1 = F(x) + C_1, \tag{20}$$

where C_1 is an arbitrary constant of integration and $F(x) = \int f(x)\, dx$ is any fixed antiderivative of $f(x)$. Observe that (20) is now a *first-order* differential equation, and is in fact of the form (6). Integrating (20) in turn, we get

$$y = \int \frac{dy}{dx}\, dx = \int F(x)\, dx + \int C_1\, dx + C_2 = \int F(x)\, dx + C_1 x + C_2,$$

where C_2 is another arbitrary constant of integration. Thus the *general* solution of the differential equation (19) involves *two* arbitrary constants, and this is a characteristic feature of the general solution of a *second-order* differential equation. Therefore, to single out a *particular* solution of (19), we must impose *two* initial conditions, since this will give us *two* algebraic equations which we can solve for the *two* constants C_1 and C_2.

  f. Find the particular solution of the differential equation

$$\frac{d^2 y}{dx^2} = x \tag{21}$$

satisfying the initial conditions

$$y\big|_{x=1} = \frac{1}{2}, \qquad y'\big|_{x=1} = -\frac{1}{2}. \tag{22}$$

SOLUTION. Note that one initial condition involves the function y, while the other involves its derivative y'. Integrating (21) twice with respect to x, we get first

$$y' = \frac{dy}{dx} = \int x\, dx + C_1 = \frac{1}{2} x^2 + C_1, \tag{23}$$

and then

$$y = \int \frac{1}{2} x^2\, dx + \int C_1\, dx + C_2 = \frac{1}{6} x^3 + C_1 x + C_2. \tag{24}$$

Substituting the second of the conditions (22) into (23) and the first into (24), we find that

$$-\frac{1}{2} = \frac{1}{2} + C_1,$$

$$\frac{1}{2} = \frac{1}{6} + C_1 + C_2.$$

Solving for C_1 and C_2, we then obtain

$$C_1 = -1, \qquad C_2 = \frac{4}{3}.$$

Thus the particular solution of the differential equation (21) satisfying the conditions (22) is just

$$y = \frac{1}{6}x^3 - x + \frac{4}{3}.$$

Instead of imposing one condition on y and the other on y' at the same point, we can impose two conditions on y at two different points. In this case, the conditions . are called *boundary conditions* rather than *initial conditions*. For example, to find the particular solution of (21) satisfying the boundary conditions

$$y|_{x=0} = 1, \qquad y|_{x=1} = 2,$$

we solve the equations

$$1 = C_2,$$

$$2 = \frac{1}{6} + C_1 + C_2$$

for C_1 and C_2. This gives

$$C_1 = \frac{5}{6}, \qquad C_2 = 1,$$

and leads to the particular solution

$$y = \frac{1}{6}x^3 + \frac{5}{6}x + 1.$$

PROBLEMS

1. Show that equation (2) is a special case of the even more general first-order differential equation $\Phi(x, y, y') = 0$, where $\Phi(x, y, z)$ is a function of three variables.
2. Find the particular solution of the differential equation $y' = -y/x$ satisfying the initial condition $y|_{x=2} = 1$.
3. Find the particular solution of the differential equation $y' + 2xy = 0$ satisfying the initial condition $y|_{x=0} = 1$.
4. Find the particular solution of the differential equations $y' = 2\sqrt{y} \ln x$ satisfying the initial condition $y|_{x=e} = 1$.
5. Show that all but one of the solutions of the differential equation $y'^2 = 4y$ are given by the formula $y = (x + C)^2$, where C is an arbitrary constant. What is the extra solution?
6. Find the general solution of the differential equation $y'' = \ln x$.

*7. By making the preliminary substitutions $y' = p$, $y'' = p(dp/dy)$, find the particular solution of the differential equation $y'' = 2y^3$ satisfying the initial conditions $y|_{x=0} = 1$, $y'|_{x=0} = 1$.

*8. A differential equation of the form

$$\frac{dy}{dx} = f\left(\frac{y}{x}\right)$$

is said to be *homogeneous*. Solve this equation by separation of variables, after making the substitution $y = ux$.

*9. Find the particular solution of the homogeneous differential equation $x + y + xy' = 0$ satisfying the initial condition $y|_{x=1} = 0$.

*10. A curve goes through the point $(-1, -1)$ and has the property that the *x*-intercept of the tangent to the curve at every point P is the square of the abscissa of P. What is the curve?

5.2 PROBLEMS OF GROWTH AND DECAY

5.21. a. Suppose the dependence of one variable, say y, on another variable, say t, is described by an "exponential law"

$$y = y_0 e^{rt}, \tag{1}$$

where $y_0 > 0$ and r are constants. Then the rate of change of y (with respect to t) is given by the derivative

$$\frac{dy}{dt} = y_0 r e^{rt}.$$

Thus y satisfies the simple differential equation

$$\frac{dy}{dt} = ry, \tag{2}$$

that is, *the rate of change of the variable y is proportional to the value of y.* The function e^{rt} is an *increasing* function of t if r is *positive*, since then $t_1 < t_2$ implies $rt_1 < rt_2$ and hence $e^{rt_1} < e^{rt_2}$; in this case, we say that y *grows exponentially* (with t), or that y is an *exponentially increasing* function of t. On the other hand, e^{rt} is a *decreasing* function of t if r is *negative*, since then $t_1 < t_2$ implies $rt_1 > rt_2$ and hence $e^{rt_1} > e^{rt_2}$; in this case, we say that y *decays* (or *falls off*) *exponentially* (with t) or that y is an *exponentially decreasing* function of t.

b. Setting $t = 0$ in (1), we find that

$$y|_{t=0} = y_0. \tag{3}$$

Thus the constant y_0 is just the *initial value* of y, that is, the value of y at the time $t = 0$. In other words, (1) is the particular solution of the differential equation (2) satisfying the initial condition (3). This can, of course, be seen directly. In fact, separating variables in (2), we get

$$\frac{dy}{y} = r \, dt,$$

so that

$$\int \frac{dy}{y} = \int r \, dt + k,$$

or

$$\ln |y| = rt + k, \tag{4}$$

where k is a constant of integration. Taking the exponential of both sides of (4), we find that

$$|y| = Ce^{rt}, \tag{5}$$

where $C = e^k > 0$. But $|y|$ can never vanish, since $Ce^{rt} > 0$. Therefore y is either positive or negative for all t. Since $y_0 > 0$, by assumption, the initial condition (3) can be satisfied only if y is positive, and then $|y| = y$, $C = y_0$, so that (5) reduces to (1).

    c.   It follows from (2) that

$$r = \frac{1}{y} \frac{dy}{dt},$$

or equivalently

$$r = \frac{d}{dt} \ln y,$$

that is, r is the *logarithmic derivative* of y (Sec. 4.53a). Thus r is not the rate of change of y, but rather the rate of change of y divided by the "current" value of y. The quantity dy/dt is called the *growth rate*, whether it be positive or negative, while r is called the *proportional growth rate*. The word "proportional" can be dropped if r is given in *percent* per unit time, since it is then clear that we can only be talking about a proportional growth rate. Sometimes r is simply called the *rate*, when there is no possibility of confusion.

5.22. Population growth

    a. Example.   A population grows exponentially at the rate r. How long does it take the population to double?

    SOLUTION.   Here r is positive and we have

$$N = N(t) = N_0 e^{rt}, \tag{6}$$

where N (for "number") is the population at time t and N_0 is the population at time $t = 0$. The function N is, of course, just the particular solution of the differential equation

$$\frac{dN}{dt} = rN \tag{7}$$

satisfying the initial condition

$$N|_{t=0} = N_0. \tag{8}$$

Let T be the *doubling time*, that is, the time at which the population is twice its initial value N_0. Then

$$N_0 e^{rT} = 2N_0,$$

or

$$e^{rT} = 2. \tag{9}$$

Note that T is independent of the population size. Taking logarithms, we get

$$rT = \ln 2,$$

or

$$T = \frac{\ln 2}{r} \approx \frac{0.693}{r}. \tag{10}$$

Thus the doubling time is inversely proportional to the proportional growth rate r, which makes sense ("the faster the growth, the shorter the doubling time"). Suppose r is measured in percent per year and t in years, as in studies of human population. Then we get the rule of thumb

$$T \approx \frac{69}{r} \text{ years.}$$

For example, the average annual growth rate of the population of Brazil during the period 1961–1968 was 3%. At this rate the population of Brazil will double in about $\frac{69}{3} = 23$ years.

   b. If T is the doubling time of a population, then the population will double *every* T years, as long as it is growing at the rate r. To see this, we merely note that

$$N(T) = e^{rT} N_0 = 2N_0,$$
$$N(2T) = e^{r2T} N_0 = (e^{rT})^2 N_0 = 2^2 N_0 = 4N_0,$$

with the help of (6) and (9), and, more generally,

$$N(nT) = e^{nrT} N_0 = (e^{rT})^n N_0 = 2^n N_0.$$

   c. The differential equation (7) merely says that the rate of change of the population is proportional to the present size of the population. This is perfectly plausible. In fact, on the one hand, we must have

$$\frac{dN}{dt} = B - D, \tag{7'}$$

where B is the *birth rate* and D the *death rate*. On the other hand, both B and D are proportional to the population size N (large cities have more maternity wards and more cemeteries than small towns). Therefore $B - D$ is also proportional to N. Comparing (7) and (7'), we find that

$$r = \frac{B - D}{N}.$$

In other words, the proportional annual growth rate of population is just the per capita excess of the birth rate over the death rate.

   d. Eventually, of course, population growth must stop, due to lack of food, spread of infectious disease, loss of fertility due to overcrowding, wars fought for dwindling resources, or whatever. It turns out that these effects of "overpopulation" are described remarkably well in many cases by introducing an extra term $-sN^2$ in the right side of (7), where s (like r) is a positive constant. The resulting "growth equation" then becomes

$$\frac{dN}{dt} = rN - sN^2, \tag{11}$$

instead of (7), subject to the same initial condition (8).

To solve (11), we separate variables and integrate, obtaining

$$\int \frac{dN}{rN - sN^2} = \int dt + c = t + c, \tag{12}$$

where c is a constant of integration. The integral on the left is not hard to evaluate. In fact, setting $\alpha = s/r$, we have

$$\int \frac{dN}{rN - sN^2} = \frac{1}{r} \int \frac{dN}{N(1 - \alpha N)} = \frac{1}{r} \int \left[\frac{1}{N} + \frac{\alpha}{1 - \alpha N} \right] dN$$

$$= \frac{1}{r} [\ln N - \ln |1 - \alpha N|] = \frac{1}{r} \ln \frac{N}{|1 - \alpha N|}.$$

Therefore (12) becomes

$$\ln \frac{N}{|1 - \alpha N|} = rt + k,$$

where $k = rc$, or

$$\frac{N}{|1 - \alpha N|} = Ce^{rt},$$

where $C = e^k$. Applying the initial condition (8), we get

$$C = \frac{N_0}{|1 - \alpha N_0|},$$

so that

$$\frac{N}{|1 - \alpha N|} = \frac{N_0}{|1 - \alpha N_0|} e^{rt},$$

where the vertical bars can now be dropped, since $1 - \alpha N$ and $1 - \alpha N_0$ have the same sign (in evaluating the integral on the left in (12), it was tacitly assumed that $1 - \alpha N \neq 0$ for all $t \geq 0$). Doing this and solving for N, we finally obtain

$$N = \frac{N_0 e^{rt}}{1 - \alpha N_0 (1 - e^{rt})},$$

or, even more simply,

$$N = \frac{N_1}{1 + \left(\dfrac{N_1}{N_0} - 1\right) e^{-rt}}, \tag{13}$$

where

$$N_1 = \frac{1}{\alpha} = \frac{r}{s}. \tag{14}$$

Note that $N_1 \neq N_0$, since $1 - \alpha N_0 \neq 0$.

Graphing N as a function of time, we get the "S-shaped" curve shown in Figure 1 for the case $N_1 = 50 N_0$. Note that the population growth is now restricted. In fact,

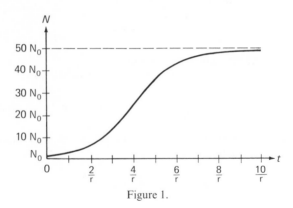

Figure 1.

if t is large enough, then $e^{-rt} \approx 0$ and (13) is close to the "stable" population level N_1 given by (14), where N_1 is independent of the size of the initial population. The validity of formula (13) has been confirmed by many observations, both of human populations and of experimental populations of bacteria, fruit flies, etc.

5.23. Radioactivity

a. Suppose it takes 2 days for 50% of the radioactivity emitted by a radio-active substance to disappear. How long does it take for 99% of the radioactivity to disappear?

SOLUTION. For simplicity, we assume that the radioactivity is entirely due to a single radioactive substance. It is known from physics that the rate of change of the mass of a substance undergoing radioactive decay is proportional at each instant of time t to the mass $m = m(t)$ of the substance actually present; the situation resembles that of a sterile population which is "dying off," except that an atom of radioactive substance, unlike a person, can have an arbitrarily large "longevity." Thus $m = m(t)$ is the particular solution of the differential equation

$$\frac{dm}{dt} = rm \tag{15}$$

satisfying the initial condition

$$m\big|_{t=0} = m_0, \tag{16}$$

where m_0 is the mass of the radioactive substance present at time $t = 0$. Since the proportional growth rate r is now negative, we replace it by $-k$, where k is positive. The differential equation (15) then becomes

$$\frac{dm}{dt} = -km. \tag{17}$$

The solution of (17), subject to the initial condition (16), is, of course, just the function

$$m = m_0 e^{-kt}, \tag{18}$$

describing the exponential decay of the amount of radioactive substance.

We now use the data of the problem to determine the number k, called the *proportional decay rate*, or simply the *decay constant*. Since 50% of the radioactivity disappears in 2 days, we have, measuring t in days,

$$m_0 e^{-2k} = \frac{1}{2} m_0.$$

This implies

$$-2k = \ln \frac{1}{2},$$

or

$$k = \frac{1}{2} \ln 2.$$

With this value of k, (18) becomes

$$m = m_0 e^{-(t/2) \ln 2}.$$

Disappearance of 99% of the radioactivity means that $m = \frac{1}{100} m_0$. Thus 99% of the radioactivity will disappear after τ days if τ satisfies the equation

$$m_0 e^{-(\tau/2) \ln 2} = \frac{1}{100} m_0,$$

or

$$-\frac{\tau}{2} \ln 2 = \ln \frac{1}{100}. \tag{19}$$

Solving (19) for τ, we get

$$\tau = 2 \frac{\ln 100}{\ln 2} = 4 \frac{\ln 10}{\ln 2} \approx 4 \frac{2.30}{0.69} \approx 13.3 \text{ days.}$$

Actually, we could have estimated the value of τ at once by the following argument: Half the radioactivity disappears in 2 days, half of what's left disappears in 2 more days, leaving one fourth the original amount after 4 days, half of what's now left disappears after another 2 days, leaving one eighth the original amount after 6 days, and so on. Hence $\frac{1}{64}$ of the original amount is left after 12 days, and $\frac{1}{128}$ is left after 14 days, so that $\frac{1}{100}$ is left after about 13 days. You will recognize this reasoning, based on repeated *halving*, as the exact analogue of the treatment of repeated *doubling*, given in Sec. 5.22b.

b. The time it takes a radioactive substance to decay to one half its original amount is called the *half-life* of the substance, and is independent of the amount originally present. The connection between the half-life T and the decay constant k is just

$$T = \frac{\ln 2}{k} \approx \frac{0.693}{k}, \tag{20}$$

as we see at once by solving the equation

$$m_0 e^{-kT} = \frac{1}{2} m_0$$

for T. Note that (20) is the same as formula (10) for the doubling time, with r replaced by k.

PROBLEMS

1. The number of bacteria in a culture doubles every hour. How long does it take a thousand bacteria to produce a million? What is the number N of bacteria in the culture at time t?

2. The world's population, equal to 3.6 billion in 1970, is growing exponentially at the rate of about 2.1% per year. Estimate the world's population in the year 1984. In the year 2001.

3. An exponentially growing population increases by 20% in 5 years. What is its doubling time?

4. One fourth of a radioactive substance disintegrates in 20 years. What is its half-life?

5. The average amount of radium in the earth's crust is about 1 atom in 10^{12}. Does it make sense to assume that this is the radium left over from a larger amount present at the time the earth was formed? The half-life of radium is 1620 years, and the age of the earth is estimated at 4.6 billion years.

6. Let $N = N(t)$, N_0 and N_1 be the same as in formula (13). Show that
   (a) If $N_0 < N_1$, $N(t)$ is increasing in $[0, \infty)$;
   (b) If $N_0 > N_1$, $N(t)$ is decreasing in $[0, \infty)$;
   (c) In both cases, $N(t) \to N_1$ as $t \to \infty$;
   (d) If $N_0 < N_1$, $N(t)$ has an inflection point at $t = t_0$, where $N(t_0) = \frac{1}{2} N_1$.

7. Suppose consumption grows exponentially at the rate of $r\%$ per year, while population grows exponentially at the rate of $s\%$ per year   How does the per capita consumption behave?

8. According to *Newton's law of cooling*, a body at temperature T cools at a rate proportional to the difference between T and the temperature of the surrounding air. Suppose the air temperature is $20°$ (Centigrade) and the body cools from $100°$ to $60°$ in 20 minutes. How long does it take the body to cool to $30°$?

9. The absorption of daylight by sea water is described by the exponential law

$$I = I_0 e^{-\mu x},$$

   where I_0 is the intensity of light at the surface of the sea and I is its intensity at the depth x. Find the constant μ, called the *absorption coefficient*, if the intensity of light at a depth of 5 meters is one thousandth of its intensity at the surface.

*10. Solve the growth equation

$$\frac{dN}{dt} = sN^2 - rN, \tag{21}$$

   differing from (11) in the sign of the right side. Show that in this case, corresponding to a birth rate proportional to the square of the population size, the population is destined for extinction if its initial size N_0 is less than $N_1 = r/s$. Show that if $N_0 > N_1$, then $N \to \infty$ as

$$t \to \frac{1}{r} \ln \frac{N_0}{N_0 - N_1}.$$

*11. Solve the growth equation

$$\frac{dN}{dt} = rN + s,$$

corresponding to a birth rate proportional to the population size, together with an "immigration" rate s.

*12. Radioactive carbon 14 ("radiocarbon"), with a half-life of 5570 years, is continually being produced in the upper atmosphere by the action of cosmic rays on nitrogen. Incorporated in carbon dioxide, the radiocarbon is mixed into the lower atmosphere, and is absorbed first by plants, during photosynthesis, and then by animals eating the plants. As long as they are alive, the plants and animals take in fresh radiocarbon, but when they die, the process ceases and the radiocarbon in their tissues slowly disintegrates, dropping to half its original amount in 5570 years. This fact leads to a method, called *radiocarbon dating*, for estimating the ages of such things as fossil organisms and old bits of wood and charcoal. For example, the age of a sliver of a mummy case can be estimated by comparing the amount of radioactivity in the sliver with the amount of radioactivity in a piece of fresh wood of the same kind and size.

Suppose a Geiger counter records m disintegrations from an old specimen of unknown age τ during the same period in which it records $n \, (>m)$ disintegrations from a similar contemporary sample. Show that

$$\tau = \frac{5570}{\ln 2} \ln \frac{n}{m} \text{ years.}$$

*13. Heartwood from a giant sequoia tree has only 75% of the radioactivity of the younger outer wood. Estimate the age of the tree.

5.3   PROBLEMS OF MOTION

5.31.   Consider the motion of a particle moving along a straight line L. As in Sec. 3.11, let s be the particle's distance at time t from some fixed reference point, where s is positive if measured in a given direction along the line and negative if measured in the opposite direction. Suppose the particle is subject to a force F, acting along the line L. Then *Newton's second law of motion* states that

$$F = ma, \tag{1}$$

where m is the particle's mass and

$$a = \frac{d^2 s}{dt^2}$$

is the particle's acceleration (Sec. 3.14a). The deceptively simple formula (1) is actually a second-order differential equation, with far-reaching physical consequences. As we now illustrate by a series of examples, once F is known, we can determine the particle's position as a function of time by solving (1), subject to appropriate initial conditions.

5.32. Examples

a.   Find the motion of a particle in the absence of any external forces.

SOLUTION.   In this case there are no forces, so that $F = 0$ in (1). It follows that

$$a = \frac{d^2 s}{dt^2} = 0, \tag{2}$$

after cancelling out the mass, which plays no role here. We now solve (2) by two consecutive integrations. Recalling that

$$a = \frac{dv}{dt},$$

where

$$v = \frac{ds}{dt}$$

is the particle's instantaneous velocity (Sec. 3.13a), we first write (2) in the form

$$a = \frac{dv}{dt} = 0. \qquad (3)$$

Integrating (3), we find that

$$v = \int \frac{dv}{dt} \, dt = \int 0 \cdot dt = C_1, \qquad (4)$$

where C_1 is a constant of integration, and hence

$$v = \frac{ds}{dt} = C_1. \qquad (5)$$

Integrating (5) in turn, we get

$$s = \int \frac{ds}{dt} \, dt = \int C_1 \, dt + C_2 = C_1 t + C_2, \qquad (6)$$

where C_2 is another constant of integration.

We must now determine the constants C_1 and C_2. This is done by taking account of the *initial conditions* of the problem, namely

$$v|_{t=0} = v_0, \qquad s|_{t=0} = s_0,$$

where v_0 and s_0 are the velocity and position of the particle at the initial time $t = 0$. Setting $t = 0$, $v = v_0$ in (4) and $t = 0$, $s = s_0$ in (6), we find at once that

$$C_1 = v_0, \qquad C_2 = s_0.$$

Therefore (4) and (6) become

$$v = v_0$$

and

$$s = v_0 t + s_0,$$

where it will be noted that s has the constant value s_0 if $v_0 = 0$. Thus we have proved *Newton's first law of motion*: Unless acted upon by an external force, a body at rest ($v_0 = 0$) remains at rest and a body in motion ($v_0 \neq 0$) continues to move with constant velocity along a straight line. It is shown in a course on mechanics that this conclusion remains true for a particle free to move in three-dimensional space, rather than just along some line L.

b.   Find the motion of a stone of mass m dropped from a point above the earth's surface.

SOLUTION.   We regard the stone as a particle, neglecting its size (Sec. 3.11). Let $s = s(t)$ be the stone's position, as measured along a vertical axis with the positive direction pointing downward and the origin at the initial position of the stone. By elementary physics, the force acting on the stone is

$$F = mg,$$

where g is the acceleration due to gravity (approximately 32 ft/sec^2) and we neglect the effect of air resistance. Thus, in this case, Newton's second law reduces to

$$a = \frac{dv}{dt} = \frac{d^2s}{dt^2} = g, \tag{7}$$

which says that the acceleration has the constant value g. Integrating (7) twice, we get first

$$v = \int \frac{dv}{dt} \, dt = \int g \, dt + C_1 = gt + C_1, \tag{8}$$

and then

$$s = \int \frac{ds}{dt} \, dt = \int (gt + C_1) \, dt + C_2 = \frac{1}{2} gt^2 + C_1 t + C_2. \tag{9}$$

This time the initial conditions are

$$v\big|_{t=0} = 0, \qquad s\big|_{t=0} = 0,$$

since the stone is "dropped" (that is, released with no initial velocity) from the point chosen as origin. Setting $t = 0$, $v = 0$ in (8) and $t = 0$, $s = 0$ in (9), we immediately get $C_1 = C_2 = 0$. Thus, finally,

$$v = gt, \tag{8'}$$

and

$$s = \frac{1}{2} gt^2, \tag{9'}$$

at least until the stone hits the ground.

    c.   Find the motion of a stone thrown vertically upward with initial velocity v_0.

SOLUTION.   We now find it more convenient to measure the stone's position along a vertical axis with the positive direction pointing *upward*. This has the effect of changing g to $-g$ in (8) and (9), since the acceleration due to gravity points *downward*. The initial conditions are now

$$v\big|_{t=0} = v_0, \qquad s\big|_{t=0} = 0.$$

Setting $t = 0$, $v = v_0$ in (8) and $t = 0$, $s = 0$ in (9), we get $C_1 = v_0$, $C_2 = 0$. Thus, in this case, (8) and (9) reduce to

$$v = v_0 - gt, \tag{8''}$$

and

$$s = v_0 t - \frac{1}{2} gt^2 \tag{9''}$$

(recall Example 3.15, where $v_0 = 96$ ft/sec).

d. Find the motion of a falling stone of mass m subject to air resistance.

SOLUTION. It is shown in physics that the effect of air resistance can be approximated by a force

$$F = -kv \qquad (k > 0),$$

proportional to the stone's velocity and acting in the direction opposite to its motion. The force acting on the stone is now the sum of two forces, its weight mg and the air resistance $-kv$. Thus, in this case, Newton's second law ($F = ma$) gives

$$m\frac{d^2 s}{dt^2} = m\frac{dv}{dt} = mg - kv, \tag{10}$$

where s is measured *downward* again, as in (7). Introducing the constants

$$\alpha = \frac{k}{m}, \qquad v_1 = \frac{g}{\alpha},$$

we can write (10) in the form

$$\frac{dv}{dt} = g - \alpha v = -\alpha(v - v_1). \tag{11}$$

Separating variables in (11) and integrating, we find that

$$\int \frac{dv}{v - v_1} = -\alpha \int dt + c,$$

where c is a constant of integration. Therefore

$$\ln |v - v_1| = -\alpha t + c,$$

or

$$v_1 - v = e^c e^{-\alpha t} = C e^{-\alpha t}, \tag{12}$$

since $v - v_1 < 0$ (why?), where $C = e^c$. Applying the initial condition $v|_{t=0} = 0$ (the stone is dropped from rest), we get $C = v_1$, so that (12) becomes

$$v = v_1(1 - e^{-\alpha t}). \tag{13}$$

The behavior of v as a function of time is shown in Figure 2. After falling for T

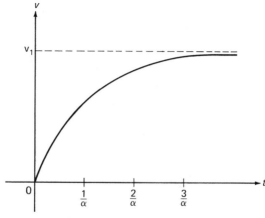

Figure 2.

seconds, where T is three or four times larger than $1/\alpha$, the stone effectively attains its *terminal velocity* v_1, which is never exceeded. Note that $v_1 = mg/k$ is proportional to the weight of the falling object, and hence is much smaller for falling feathers than for falling bricks! Nevertheless, feathers and bricks fall in exactly the same way *in a vacuum*.

To find the stone's equation of motion, we integrate (13), obtaining

$$s = \int v \, dt = v_1 \int (1 - e^{-\alpha t}) \, dt + C = v_1 \left(t + \frac{e^{-\alpha t}}{\alpha} \right) + C.$$

Applying the initial condition $s|_{t=0} = 0$, we find that $C = -v_1/\alpha$. Therefore

$$s = v_1 t - \frac{v_1}{\alpha} (1 - e^{-\alpha t}),$$

at least until the stone hits the ground.

5.33. Work and energy

a. Suppose a particle of mass m, moving along a straight line L, is acted upon by a force $F = F(s)$ which is a continuous function of its position s. Then, according to Newton's second law,

$$ma = m \frac{dv}{dt} = F(s),$$

or, by the chain rule,

$$m \frac{dv}{ds} \frac{ds}{dt} = mv \frac{dv}{ds} = F(s), \tag{14}$$

if we think of the velocity v as a function of s rather than t. Let

$$v_0 = v(s_0), \qquad v_1 = v(s_1)$$

be the particle's velocity at two different positions s_0 and s_1. Then, integrating (14) with respect to s from s_0 to s_1, we get

$$\int_{s_0}^{s_1} mv \frac{dv}{ds} \, ds = \int_{s_0}^{s_1} mv \, dv = \int_{s_0}^{s_1} F(s) \, ds,$$

or

$$\left[\frac{1}{2} mv^2 \right]_{s_0}^{s_1} = \frac{1}{2} mv_1^2 - \frac{1}{2} mv_0^2 = \int_{s_0}^{s_1} F(s) \, ds. \tag{15}$$

In other words, as a result of the action of the force, the quantity $T = \frac{1}{2}mv^2$, called the *kinetic energy* of the particle, increases by an amount

$$W = \int_{s_0}^{s_1} F(s) \, ds, \tag{16}$$

called the *work* done by the force on the particle in moving it from s_0 to s_1.

b. In the absence of any force, $F \equiv 0$ and the work (16) vanishes. Then (15) reduces to

$$\frac{1}{2} mv_1^2 = \frac{1}{2} mv_0^2, \tag{17}$$

so that the kinetic energy remains unchanged, or, in the language of physics, is *conserved*. If $F \equiv$ constant, (16) becomes

$$W = \int_{s_0}^{s_1} F(s)\, ds = F \int_{s_0}^{s_1} ds = F(s_1 - s_0).$$

Thus, in this case, the work equals the product of the force F and the "displacement" $s_1 - s_0$, as taught in elementary physics.

    c. If $F = mg$, $s_0 = 0$, $v_0 = 0$, we have the problem of the falling stone, as in Example 5.32b. Then (16) gives $W = mgs_1$ and (15) reduces to

$$\frac{1}{2} mv^2 = mgs,$$

after dropping the subscript 1 twice. Solving this equation for v, we get

$$v = \sqrt{2gs}.$$

The same result can be obtained by eliminating t from formulas (8') and (9'), but here we have used the concepts of work and kinetic energy to find the connection between the stone's velocity and its position without bothering to express either as a function of time.

    d. Now let $V(s)$ be any antiderivative of the function $-F(s)$, where the existence of $V(s)$ is guaranteed by the assumed continuity of $F(s)$ and Theorem 4.23a. Then it follows from (15) and the fundamental theorem of calculus that

$$\frac{1}{2} mv_1^2 - \frac{1}{2} mv_0^2 = -V(s) \Big|_{s_0}^{s_1} = V(s_0) - V(s_1),$$

or equivalently

$$\frac{1}{2} mv_1^2 + V(s_1) = \frac{1}{2} mv_0^2 + V(s_0). \tag{18}$$

The function $V(s)$ is called the *potential energy* (of the particle), and the sum of the kinetic energy $T = \frac{1}{2}mv^2$ and the potential energy $V = V(s)$ is called the *total energy* $E = T + V$. Thus equation (18) says that the total energy remains unchanged, or synonymously, is *conserved*, in the presence of any force $F = F(s)$. In the absence of any force, $F \equiv 0$, $V \equiv$ constant, and then formula (18) for the conservation of the total energy reduces to formula (17) for the conservation of the kinetic energy. Note that the potential energy V, being an antiderivative, is defined only to within an arbitrary "additive constant," and hence the same is true of the total energy $E = T + V$. This leads to no difficulties, since formula (18) remains valid if we replace $V(s)$ by $V(s) + C$, where C is an arbitrary constant.

    e. If $F = -mg$ and $s_0 = 0$, $v_0 \neq 0$, we have the problem of the stone thrown upward with initial velocity v_0, as in Example 5.32c. In this case, $V = mgs$ is an antiderivative of $-F = mg$, and (18) takes the form

$$\frac{1}{2} mv^2 + mgs = \frac{1}{2} mv_0^2, \tag{19}$$

after dropping the subscript 1 in two places. To find the maximum height reached by the stone, we set $v = 0$ in (19) and solve for s, obtaining

$$s = \frac{v_0^2}{2g}.$$

The same result can be obtained by solving (8'') for the time t at which v vanishes, and then substituting this value of t into (9'').

f. Example. With what velocity v_0 must a rocket be fired vertically upward in order to completely escape the earth's gravitational attraction?

SOLUTION. According to *Newton's law of gravitation*, the force attracting the rocket back to earth is given by the "inverse square law"

$$F = F(s) = -\frac{kMm}{s^2}, \tag{20}$$

where k is a positive constant, M is the mass of the earth, m is the mass of the rocket, and s is the distance between the rocket (regarded as a particle) and the center of the earth. Here, of course, we choose the s-axis vertically upward along a line going through the center of the earth, and the minus sign in (20) expresses the fact that the force of gravitation is *attractive*, pulling the rocket back to earth.

The work done on the rocket by the earth's gravitational pull as the rocket leaves the surface of the earth and goes off to a remote point in outer space is given by the integral

$$W = \int_{s_0}^{s_1} F(s)\, ds = -\int_{s_0}^{s_1} \frac{kMm}{s^2}\, ds = \left[\frac{kMm}{s}\right]_{s_0}^{s_1} = \frac{kMm}{s_1} - \frac{kMm}{s_0},$$

where s_0 equals R, the radius of the earth, and s_1 is a very large number. Therefore

$$W = -\frac{kMm}{R}, \tag{21}$$

after dropping the negligibly small number kMm/s_1. The work W equals the change

$$\frac{1}{2}mv_1^2 - \frac{1}{2}mv_0^2 \tag{22}$$

in the rocket's kinetic energy in going from the earth's surface to outer space. Since we are looking for the smallest value of the rocket's initial velocity that will allow it to escape the earth's gravitational pull, we choose $v_1 = 0$ as the rocket's final velocity, so that the rocket arrives in outer space with its initial velocity v_0 completely lost. Equating (21) and (22), with $v_1 = 0$, we get

$$\frac{1}{2}mv_0^2 = \frac{kMm}{R}.$$

Therefore v_0 is given by the formula

$$v_0 = \sqrt{\frac{2kM}{R}}, \tag{23}$$

and is independent of the rocket's mass m.

To evaluate (23), we observe that the force acting on the rocket at the earth's surface is $-kMm/R^2$ by (20) and $-mg$ in terms of the constant g, the "acceleration due to gravity," which figures in terrestrial problems involving gravitation. Equating these two expressions, we find that the "universal gravitational constant" k equals

$$k = \frac{gR^2}{M}. \tag{24}$$

Substituting this into (23), we get

$$v_0 = \sqrt{\frac{2gR^2M}{MR}} = \sqrt{2gR}. \tag{25}$$

Since, to a good approximation, $R = 4000$ miles and $g = 32$ ft/sec^2, we finally have

$$v_0 = \sqrt{\frac{2 \cdot 32 \cdot 4000}{5280}} \text{ mi/sec} \approx 7.0 \text{ mi/sec}$$

(there are 5280 feet in a mile). In the language of rocketry, the quantity v_0 is called the earth's *escape velocity*. A rocket fired upward with a velocity less than v_0 must eventually "fall" back to earth, unless it is "captured" by the gravitational attraction of some other celestial body.

PROBLEMS

1. Find the equation of motion $s = s(t)$ of a particle of mass m acted upon by a constant force F, given that the particle is initially at rest at the point $s = 0$.

2. A particle of mass m moves under the action of a constant force F. Suppose the particle's position at time $t = t_0$ is $s = s_0$. What velocity v_0 must the particle have at time $t = t_0$ in order to arrive at the point $s = s_1$ at time $t = t_1$?

3. Suppose a particle of mass m is subject to a force $F = kt$, proportional to the time that has elapsed since the onset of the motion. Find the resulting equation of motion, assuming that the particle starts from the point $s = 0$ with initial velocity v_0.

4. With what velocity must a stone be thrown vertically upward from ground level to reach a maximum height of 64 feet? How many seconds after it is thrown will the stone hit the ground?

5. The acceleration due to gravity on the moon is about 5 ft/sec^2, as compared with 32 ft/sec^2 on the earth. Suppose a man can jump 5 feet high on the earth. How high can he jump on the moon?

6. Which has more kinetic energy, a one-ounce bullet going 500 mi/hr or a ten-ton truck going 1 mi/hr? What happens if the bullet goes 600 mi/hr?

7. According to *Hooke's law*, the tension in a stretched spring equals ks, where k is a positive constant and s is the length of the spring minus its unstretched length. Show that the potential energy V of the stretched spring equals $\frac{1}{2}ks^2$.

8. A particle is attracted to each of two fixed points with a force proportional to the distance between the particle and the point. How much work is done in moving the particle from one point to another along the line connecting them? Assume that the constant of proportionality k is the same for both points.

9. Show that to reach an altitude of h miles, a rocket must be fired vertically upward with velocity

$$v_0 = \sqrt{\frac{2gRh}{R + h}},$$

where g is the acceleration due to gravity and R is the earth's radius. Verify that this expression approaches the value (25) as $h \to \infty$.

10. If a rocket is fired vertically upward with a velocity of 1 mi/sec, how high will it rise?

11. Estimate the moon's escape velocity, given that the moon has approximately $\frac{3}{11}$ the radius and $\frac{1}{81}$ the mass of the earth.

*12. Two men stand on the edge of a roof h feet above the ground. The first man throws a stone downward with velocity v_0 ft/sec, while the second simultaneously throws another stone upward with the same velocity. Show that both stones hit the ground with the same velocity v_1, but naturally at different times t_1 and t_2. Find v_1, t_1, t_2 and $\Delta t = t_2 - t_1$.

*13. A spider hangs from the ceiling by a single strand of web. Suppose the spider's weight doubles the unstretched length of the strand, stretching it from s to $2s$. Show that to climb back to the ceiling, the spider need only do 75% of the work required to climb an inelastic strand of length $2s$.

Chapter **6**

FUNCTIONS OF SEVERAL VARIABLES

6.1 FROM TWO TO n DIMENSIONS

6.11. Rectangular coordinates in space

a. Rectangular coordinates in space are the natural extension of rectangular coordinates in the plane (Sec. 1.7). Suppose we construct three mutually perpendicular lines Ox, Oy and Oz, known as the *coordinate axes*, intersecting in a point O, called the *origin* (see Figure 1). Just as in the plane, each line is regarded as extending indefinitely in both directions, and each is equipped with a *positive direction*, as indicated by the arrowheads in the figure. The coordinate axes Ox, Oy and Oz are called the *x-axis*, the *y-axis* and the *z-axis*, respectively. These axes determine three

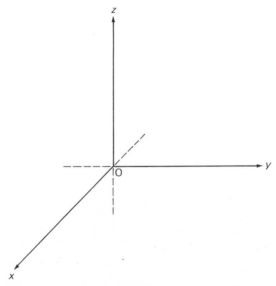

Figure 1.

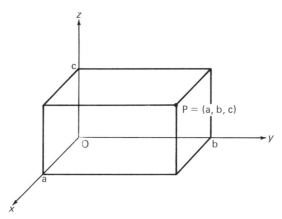

Figure 2.

mutually perpendicular *coordinate planes*, the *xy-plane* containing the *x* and *y*-axes, the *yz-plane* containing the *y* and *z*-axes, and the *xz-plane* containing the *x* and *z*-axes. In Figure 1 the *yz*-plane is the plane of the paper, and the positive *x*-axis points out from the paper at right angles to the *yz*-plane.

    b. We now associate three numbers with any given point *P* in space, by making the following construction, analogous to the construction in Sec. 1.72: Through the point *P* we draw three planes, one perpendicular to the *x*-axis, another perpendicular to the *y*-axis, and a third perpendicular to the *z*-axis. Suppose that, as in Figure 2, the first plane intersects the *x*-axis in the point with coordinate *a*, the second plane intersects the *y*-axis in the point with coordinate *b*, and the third plane intersects the *z*-axis in the point with coordinate *c*. Then the numbers *a*, *b* and *c* are called the *rectangular coordinates*, or simply the *coordinates*, of the point *P*. More exactly, *a* is called the *x-coordinate* of *P*, *b* is called the *y-coordinate* of *P*, and *c* is called the *z-coordinate* of *P*. Figure 2 is, of course, just the three-dimensional analogue of Figure 10, p. 21.

    c. The point *P* with *a*, *b* and *c* as its *x*, *y* and *z*-coordinates may also be denoted by (a, b, c). The symbol (a, b, c) is called an *ordered triple*, and is a special kind of three-element set of real numbers, namely one in which *the order of the elements matters*. More generally, an *n*-element set of real numbers in which the order of the elements matters is called an *ordered n-tuple*, and is denoted by $(a_1, a_2, \ldots, a_n)$.

    d. Example. Find the distance $|P_1 P_2|$ between two points $P_1 = (x_1, y_1, z_1)$ and $P_2 = (x_2, y_2, z_2)$ in space.

    SOLUTION. Consulting Figure 3, which generalizes Figure 11, p. 22, we find that

$$|P_1 P_2|^2 = |P_1 Q|^2 + |Q P_2|^2,$$

by the Pythagorean theorem. Therefore

$$|P_1 P_2| = \sqrt{|AB|^2 + |Q P_2|^2}, \tag{1}$$

where we use the fact that $|P_1 Q| = |AB|$. But $A = (x_1, y_1)$, $B = (x_2, y_2)$, regarded as points in the *xy*-plane, and hence

$$|AB|^2 = (x_1 - x_2)^2 + (y_1 - y_2)^2, \tag{2}$$

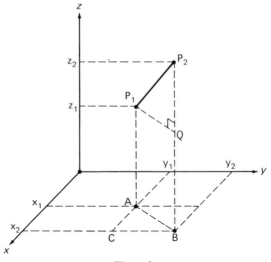

Figure 3.

by the formula for the distance between two points in the plane (Sec. 1.74). Substituting (2) into (1), and noting that $|QP_2| = |z_1 - z_2|$, we finally get

$$|P_1P_2| = \sqrt{(x_1 - x_2)^2 + (y_1 - y_2)^2 + (z_1 - z_2)^2}.$$

For example, the distance between the points $P_1 = (3, 1, 9)$ and $P_2 = (-1, 4, -3)$ is

$$|P_1P_2| = \sqrt{(3 + 1)^2 + (1 - 4)^2 + (9 + 3)^2} = \sqrt{4^2 + 3^2 + 12^2} = \sqrt{169} = 13.$$

6.12. By *n-dimensional space*, or simply *n-space*, we mean the set, denoted by R_n, of all ordered n-tuples $(x_1, x_2, \ldots, x_n)$ of real numbers. If $n = 1$, we have the real number system $R_1 = R$. Thus one-space R_1 is the line, two-space R_2 is the plane, and three-space R_3 is ordinary three-dimensional space. Here, of course, we rely on the one-to-one correspondence between R and the points of the line (Sec. 1.36a), and the analogous one-to-one correspondence between R_2 and the plane (Sec. 1.72), and between R_3 and space (see Prob. 1). The elements of R_n are called "points," just as in the case of one, two and three dimensions.

The *distance* between two points $P_1 = (a_1, a_2, \ldots, a_n)$ and $P_2 = (b_1, b_2, \ldots, b_n)$ in n-space is defined by

$$|P_1P_2| = \sqrt{(a_1 - b_1)^2 + (a_2 - b_2)^2 + \cdots + (a_n - b_n)^2},$$

or, more concisely, by

$$|P_1P_2| = \sqrt{\sum_{i=1}^{n} (a_i - b_i)^2}. \tag{3}$$

When we set n equal to 1, 2 and 3 in formula (3), it reduces in turn to the formula for the distance between two points on the line, in the plane, and in space (check this).

6.13. By a (numerical) *function of n variables* we simply mean a function $f(x_1, x_2, \ldots, x_n)$ whose domain is some subset of n-space. For simplicity, we will usually restrict the number of independent variables $x_1, x_2, \ldots, x_n$ to two or three. Other things being equal, we write x, y for x_1, x_2 if $n = 2$ and x, y, z for x_1, x_2, x_3

if $n = 3$, to preserve the standard labelling of the coordinate axes in the plane and in space.

When dealing with functions of several variables $x_1, x_2, \ldots, x_n$, we often want to be vague about the actual number of variables. We then write $f(P)$ instead of $f(x_1, x_2, \ldots, x_n)$, where $P = (x_1, x_2, \ldots, x_n)$ is a variable point of *n*-space.

6.14. Next we generalize the considerations of Sec. 2.3 to the case of three dimensions. By the *solution set* of the equation

$$F(x, y, z) = 0, \tag{4}$$

where $F(x, y, z)$ is a function of three variables, we mean the set S of all ordered triples (x, y, z) for which (4) holds. Suppose we introduce a three-dimensional system of rectangular coordinates, by setting up perpendicular axes Ox, Oy and Oz, as in Sec. 6.11a, and then plot all the elements of S as points in space. These points make up a "three-dimensional picture," called the *graph of S*, or, equivalently, the *graph of equation* (4). The same technique can be applied to a function

$$z = f(x, y) \tag{5}$$

of two variables. Let S be the set of all ordered triples (x, y, z) for which (5) holds. Then, plotting all the elements of S as points in space, we get a "picture," called the *graph of S*, or, equivalently, the *graph of the function* (5). Note that (5) is a special case of (4), corresponding to the choice $F(x, y, z) = z - f(x, y)$.

The graph of an equation (4) or function (5) typically looks like a "surface," possibly made up of several "pieces." We will often refer to these graphs as "the surface $F(x, y, z) = 0$," or "the surface $z = f(x, y)$."

6.15. Examples

a. Graph the equation

$$x^2 + y^2 + z^2 = 1. \tag{6}$$

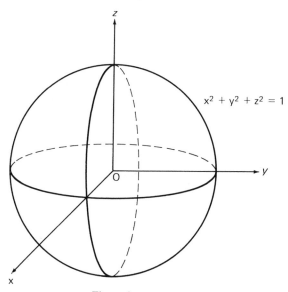

$$x^2 + y^2 + z^2 = 1$$

Figure 4.

SOLUTION. Since $x^2 + y^2 + z^2$ is the square of the distance between the point (x, y, z) and the origin $O = (0, 0, 0)$, the point (x, y, z) belongs to the graph of (6) when the distance between (x, y, z) and O equals 1, and only then. Therefore the graph of (6) is the sphere of unit radius with its center at O, shown in Figure 4.

b. Graph the function

$$z = x^2 + y^2. \tag{7}$$

SOLUTION.   Here every point of the circle

$$x^2 + y^2 = C \qquad (C \geqslant 0)$$

corresponds to the same value of z, namely C. Equivalently, every plane $z = C$ $(C \geqslant 0)$ parallel to the xy-plane intersects the graph of (7) in a circle, namely the circle of radius $\sqrt{C}$ (why the square root?) with its center on the z-axis. The value $C = 0$ gives rise to the "degenerate circle" $x^2 + y^2 = 0$, consisting of the single point $O = (0, 0, 0)$. Thus the graph of (7) is the surface shown in Figure 5. This surface intersects the xz-plane $(y = 0)$ in the parabola $z = x^2$ and the yz-plane $(x = 0)$ in the parabola $z = y^2$, as we find by substituting first $y = 0$ and then $x = 0$ into (7). It is easy to see (how?) that the surface (7) is "generated" (that is, "swept out") by rotating either of these parabolas about the z-axis. For this reason, the surface (7) is called a *paraboloid of revolution*.

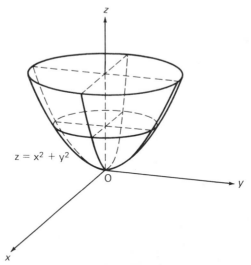

Figure 5.

c. Graph the function

$$z = 1 - \sqrt{x^2 + y^2} \qquad (x^2 + y^2 \leqslant 1). \tag{8}$$

SOLUTION.   The graph of (8) is the right circular cone shown in Figure 6, with its vertex at the point $(0, 0, 1)$. How is this deduced from (8)?

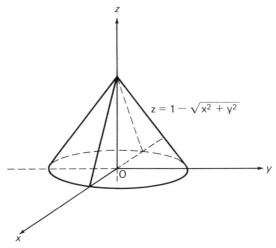

$$z = 1 - \sqrt{x^2 + y^2}$$

Figure 6.

6.16. Regions and neighborhoods

a.   In general, the domain D of a function $z = f(x, y)$ can be an arbitrary set of points in the xy-plane, but in the simple cases considered here D will always be either the whole plane, or a subset of the plane bounded by one or more curves, parts or all of which make up the *boundary* of D. Suppose D is *connected*, in the sense that any point of D can be joined to any other point of D by a curve which never leaves D. (For example, the map of Ohio is connected, but not the map of Hawaii.) Then D is said to be a *region*. A region is said to be *closed* if it contains its boundary and *open* if it does not. More generally, a region may contain some but not all of its boundary points. A region is said to be *finite* if it lies entirely inside some circle $x^2 + y^2 = r^2$ of sufficiently large radius r; otherwise the region is said to be *infinite*. The symbol R will henceforth be used to denote a *region*, rather than the real number system.

b.   The domain of a function $z = f(x, y)$ is understood to be the *largest* set of points (x, y) for which the function is defined, just as in the case of one variable (Sec. 2.15a). Thus the domain of the function

$$z = \sqrt{1 - x^2 - y^2}$$

is the set of all points (x, y) such that the square root makes sense, that is, such that $x^2 + y^2 \leqslant 1$ (what is the graph of the function?). This is the finite region consisting of the "unit circle"

$$x^2 + y^2 = 1 \tag{9}$$

and its interior. This region is closed, since it contains its boundary, namely the circle (9). We can talk about "the region $x^2 + y^2 \leqslant 1$," just as we talk about "the interval $-1 \leqslant x \leqslant 1$."

On the other hand, the domain of the function

$$z = \ln (x + y) \tag{10}$$

is the set of all points (x, y) such that $x + y > 0$, or equivalently $y > -x$. Thus the domain of (10) is the infinite region R consisting of all points lying to the right of the line $y = -x$ (sketch a figure). The line $y = -x$ is the boundary of R, but is not contained in R, since ln 0 is meaningless. Therefore R is an open region.

   c.   Given any fixed point $P_0 = (a, b)$ in the plane, let N be the set of all points $P = (x, y)$ such that

$$|PP_0| < \delta, \tag{11}$$

where δ is a positive number and $|PP_0|$ is the distance between P and P_0. In terms of coordinates, N is the set of all points (x, y) such that

$$\sqrt{(x - a)^2 + (y - b)^2} < \delta, \tag{12}$$

that is, the interior of the *circle*

$$(x - a)^2 + (y - b)^2 = \delta^2 \tag{13}$$

of radius δ with its center at (a, b). Since N does not contain the circle (13) itself, N is an open region. A region of this kind is called a *neighborhood* of the point $P_0 = (a, b)$. Clearly N is the two-dimensional generalization of the one-dimensional neighborhood $|x - a| < \delta$ (Sec. 1.63b).

   The merit of the inequality (11), as opposed to (12), is that it leaves the number of independent variables unspecified. If $n = 3$, we write

$$P_0 = (a, b, c), \qquad P = (x, y, z).$$

Then the set N of all points P satisfying (11) is the interior of the *sphere*

$$(x - a)^2 + (y - b)^2 + (z - c)^2 = \delta^2$$

of radius δ with its center at (a, b, c). We again call N a neighborhood, this time of the point $P_0 = (a, b, c)$ in three-space. Similarly, the set N of all points $P = (x_1, x_2, \ldots, x_n)$ in n-space satisfying the inequality (11) is the "interior" of the "n-dimensional sphere"

$$\sum_{i=1}^{n} (x - x_i)^2 = \delta^2$$

of radius δ with its center at $(a_1, a_2, \ldots, a_n)$. Naturally we again call N a neighborhood, this time of the point $P_0 = (a_1, a_2, \ldots, a_n)$ in n-space.

   By a *deleted neighborhood* of a point P_0 we mean any neighborhood of P_0 with the point P_0 itself excluded.

PROBLEMS

1.   We have shown how to find the coordinates of a given point in R_3. How does one find the point in R_3 with given coordinates?
2.   If a cube has the points $(1, 1, 1)$, $(1, -1, -1)$, $(-1, 1, -1)$ and $(-1, -1, -1)$ as four of its vertices, what are the other four?
3.   Find the distance from the origin to the point
     (a) $(4, -2, -4)$;   (b) $(-4, 12, 6)$;   (c) $(12, 16, -15)$.

4. Verify that the points $(3, -1, 6)$, $(-1, 7, -2)$ and $(1, -3, 2)$ are the vertices of a right triangle.

5. Which of the points $(\sqrt{2}, \sqrt{2}, 0)$, $(\sqrt{2}, \sqrt{2}, 1)$, $(\sqrt{2}, \sqrt{2}, \sqrt{2})$, $(1, \sqrt{3}, 0)$, $(0, \sqrt{3}, 1)$, $(-1, -\sqrt{2}, 1)$ lie on the surface of the sphere of radius 2 with its center at the origin?

6. Show that the sphere $x^2 + y^2 + z^2 - 4x - 6y - 2z + 13 = 0$ is tangent to the xy-plane.

7. *By a surface of revolution* we mean any geometrical figure in R_3 generated by rotating a plane curve about a straight line lying in its plane. Can the same surface of revolution be generated by rotating a given plane curve about two different axes?

8. Describe the surface of revolution obtained by rotating the line $x = 0$, $y = a$ about
   (a) The y-axis;   (b) The z-axis.

9. Find the piecewise linear curves in which the cone (8) intersects
   (a) The xz-plane;   (b) The yz-plane.

10. Justify thinking of intervals as "one-dimensional regions."

11. Write inequalities describing the finite open region bounded by the lines $x = \pm 1$, $y = \pm 1$.

*12. Find the two points of the x-axis at distance 12 from the point $(-3, 4, 8)$.

*13. Find the distance between the two points $P_1 = (1, 1, 1, 1)$ and $P_2 = (0, 2, 0, 2)$ of R_4. Which is closer to the origin of four-space?

*14. What is the surface $x^2 + y^2 = z^2$?

*15. Why is the domain of the function $f(x, y) = \ln(x^2 - y^2)$ not a region?

6.2 LIMITS AND DIFFERENTIATION

6.21. Limits and continuity

a. To define the limit of a function of several variables, we need only make slight changes in the definition of the limit of a function of a single variable, given in Sec. 2.44a. Thus we say that a (numerical) function $f(P)$ of several variables, defined in a deleted neighborhood of a point P_0, *approaches the limit A as P approaches P_0*, or that *$f(P)$ has the limit A at P_0*, if $f(P)$ gets "closer and closer" to A as P gets "closer and closer" to P_0 without ever coinciding with P_0. This fact is expressed by writing

$$\lim_{P \to P_0} f(P) = A \qquad (1)$$

or

$$f(P) \to A \quad \text{as} \quad P \to P_0.$$

In the "ε, δ language" introduced in Sec. 2.44b, (1) means that, given any $\varepsilon > 0$, we can find a number $\delta > 0$ such that $|f(P) - A| < \varepsilon$ whenever $0 < |PP_0| < \delta$. As in Sec. 2.44d, it is often convenient to talk about having a limit without specifying what the limit is. Thus we say that a function $f(P)$ *has a limit at P_0* if there is some number A such that $f(P) \to A$ as $P \to P_0$.

b. In two dimensions, we have $f(P) = f(x, y)$ and $P_0 = (a, b)$, say. We can then write the limit (1) more explicitly as a "double limit"

$$\lim_{\substack{x \to a \\ y \to b}} f(x, y) = A. \tag{2}$$

Do not make the mistake of confusing (2) with the "iterated limit"

$$\lim_{x \to a} \{\lim_{y \to b} f(x, y)\},$$

which means something quite different.

c. Just as in the case of a function of one variable (Sec. 2.63a), a function $f(P)$ of several variables, defined in a neighborhood of a point P_0, is said to be *continuous at P_0* if

$$\lim_{P \to P_0} f(P) = f(P_0).$$

If a function $f(P)$ is continuous at every point of a region R, we say that $f(P)$ is *continuous in R*. When we call a function continuous, without further qualification, we always mean continuous at some point or in some region, where the context makes it clear just what is meant.

6.22. Partial derivatives

a. Let f be a function of n variables defined in a neighborhood of a point $(x_1, x_2, \ldots, x_n)$. Then by the *partial derivative of f with respect to x_i at $(x_1, x_2, \ldots, x_n)$*, denoted by the expression

$$\frac{\partial f(x_1, x_2, \ldots, x_n)}{\partial x_i}, \tag{3}$$

we mean the limit

$$\lim_{\Delta x_i \to 0} \frac{f(x_1, \ldots, x_i + \Delta x_i, \ldots, x_n) - f(x_1, \ldots, x_i, \ldots, x_n)}{\Delta x_i},$$

where x_i is given an increment Δx_i, *but all the other variables are held fixed*, provided this limit exists and is finite. Clearly there are n such partial derivatives, corresponding to the n subscripts $1, 2, \ldots, n$. Thinking of the symbol $\partial/\partial x_i$ as a single entity, whose effect is to form the partial derivative with respect to x_i of any function written after it, we can also write (3) as

$$\frac{\partial}{\partial x_i} f(x_1, x_2, \ldots, x_n).$$

The symbol ∂ is still pronounced "dee," even though we are now dealing with a "curved dee."

b. To evaluate (3), we need only treat all the independent variables except x_i as if they are constants. Thus no extra technique is required to calculate partial derivatives. For example, the function

$$f(x, y) = xe^{xy}$$

of two variables has two partial derivatives, namely

$$\frac{\partial f(x, y)}{\partial x} = \frac{\partial}{\partial x}(xe^{xy}) = e^{xy} + xye^{xy},$$

calculated by regarding y as a constant and then treating $\partial/\partial x$ like d/dx, and

$$\frac{\partial f(x, y)}{\partial y} = \frac{\partial}{\partial y}(xe^{xy}) = x^2e^{xy},$$

calculated by regarding x as a constant and treating $\partial/\partial y$ like d/dy.

    c. There are other ways of writing (3). If $u = f(x_1, x_2, \ldots, x_n)$, we can abbreviate (3) to $\partial f/\partial x_i$ or $\partial u/\partial x_i$. We can also write (3) as

$$f_{x_i}(x_1, x_2, \ldots, x_n), \tag{4}$$

or simply f_{x_i} or u_{x_i}, where the subscript x_i calls for (partial) differentiation with respect to x_i. We can go a step further, and write just

$$f_i(x_1, x_2, \ldots, x_n),$$

instead of (4), or simply f_i or u_i, where the subscript i calls for differentiation with respect to the ith argument, whatever it be called. For example, if

$$u = f(r, s, t) = rs^2 \ln t,$$

then $f_t(r, s, t)$, f_t, u_t, f_3, and u_3 are all ways of writing the same partial derivative

$$\frac{\partial}{\partial t}(rs^2 \ln t) = \frac{rs^2}{t}.$$

    d. Partial derivatives of higher order are defined in the natural way. For example, if $z = f(x, y)$, we have four *second partial derivatives*, two of the form

$$\frac{\partial^2 z}{\partial x^2} = \frac{\partial}{\partial x}\left(\frac{\partial z}{\partial x}\right), \qquad \frac{\partial^2 z}{\partial y^2} = \frac{\partial}{\partial y}\left(\frac{\partial z}{\partial y}\right),$$

and two "mixed" derivatives of the form

$$\frac{\partial^2 z}{\partial x\, \partial y} = \frac{\partial}{\partial x}\left(\frac{\partial z}{\partial y}\right), \qquad \frac{\partial^2 z}{\partial y\, \partial x} = \frac{\partial}{\partial y}\left(\frac{\partial z}{\partial x}\right),$$

one obtained by differentiating z first with respect to y and then with respect to x, the other obtained by carrying out the differentiations in the opposite order. Similarly, if $u = f(x, y, z)$, then

$$\frac{\partial^3 u}{\partial x\, \partial y\, \partial z} = \frac{\partial}{\partial x}\left(\frac{\partial^2 u}{\partial y\, \partial z}\right), \qquad \frac{\partial^4 u}{\partial x^2\, \partial y\, \partial z} = \frac{\partial}{\partial x}\left(\frac{\partial^3 u}{\partial x\, \partial y\, \partial z}\right) = \frac{\partial^2}{\partial x^2}\left(\frac{\partial^2 u}{\partial y\, \partial z}\right),$$

and so on.

    e. Example. If

$$f(x, y, z) = xy \ln z,$$

then

$$\frac{\partial f}{\partial x} = y \ln z, \qquad \frac{\partial f}{\partial y} = x \ln z, \qquad \frac{\partial f}{\partial z} = \frac{xy}{z},$$

$$\frac{\partial^2 f}{\partial x^2} = \frac{\partial}{\partial x}(y \ln z) = 0,$$

$$\frac{\partial^2 f}{\partial y^2} = \frac{\partial}{\partial y}(x \ln z) = 0,$$

$$\frac{\partial^2 f}{\partial z^2} = \frac{\partial}{\partial z}\left(\frac{xy}{z}\right) = -\frac{xy}{z^2},$$

$$\frac{\partial^2 f}{\partial x \, \partial y} = \frac{\partial^2 f}{\partial y \, \partial x} = \frac{\partial}{\partial x}(x \ln z) = \frac{\partial}{\partial y}(y \ln z) = \ln z,$$

$$\frac{\partial^2 f}{\partial y \, \partial z} = \frac{\partial^2 f}{\partial z \, \partial y} = \frac{\partial}{\partial y}\left(\frac{xy}{z}\right) = \frac{\partial}{\partial z}(x \ln z) = \frac{x}{z},$$

$$\frac{\partial^2 f}{\partial x \, \partial z} = \frac{\partial^2 f}{\partial z \, \partial x} = \frac{\partial}{\partial x}\left(\frac{xy}{z}\right) = \frac{\partial}{\partial z}(y \ln z) = \frac{y}{z}.$$

Note that the value of each mixed partial derivative is independent of the order of differentiation. It is not hard to show that this is always true for a function with continuous first and second partial derivatives, but the proof will not be given here.

6.23. Differentiable functions and differentials

a. It will be recalled from Sec. 2.55a that if f is a function of a single variable, and if f has a derivative $f'(x)$ at a point x, then the expression

$$\alpha(\Delta x) = \frac{\Delta f(x) - f'(x) \, \Delta x}{\Delta x},$$

where $\Delta f(x) = f(x + \Delta x) - f(x)$ is the increment of f at x, approaches zero as $\Delta x \to 0$. Therefore, if f has a derivative $f'(x)$ at the point x, we can write the increment $\Delta f(x)$ in the form

$$\Delta f(x) = f'(x) \, \Delta x + \alpha(\Delta x) \, \Delta x,$$

where

$$\lim_{\Delta x \to 0} \alpha(\Delta x) = 0. \tag{5}$$

We can also write the increment as

$$\Delta f(x) = df(x) + \alpha(\Delta x) \, \Delta x,$$

in terms of $df(x) = f'(x) \, \Delta x$, the differential of f at the point x.

Conversely, if $\Delta f(x)$ can be represented in the form

$$\Delta f(x) = A \, \Delta x + \alpha(\Delta x) \, \Delta x, \tag{6}$$

where A is a constant and (5) holds, then the derivative $f'(x)$ exists and equals A. To see this, we need only divide (6) by Δx and take the limit as $\Delta x \to 0$, obtaining

$$f'(x) = \lim_{\Delta x \to 0} \frac{\Delta f(x)}{\Delta x} = A + \lim_{\Delta x \to 0} \alpha(\Delta x) = A.$$

Thus, in the case of a function f of one variable, having a differential $df(x)$ and having a derivative $f'(x)$ are equivalent properties, both described by saying that f is *differentiable* at x.

    b.   In the case of a function of several variables, the situation is rather different. For such a function, having a differential (in a sense to be defined in a moment for a function of two variables) and having partial derivatives are *not* equivalent properties. Indeed, the mere fact of having partial derivatives does not guarantee having a differential (it does, though, if the partial derivatives are *continuous*). On the other hand, having a differential does guarantee having partial derivatives. Thus, having a differential is a "stronger" requirement than having partial derivatives, and we will use the word "differentiable" exclusively in the sense of having a differential. These things are worth knowing, but, with the exception of Theorem 6.23d, we omit the proofs, which are tedious and rather technical. They can be found in a more advanced course, together with the proof alluded to at the end of Sec. 6.22e.

    c. Definition.   Given a function $z = f(x, y)$ of two variables, by the *increment of f at the point* (x, y) we mean the expression

$$\Delta f(x, y) = f(x + \Delta x, y + \Delta y) - f(x, y). \tag{7}$$

Suppose that for every point $(x + \Delta x, y + \Delta y)$ in some neighborhood of (x, y), we can write $\Delta f(x, y)$ in the form

$$\Delta f(x, y) = A \, \Delta x + B \, \Delta y + \alpha(\Delta x, \Delta y) \, \Delta x + \beta(\Delta x, \Delta y) \, \Delta y, \tag{8}$$

analogous to (6), where A and B are constants, and

$$\lim_{\substack{\Delta x \to 0 \\ \Delta y \to 0}} \alpha(\Delta x, \Delta y) = 0, \qquad \lim_{\substack{\Delta x \to 0 \\ \Delta y \to 0}} \beta(\Delta x, \Delta y) = 0. \tag{9}$$

Then we say that f *has a differential* or is *differentiable at* (x, y), and the expression $A \, \Delta x + B \, \Delta y$ in (8) is called the *(total) differential of f at* (x, y), denoted by $df(x, y)$, so that

$$df(x, y) = A \, \Delta x + B \, \Delta y. \tag{10}$$

For brevity, we will often write Δf and df for $\Delta f(x, y)$ and $df(x, y)$.

    d. Theorem.   *If f is differentiable at* (x, y), *then*

(1) *f is continuous at* (x, y);
(2) *f has partial derivatives f_x and f_y at* (x, y);
(3) *The increment Δf can be written in the form*

$$\Delta f(x, y) = f_x(x, y) \, \Delta x + f_y(x, y) \, \Delta y + \alpha(\Delta x, \Delta y) \, \Delta x + \beta(\Delta x, \Delta y) \, \Delta y; \tag{11}$$

(4) *The differential df can be written in the form*

$$df(x, y) = f_x(x, y) \, \Delta x + f_y(x, y) \, \Delta y. \tag{12}$$

    Proof.   It follows at once from (8) and (9) that

$$\lim_{\substack{\Delta x \to 0 \\ \Delta y \to 0}} \Delta f(x, y) = 0,$$

and therefore

$$\lim_{\substack{\Delta x \to 0 \\ \Delta y \to 0}} f(x + \Delta x, y + \Delta y) = f(x, y),$$

which expresses the continuity of f at (x, y) in increment notation. Setting $\Delta y = 0$ in (8), we get

$$f(x + \Delta x, y) - f(x, y) = A\,\Delta x + \alpha(\Delta x, 0)\,\Delta x,$$

with the help of (7), and hence

$$f_x(x, y) = \lim_{\Delta x \to 0} \frac{f(x + \Delta x, y) - f(x, y)}{\Delta x} = A + \lim_{\Delta x \to 0} \alpha(\Delta x, 0) = A$$

(see Prob. 1), that is, the partial derivative $f_x(x, y)$ exists and equals A. Similarly, setting $\Delta x = 0$ in (8), we get

$$f(x, y + \Delta y) - f(x, y) = B\,\Delta y + \beta(0, \Delta y)\,\Delta y,$$

and hence

$$f_y(x, y) = \lim_{\Delta y \to 0} \frac{f(x, y + \Delta y) - f(x, y)}{\Delta y} = B + \lim_{\Delta y \to 0} \beta(0, \Delta y) = B,$$

so that $f_y(x, y)$ exists and equals B. Replacing A and B in (8) and (10) by $f_x(x, y)$ and $f_y(x, y)$, we get (11) and (12). □

  e.   It follows from (12) that

$$dx = \frac{\partial x}{\partial x}\,\Delta x + \frac{\partial x}{\partial y}\,\Delta y = 1 \cdot \Delta x + 0 \cdot \Delta y = \Delta x,$$

$$dy = \frac{\partial y}{\partial x}\,\Delta x + \frac{\partial y}{\partial y}\,\Delta y = 0 \cdot \Delta x + 1 \cdot \Delta y = \Delta y.$$

In other words, the increments and the differentials of the *independent* variables are equal (recall Sec. 2.56a). Thus we can write (12) in the form

$$df(x, y) = f_x(x, y)\,dx + f_y(x, y)\,dy,$$

or, even more concisely, as

$$df = \frac{\partial f}{\partial x}\,dx + \frac{\partial f}{\partial y}\,dy. \tag{13}$$

The generalization of (13) to the case of a function $f(x_1, x_2, \ldots, x_n)$ of n variables is just

$$df = \frac{\partial f}{\partial x_1}\,dx_1 + \frac{\partial f}{\partial x_2}\,dx_2 + \cdots + \frac{\partial f}{\partial x_n}\,dx_n, \tag{13'}$$

and is proved in much the same way (we omit the details).

  f. Example.   Estimate the quantity

$$Q = \sqrt{(1.98)^2 + (4.02)^2 + (3.96)^2}.$$

SOLUTION.   If

$$f(x, y, z) = \sqrt{x^2 + y^2 + z^2},$$

then

$$Q = f(x + \Delta x, y + \Delta y, z + \Delta z),$$

where

$$x = 2, \qquad y = 4, \qquad z = 4,$$
$$\Delta x = -0.02, \qquad \Delta y = 0.02, \qquad \Delta z = -0.04.$$

Therefore

$$Q = f(x, y, z) + \Delta f(x, y, z) \approx f(x, y, z) + df(x, y, z),$$

where we have approximated the increment Δf by the differential df. Using (13′) with $n = 3$, we have

$$df = \frac{\partial f}{\partial x} dx + \frac{\partial f}{\partial y} dy + \frac{\partial f}{\partial z} dz = \frac{x \, dx + y \, dy + z \, dz}{\sqrt{x^2 + y^2 + z^2}},$$

and therefore

$$Q \approx \sqrt{2^2 + 4^2 + 4^2} + \frac{2(-0.02) + 4(0.02) + 4(-0.04)}{\sqrt{2^2 + 4^2 + 4^2}} = 6 - \frac{0.12}{6} = 5.98.$$

An exact calculation shows that

$$Q = \sqrt{35.7624} = 5.9802$$

to four decimal places.

PROBLEMS

1. Show that formula (2) implies

$$\lim_{x \to a} f(x, b) = A, \qquad \lim_{y \to b} f(a, y) = A.$$

2. Show that

$$\lim_{\substack{x \to 0 \\ y \to 0}} \frac{x + y}{x - y}$$

   does not exist.

3. Does $(x, y) \to (a, b)$ imply $x \to a$, $y \to b$, and conversely? Justify your answer.

4. Suppose $f(x, y)$ is independent of y, so that $f(x, y) \equiv g(x)$, and suppose $g(x)$ is continuous at the point a. Show that $f(x, y)$ is continuous at the point (a, b) for arbitrary b.

5. Suppose $f(x, y)$ is independent of x, so that $f(x, y) \equiv h(y)$, and suppose $h(y)$ is continuous at the point b. Show that $f(x, y)$ is continuous at the point (a, b) for arbitrary a.

6. Let $g(x)$ be continuous at the point a, while $h(y)$ is continuous at the point b. Show that the functions $f(x) \pm g(y)$, $f(x)g(y)$ and $f(x)/g(y)$ are all continuous at (a, b), provided that $g(b) \neq 0$ in the last case.

7. Use continuity to evaluate

   (a) $\displaystyle\lim_{\substack{x \to 1 \\ y \to 1}} \frac{xy}{x^2 + y^2};$ (b) $\displaystyle\lim_{\substack{x \to 0 \\ y \to 0}} e^{xy};$ (c) $\displaystyle\lim_{\substack{x \to a \\ y \to 2a}} \ln \frac{x}{y}.$

8. Define and evaluate the limit

$$A = \lim_{\substack{x \to 0 \\ y \to 1 \\ z \to e}} \ln \frac{z}{\sqrt{x^2 + y^2}}.$$

9. Where is the function $f(x, y) = 1/\sqrt{xy}$ continuous?
10. Where does the function

$$f(x, y, z) = \frac{xyz}{1 - x^2 - y^2}$$

fail to be continuous?

11. Find $\partial z/\partial x$ and $\partial z/\partial y$ if

(a) $z = x^2 y^3 + x^3 y^2$;   (b) $z = \dfrac{x + y}{x - y}$;   (c) $z = e^{-x/y}$.

12. Find $\partial u/\partial x$, $\partial u/\partial y$ and $\partial u/\partial z$ if

(a) $u = e^{xyz}$;   (b) $u = \ln \dfrac{xy}{z}$;   (c) $u = (xy)^z$.

13. Find

(a) $\dfrac{\partial^3 u}{\partial x^2 \, \partial y}$ if $u = (x + y) \ln (xy)$;   (b) $\dfrac{\partial^3 u}{\partial x \, \partial y \, \partial z}$ if $u = e^{xyz}$.

14. Verify by direct calculation that $\partial^2 z/\partial x \, \partial y = \partial^2 z/\partial y \, \partial x$ if
(a) $z = \ln (x + y)$;   (b) $z = \ln (xy)$.
15. Verify directly from the definition of the differential that each of the following functions is differentiable in the whole xy-plane:
(a) $f(x, y) = x^2 + y^2$;   (b) $f(x, y) = xy$.
16. Find the (total) differential of
(a) $z = xy - x^2 y^3 + x^3 y$;   (b) $z = y^x$.
17. Use differentials to estimate
(a) $\sqrt{(2.97)^2 + (4.05)^2}$;   (b) $(1.002)(2.003)^2(2.999)^3$.
*18. Verify that the function

$$u = \ln \frac{1}{\sqrt{x^2 + y^2}} \qquad (x^2 + y^2 \neq 0)$$

satisfies the equation

$$\frac{\partial^2 u}{\partial x^2} + \frac{\partial^2 u}{\partial y^2} = 0,$$

known as *Laplace's equation*.

    Comment. An equation like this, involving one or more partial derivatives of a function, is called a *partial* differential equation, as opposed to the *ordinary* differential equations considered in Chapter 5.
*19. Is the function $f(x, y) = \sqrt{x^2 + y^2}$ differentiable at the origin?
*20. According to the *perfect-gas law*, the pressure p, the volume V and the tem-

perature T (in degrees Kelvin) of a confined gas are related by the formula $pV = kT$, where k is a constant of proportionality. Show that

$$\frac{\partial V}{\partial T}\frac{\partial T}{\partial p}\frac{\partial p}{\partial V} = -1.$$

Comment. This formula should cure you of any temptation to treat partial derivatives like fractions.

6.3 THE CHAIN RULE

6.31. a. We now generalize Theorem 2.82a on the derivative of a composite function to the case of a function of n variables. For simplicity, we choose $n = 2$, just as in Theorem 6.23d, but the result is readily extended to the case $n > 2$.

THEOREM **(Chain rule).** *Suppose x and y are functions of a single variable, both differentiable at t, and suppose f is a function of two variables, differentiable at $(x(t), y(t))$. Then the composite function F, defined by $F(t) \equiv f(x(t), y(t))$, is differentiable at t, with derivative*

$$F'(t) = f_x(x(t), y(t))x'(t) + f_y(x(t), y(t))y'(t). \tag{1}$$

Proof. The proof is the natural generalization of the proof of Theorem 2.82a. Let $z = f(x, y)$. Since x and y are differentiable at t, then, as in Sec. 6.23a,

$$\begin{aligned}
\Delta x &= x(t + \Delta t) - x(t) = [x'(t) + \lambda(\Delta t)]\,\Delta t, \\
\Delta y &= y(t + \Delta t) - y(t) = [y'(t) + \mu(\Delta t)]\,\Delta t,
\end{aligned} \tag{2}$$

where $\lambda(\Delta t) \to 0$, $\mu(\Delta t) \to 0$ as $\Delta t \to 0$. Moreover, since f is differentiable at $(x(t), y(t))$, then, by Theorem 6.23d,

$$\begin{aligned}
\Delta z &= f(x + \Delta x, y + \Delta y) - f(x, y) \\
&= [f_x(x, y) + \alpha(\Delta x, \Delta y)]\,\Delta x + [f_y(x, y) + \beta(\Delta x, \Delta y)]\,\Delta y,
\end{aligned}$$

where we temporarily drop the argument t in many places, and $\alpha(\Delta x, \Delta y)$, $\beta(\Delta x, \Delta y)$ satisfy the conditions (9), p. 219. Substituting the expressions for Δx and Δy into the formula for Δz, we get

$$\begin{aligned}
\Delta z = &[f_x(x, y) + \alpha(\Delta x, \Delta y)][x'(t) + \lambda(\Delta t)]\,\Delta t \\
&+ [f_y(x, y) + \beta(\Delta x, \Delta y)][y'(t) + \mu(\Delta t)]\,\Delta t.
\end{aligned} \tag{3}$$

It follows from the expressions for Δx and Δy (or from the continuity of x and y at t) that $\Delta t \to 0$ implies $\Delta x \to 0$, $\Delta y \to 0$, so that $\Delta t \to 0$ implies not only $\lambda(\Delta t) \to 0$, $\mu(\Delta t) \to 0$, but also $\alpha(\Delta x, \Delta y) \to 0$, $\beta(\Delta x, \Delta y) \to 0$. Therefore, dividing (3) by Δt and taking the limit as $\Delta t \to 0$, we find that

$$\begin{aligned}
\lim_{\Delta t \to 0} \frac{\Delta z}{\Delta t} = &\lim_{\Delta t \to 0}\,[f_x(x, y) + \alpha(\Delta x, \Delta y)] \lim_{\Delta t \to 0}\,[x'(t) + \lambda(\Delta t)] \\
&+ \lim_{\Delta t \to 0}\,[f_y(x, y) + \beta(\Delta x, \Delta y)] \lim_{\Delta t \to 0}\,[y'(t) + \mu(\Delta t)] \\
= &f_x(x(t), y(t))x'(t) + f_y(x(t), y(t))y'(t),
\end{aligned} \tag{4}$$

where we reinstate the argument t in four places. On the other hand,

$$\Delta z = f(x + \Delta x, y + \Delta y) - f(x, y)$$
$$= f(x(t) + x(t + \Delta t) - x(t), y(t) + y(t + \Delta t) - y(t)) - f(x(t), y(t))$$
$$= f(x(t + \Delta t), y(t + \Delta t)) - f(x(t), y(t)) = F(t + \Delta t) - F(t),$$

which implies

$$\lim_{\Delta t \to 0} \frac{\Delta z}{\Delta t} = \lim_{\Delta t \to 0} \frac{F(t + \Delta t) - F(t)}{\Delta t} = F'(t). \tag{5}$$

Comparing (4) and (5), we get (1). $\square$

b.   Formula (1) can be written more concisely as

$$\frac{dF}{dt} = \frac{\partial f}{\partial x} \frac{dx}{dt} + \frac{\partial f}{\partial y} \frac{dy}{dt}. \tag{6}$$

We can simplify (6) even further by changing F to f:

$$\frac{df}{dt} = \frac{\partial f}{\partial x} \frac{dx}{dt} + \frac{\partial f}{\partial y} \frac{dy}{dt}. \tag{7}$$

After all, since f is a function of two variables, the fact that we write an *ordinary* derivative df/dt on the left in (7) means that each argument of f is being thought of as a function of a single variable, namely t. With this understanding, we can do without the extra symbol F, which was introduced only to make the distinction between $f(x, y)$ and $f(x(t), y(t))$ more explicit.

c.   The case where x and y are functions of *several* variables presents no difficulties. For example, suppose $x = x(t, u)$, $y = y(t, u)$, where x and y are differentiable functions of two variables. If u is held fixed, x and y reduce to functions of a single variable t, and we can apply the theorem without further ado, obtaining

$$\frac{\partial f}{\partial t} = \frac{\partial f}{\partial x} \frac{\partial x}{\partial t} + \frac{\partial f}{\partial y} \frac{\partial y}{\partial t}, \tag{8}$$

where all three ordinary derivatives in (7) now become partial derivatives. Similarly, holding t fixed, we get

$$\frac{\partial f}{\partial u} = \frac{\partial f}{\partial x} \frac{\partial x}{\partial u} + \frac{\partial f}{\partial y} \frac{\partial y}{\partial u}. \tag{8'}$$

Here again we might have introduced a composite function F, defined by

$$F(t, u) \equiv f(x(t, u), y(t, u)),$$

but it is simpler to regard $f(x, y)$ and $f(x(t, u), y(t, u))$ as being the same function, written in terms of different independent variables.

d.   The generalization of formula (7) to the case of a function $f(x_1, x_2, \ldots, x_n)$ whose n arguments depend on a single new independent variable t is given by

$$\frac{df}{dt} = \frac{\partial f}{\partial x_1} \frac{dx_1}{dt} + \frac{\partial f}{\partial x_2} \frac{dx_2}{dt} + \cdots + \frac{\partial f}{\partial x_n} \frac{dx_n}{dt} = \sum_{i=1}^{n} \frac{\partial f}{\partial x_i} \frac{dx_i}{dt}, \tag{9}$$

and is proved in much the same way (we omit the lengthy details). Similarly, the generalization of (8) and (8') to the case of a function $f(x_1, x_2, \ldots, x_n)$ whose n arguments depend on m new independent variables $t_1, t_2, \ldots, t_m$ is given by

$$\frac{\partial f}{\partial t_j} = \frac{\partial f}{\partial x_1}\frac{\partial x_1}{\partial t_j} + \frac{\partial f}{\partial x_2}\frac{\partial x_2}{\partial t_j} + \cdots + \frac{\partial f}{\partial x_n}\frac{\partial x_n}{\partial t_j} = \sum_{i=1}^{n} \frac{\partial f}{\partial x_i}\frac{\partial x_i}{\partial t_j} \qquad (j = 1, 2, \ldots, m) \quad (10)$$

The last two formulas are the "master" chain rules, of which all previous versions (including the first formula in Sec. 2.82c) are merely special cases. Note the following common features of (9) and (10):

(1) The right side contains n terms, one for each "intermediate" variable $x_1, x_2, \ldots, x_n$;
(2) Each of these terms is a product of two derivatives, with the intermediate variable appearing in the denominator of one derivative and in the numerator of the other.

6.32. Examples

a. Let f be any differentiable function of two variables. Prove that the function $u = f(x - y, y - z)$ satisfies the partial differential equation

$$\frac{\partial u}{\partial x} + \frac{\partial u}{\partial y} + \frac{\partial u}{\partial z} = 0. \tag{11}$$

SOLUTION. Let $s = x - y, t = y - z$. Then, by the chain rule,

$$\frac{\partial u}{\partial x} = \frac{\partial u}{\partial s}\frac{\partial s}{\partial x} + \frac{\partial u}{\partial t}\frac{\partial t}{\partial x} = \frac{\partial u}{\partial s},$$

$$\frac{\partial u}{\partial y} = \frac{\partial u}{\partial s}\frac{\partial s}{\partial y} + \frac{\partial u}{\partial t}\frac{\partial t}{\partial y} = -\frac{\partial u}{\partial s} + \frac{\partial u}{\partial t},$$

$$\frac{\partial u}{\partial z} = \frac{\partial u}{\partial s}\frac{\partial s}{\partial z} + \frac{\partial u}{\partial t}\frac{\partial t}{\partial z} = -\frac{\partial u}{\partial t},$$

and adding these three equations, we get (11). Here, of course, we have used the formulas

$$\frac{\partial s}{\partial x} = \frac{\partial}{\partial x}(x - y) = 1, \qquad \frac{\partial t}{\partial x} = \frac{\partial}{\partial x}(y - z) = 0,$$

$$\frac{\partial s}{\partial y} = \frac{\partial}{\partial y}(x - y) = -1, \qquad \frac{\partial t}{\partial y} = \frac{\partial}{\partial y}(y - z) = 1,$$

$$\frac{\partial s}{\partial z} = \frac{\partial}{\partial z}(x - y) = 0, \qquad \frac{\partial t}{\partial z} = \frac{\partial}{\partial z}(y - z) = -1.$$

b. Given a function $F(x, y)$ of two variables, suppose the equation

$$F(x, y) = 0 \tag{12}$$

defines y as an "implicit" function of x, in the sense that there exists a function $y = y(x)$ such that

$$F(x, y(x)) = 0$$

for all x in some interval I. Then, assuming that the functions $F(x, y)$ and $y(x)$ are both differentiable, we can use the chain rule to differentiate (12) with respect to x:

$$\frac{dF(x, y)}{dx} = F_x(x, y)\frac{dx}{dx} + F_y(x, y)\frac{dy}{dx} = F_x(x, y) + F_y(x, y)\frac{dy}{dx} = 0. \quad (13)$$

Solving (13) for dy/dx, we get

$$y' = \frac{dy}{dx} = -\frac{F_x(x, y)}{F_y(x, y)}, \quad (14)$$

provided that $F_y(x, y) \neq 0$. This is, of course, just a more official version of the technique of implicit differentiation introduced in Sec. 2.83d. For example, the equation

$$x^2 - xy + y^3 = 1$$

considered there is of the form (12) if

$$F(x, y) = x^2 - xy + y^3 - 1.$$

We then have

$$F_x(x, y) = 2x - y, \qquad F_y(x, y) = -x + 3y^2,$$

so that (14) becomes

$$y' = \frac{2x - y}{x - 3y^2},$$

which is precisely formula (12), p. 86.

   c.   The technique of implicit differentiation can also be used to calculate partial derivatives. Thus, given a function $F(x, y, z)$ of three variables, suppose the equation

$$F(x, y, z) = 0 \quad (15)$$

defines z as an "implicit" function of x and y, in the sense that there exists a function $z = z(x, y)$ such that

$$F(x, y, z(x, y)) = 0$$

for all (x, y) in some region R. Then, assuming that the functions $F(x, y, z)$ and $z(x, y)$ are both differentiable, we can use the chain rule to differentiate (15) with respect to x and y (dropping arguments for simplicity):

$$\frac{\partial F}{\partial x} = F_x\frac{\partial x}{\partial x} + F_y\frac{\partial y}{\partial x} + F_z\frac{\partial z}{\partial x} = F_x + F_z\frac{\partial z}{\partial x} = 0, \quad (16)$$

$$\frac{\partial F}{\partial y} = F_x\frac{\partial x}{\partial y} + F_y\frac{\partial y}{\partial y} + F_z\frac{\partial z}{\partial y} = F_y + F_z\frac{\partial z}{\partial y} = 0, \quad (16')$$

since

$$\frac{\partial x}{\partial x} = \frac{\partial y}{\partial y} = 1, \qquad \frac{\partial x}{\partial y} = \frac{\partial y}{\partial x} = 0.$$

Solving (16) and (16′) for $\partial z/\partial x$ and $\partial z/\partial y$, we get

$$z_x = \frac{\partial z}{\partial x} = -\frac{F_x(x, y, z)}{F_z(x, y, z)}, \qquad z_y = \frac{\partial z}{\partial y} = -\frac{F_y(x, y, z)}{F_z(x, y, z)}, \tag{17}$$

provided, of course, that $F_z(x, y, z) \neq 0$.

   d.   Given that

$$e^{-xy} - 2z + e^z = 0, \tag{18}$$

find $\partial z/\partial x$ and $\partial z/\partial y$.

   SOLUTION.   Here $F(x, y, z)$ is just the left side of (18). Therefore, by (17),

$$z_x = \frac{\partial z}{\partial x} = \frac{ye^{-xy}}{e^z - 2}, \qquad z_y = \frac{\partial z}{\partial y} = \frac{xe^{-xy}}{e^z - 2},$$

provided that $z \neq \ln 2$.

PROBLEMS

1. Find dz/dt by both the chain rule and by direct substitution if
    (a) $z = x^2 + xy^2$, $x = e^t$, $y = 1/t$;
    (b) $z = e^{u+v} \ln (u + v)$, $u = 2t^2$, $v = 1 - 2t^2$
2. Find $\partial z/\partial x$ and $\partial z/\partial y$ in two different ways if
    (a) $z = u + v^2$, $u = x^2$, $v = \ln (x + y)$;   (b) $z = \ln \dfrac{s}{t}$, $s = \dfrac{x}{y}$, $t = \dfrac{y}{x}$.

3. Given that $z = f(x, y)$, express $\partial z/\partial x$ and $\partial z/\partial y$ in terms of $\partial z/\partial u$ and $\partial z/\partial v$ if
    (a) $u = px + qy$, $v = rx + sy$;   (b) $u = xy$, $v = y/x$.
4. Given that

$$z^2 - 2xyz + 3 = 0,$$

    find $z_x = \partial z/\partial x$ and $z_y = \partial z/\partial y$. Evaluate these derivatives at the point $x = 1$, $y = 2$.
5. A function $f(x, y)$ with domain D is said to be *homogeneous of degree k* if $(x, y) \in D$ implies $(tx, ty) \in D$ for all $t > 0$ and if

$$f(tx, ty) = t^k f(x, y) \tag{19}$$

    for all $(x, y) \in D$ and $t > 0$. Show that each of the following functions is homo-geneous, and find its degree:
    (a) $x^2 + xy + y^2$;   (b) $\sqrt{x^2 + y^2}$;   (c) $\sqrt{x - y}$;

    (d) $\ln \dfrac{x}{y}$;   (e) $\dfrac{1}{x + y}$.

6. Show that if $f(x, y)$ is homogeneous of degree k and differentiable at every point of its domain, then

$$xf_x(x, y) + yf_y(x, y) = kf(x, y), \tag{20}$$

    a result known as *Euler's theorem on homogeneous functions*. Verify by direct calculation that each of the functions in the preceding problem satisfies (20).
*7. Let

$$x^2 + 2y^2 + 3z^2 + xy - z - 9 = 0.$$

Use repeated implicit differentiation to find $\partial^2 z/\partial x^2$, $\partial^2 z/\partial x \partial y$, $\partial^2 z/\partial y \partial x$ and $\partial^2 z/\partial y^2$ at the point $x = 1$, $y = -2$, $z = 1$.

*8. Show that if f is continuous in $[a, b]$ and if $a \leqslant u(x) \leqslant v(x) \leqslant b$, then

$$\frac{d}{dx} \int_{u(x)}^{v(x)} f(t)\, dt = f(v(x)) \frac{dv}{dx} - f(u(x)) \frac{du}{dx},$$

provided that u and v are differentiable.

6.4 EXTREMA IN n DIMENSIONS

6.41. a. In this section we favor the notation $X = (x_1, x_2, \ldots, x_n)$ for a variable point in n-space, reserving the symbol P for a *fixed* point of n-space. Global and local extrema of a function $f(X) = f(x_1, x_2, \ldots, x_n)$ of n variables are defined exactly as in Secs. 3.32a and 3.61b, with x, p and q replaced by n-dimensional points X, P and Q, respectively. As in Sec. 3.62b, if $f(X)$ has a local extremum at P, the extremum is said to be *strict* if $f(X) \neq f(P)$ for all $X \neq P$ in some neighborhood of P, that is, for all X in some *deleted* neighborhood of P (Sec. 6.16c).

In R_n we have the following analogue of Theorem 3.32c, which we state without proof:

THEOREM. *If f is continuous in a finite closed region R, then R contains points P and Q such that*

$$f(Q) \leqslant f(X) \leqslant f(P)$$

for all $X \in R$. In other words, f has both a maximum and a minimum in R, at the points P and Q, respectively.

For example, the function

$$f(x, y) = \sqrt{1 - x^2 - y^2}, \tag{1}$$

whose graph is the upper half of the sphere shown in Figure 4, p. 212, has a global minimum, equal to 0, at every point of the circle $x^2 + y^2 = 1$, but no local minima (why not?), and both a global maximum and a strict local maximum, equal to 1, at the origin $O = (0, 0)$. As another example, the function

$$f(x, y) = x^2 + y^2, \tag{2}$$

whose graph is the paraboloid of revolution shown in Figure 5, p. 212, has both a global minimum and a strict local minimum, equal to 0, at the origin O, but no global or local maxima. This can be seen by inspection of Figures 4 and 5, and will be verified in Sec. 6.42 by using partial differentiation.

b. There is a natural generalization of Theorem 3.63a for functions of several variables:

THEOREM. *If $f(X) = f(x_1, x_2, \ldots, x_n)$ has a local extremum at a point $P = (a_1, a_2, \ldots, a_n)$, then either $f(X)$ is nondifferentiable at P, or the partial derivatives of $f(X)$ all vanish at P:*

$$\frac{\partial f(P)}{\partial x_1} = \frac{\partial f(P)}{\partial x_2} = \cdots = \frac{\partial f(P)}{\partial x_n} = 0. \tag{3}$$

Proof. Obviously, $f(X)$ is either nondifferentiable at P or differentiable at P.

In the latter case, the partial derivatives of $f(X)$ at P all exist, by the natural generalization of Theorem 6.23d to the case of n variables. But then all n functions

$$f(x_1, a_2, \ldots, a_n), \qquad f(a_1, x_2, \ldots, a_n), \ldots, \qquad f(a_1, a_2, \ldots, x_n) \qquad (4)$$

of a single variable are differentiable, the first at a_1, the second at a_2, and so on, with derivatives

$$f'(x_1, a_2, \ldots, a_n)\Big|_{x_1 = a_1} = \frac{\partial f(P)}{\partial x_1},$$

$$f'(a_1, x_2, \ldots, a_n)\Big|_{x_2 = a_2} = \frac{\partial f(P)}{\partial x_2}, \qquad (5)$$

$$\cdot \quad \cdot \quad \cdot \quad \cdot \quad \cdot \quad \cdot \quad \cdot \quad \cdot$$

$$f'(a_1, a_2, \ldots, x_n)\Big|_{x_n = a_n} = \frac{\partial f(P)}{\partial x_n}.$$

The first of the functions (4) has a local extremum at a_1, the second has a local extremum at a_2, and so on (why?). Therefore, applying Theorem 3.63a n times, we find that the left sides of all n equations (5) vanish. But then (3) holds.

Note that the theorem reduces to Theorem 3.63a if $n = 1$. $\square$

    c. By analogy with Sec. 3.63c, by a *critical point* of a function f of n variables we mean either a point where f is nondifferentiable or a point where the condition (3) holds, and by a *stationary point* of f we mean a point where (3) holds. Thus a critical point of f is either a point where f is nondifferentiable or a stationary point of f. According to the theorem, if f has a local extremum at P, then P is a critical point of f. On the other hand, just as in the case of one variable (Sec. 3.63c), if P is a critical point of f, there is no necessity for f to have a local extremum at P. For example, consider the function

$$f(x,y) = x^2 - y^2. \qquad (6)$$

The partial derivatives

$$\frac{\partial f}{\partial x} = 2x, \qquad \frac{\partial f}{\partial y} = -2y$$

vanish at the origin $O = (0, 0)$, which is therefore a critical point of f, in fact a stationary point. But f does not have a local extremum at O. In fact,

$$\frac{\partial^2 f}{\partial x^2} = 2, \qquad \frac{\partial^2 f}{\partial y^2} = -2,$$

so that, by the second derivative test for functions of a single variable (Theorem 3.65a), the function $f(x, 0)$ has a local minimum at O, while the function $f(0, y)$ has a local maximum at O, and this obviously prevents $f(x, y)$ from having either a local minimum or a local maximum at O.

    Thus what we really want are conditions on a function f which *compel* f to have a local extremum at a given point P. For the case of a function of two variables such conditions are given by the following two-dimensional generalization of Theorem 3.65a, which we state without proof:

THEOREM. *Suppose $f(x, y)$ has continuous second partial derivatives in a neighborhood of a critical point $P = (a, b)$, and let*

$$A = \frac{\partial^2 f(P)}{\partial x^2}, \qquad B = \frac{\partial^2 f(P)}{\partial x\, \partial y}, \qquad C = \frac{\partial^2 f(P)}{\partial y^2}, \qquad D = AC - B^2.$$

Then $f(x, y)$ has a strict local maximum at P if $D > 0$, $A < 0$, and a strict local minimum at P if $D > 0$, $A > 0$, but no extremum at P if $D < 0$.

If $D = 0$, there may or may not be an extremum at P. For example, $D = 0$ at the origin $O = (0, 0)$ if $f(x, y) = x^2 + y^4$ or if $f(x, y) = x^2 + y^3$, but in the first case the function $f(x, y)$ clearly has a local minimum at O, since it vanishes at O and is positive everywhere else, while in the second case $f(x, y)$ has no extremum at O, since $f(0, y) = y^3$ is increasing in the interval $-\infty < y < \infty$.

6.42. Examples

a. The function (1) is differentiable for $x^2 + y^2 < 1$, with partial derivatives

$$\frac{\partial f}{\partial x} = -\frac{x}{\sqrt{1 - x^2 - y^2}}, \qquad \frac{\partial f}{\partial y} = -\frac{y}{\sqrt{1 - x^2 - y^2}}.$$

Therefore, by Theorem 6.41b, the only critical point of f in the open disk $x^2 + y^2 < 1$ is at the point $O = (0, 0)$, where these partial derivatives vanish. Moreover, as you can easily verify,

$$A = \frac{\partial^2 f(0, 0)}{\partial x^2} = -1, \quad B = \frac{\partial^2 f(0, 0)}{\partial x\, \partial y} = 0, \quad C = \frac{\partial^2 f(0, 0)}{\partial y^2} = -1, \quad D = AC - B^2 = 1.$$

It follows from Theorem 6.41c that f has a strict local maximum, equal to 1, at the point O, as already observed.

b. The function (2) is differentiable in the whole xy-plane, with partial derivatives

$$\frac{\partial f}{\partial x} = 2x, \qquad \frac{\partial f}{\partial y} = 2y.$$

Again the only critical point of f, this time in the whole plane, is at the origin $O = (0, 0)$. We now have

$$A = \frac{\partial^2 f(0, 0)}{\partial x^2} = 2, \quad B = \frac{\partial^2 f(0, 0)}{\partial x\, \partial y} = 0, \quad C = \frac{\partial^2 f(0, 0)}{\partial y^2} = 2, \quad D = AC - B^2 = 4.$$

Therefore, by Theorem 6.41c, f has a strict local minimum, equal to 0, at the origin. Actually, this fact is obvious without making any calculations at all, since the function f is positive everywhere except at the origin, where it vanishes.

c. Inspection of Figure 6, p. 213, shows that the function

$$f(x, y) = 1 - \sqrt{x^2 + y^2}$$

has both a global maximum and a strict local maximum, equal to 1, at the origin $O = (0, 0)$. Therefore, by Theorem 6.41b, O is a critical point of f, but this time not because O is a stationary point of f, but because f is nondifferentiable at O. In fact, the derivative

$$\frac{\partial f(0,0)}{\partial x} = \lim_{\Delta x \to 0} \frac{f(0 + \Delta x, 0) - f(0,0)}{\Delta x} = -\lim_{\Delta x \to 0} \frac{|\Delta x|}{\Delta x}$$

fails to exist, for the reason given in Sec. 2.45e, and the same is true of the derivative $\partial f(0,0)/\partial y$.

d. Find the local extrema of the function

$$f(x, y) = x^3 + y^3 - 3xy.$$

SOLUTION. Solving the system

$$\frac{\partial f}{\partial x} = 3x^2 - 3y = 0, \qquad \frac{\partial f}{\partial y} = 3y^2 - 3x = 0,$$

we find that f has two critical points, namely $(0,0)$ and $(1,1)$. Since

$$A = \frac{\partial^2 f}{\partial x^2} = 6x, \qquad B = \frac{\partial^2 f}{\partial x\, \partial y} = -3, \qquad C = \frac{\partial^2 f}{\partial y^2} = 6y,$$

we have

$$A = 0, \qquad B = -3, \qquad C = 0, \qquad D = AC - B^2 = -9$$

at $(0,0)$, and

$$A = 6, \qquad B = -3, \qquad C = 6, \qquad D = AC - B^2 = 27$$

at $(1,1)$. It follows from Theorem 6.41c that f has no extremum at $(0,0)$ and a strict local minimum, equal to -1, at $(1,1)$.

e. Find the brick of largest volume with a given surface area $2c$.

SOLUTION. Let x be the length, y the width and z the height of the brick. Then the brick has volume

$$V = xyz \tag{7}$$

and surface area

$$2(xy + xz + yz) = 2c$$

(there are two faces of area xy, two of area xz and two of area yz). Thus our problem is to find the largest value of (7), subject to the "side condition" or "constraint"

$$xy + xz + yz - c = 0. \tag{8}$$

Suppose we solve (8) for z, obtaining

$$z = \frac{c - xy}{x + y}. \tag{9}$$

Substituting (9) into (7), we then get

$$V = xy \frac{c - xy}{x + y}, \tag{10}$$

and we can solve the problem by maximizing (10) as a function of the *two* variables x and y. This approach leads to no particular difficulties (see Probs. 9 and 10), but we now solve the problem by another method, which is both more elegant and of interest in its own right.

Suppose we multiply the constraint (8) by a new variable λ, called a *Lagrange multiplier*, and add the result to (7), obtaining a new function

$$V^* = xyz + \lambda(xy + xz + yz - c) \tag{11}$$

of *four* variables x, y, z and λ. We then look for local extrema of V^*, by the usual technique of setting the partial derivatives of V^* equal to zero:

$$\frac{\partial V^*}{\partial x} = yz + \lambda(y + z) = 0,$$

$$\frac{\partial V^*}{\partial y} = xz + \lambda(x + z) = 0,$$

$$\frac{\partial V^*}{\partial z} = xy + \lambda(x + y) = 0, \tag{12}$$

$$\frac{\partial V^*}{\partial \lambda} = xy + xz + yz - c = 0.$$

The fact that we get back the constraint (8) as the last of these equations is, of course, essential to the success of the method. Suppose the four equations (12) can be solved for λ as a function of x, y and z. Then, substituting λ back into the first three equations, we get three equations in x, y and z, whose solutions are just the values of x, y and z for which the original function V achieves its local extrema, subject to the constraint (8). To see this, we merely observe that if the equations (12) hold, then the constraint (8) is automatically satisfied, so that the term in parentheses in (11) vanishes, and the "unconstrained extrema" of V^* reduce to the "constrained extrema" of V.

We now solve the equations (12) for λ. To this end, we multiply the first equation by x, the second by y, and the third by z. We then add the results and invoke the fourth equation, namely the constraint, obtaining

$$3xyz + 2\lambda(xy + xz + yz) = 3xyz + 2\lambda c = 0,$$

which determines λ as a function of x, y and z:

$$\lambda = -\frac{3xyz}{2c}. \tag{13}$$

Substituting (13) back into the first three equations (12), we get

$$yz\left[1 - \frac{3x}{2c}(y + z)\right] = 0,$$

$$xz\left[1 - \frac{3y}{2c}(x + z)\right] = 0,$$

$$xy\left[1 - \frac{3z}{2c}(x + y)\right] = 0,$$

or equivalently,

$$\frac{3x}{2c}(y + z) = 1, \qquad \frac{3y}{2c}(x + z) = 1, \qquad \frac{3z}{2c}(x + y) = 1, \tag{14}$$

since x, y and z are nonzero, because of their physical meaning. The first and second of the equations (14) imply $x(y + z) = y(x + z)$ and hence $x = y$, while the second and third imply $y(x + z) = z(x + y)$ and hence $y = z$, so that

$$x = y = z. \tag{15}$$

Substituting (15) into the constraint (8), we get $3x^2 = c$. It follows that

$$x = y = z = \sqrt{\frac{c}{3}}. \tag{16}$$

It is "physically obvious" that the function V must have a maximum, rather than a minimum, at the point (16), since there are long, skinny bricks with surface area $2c$ which have "arbitrarily small" volume (see Prob. 5). Thus, finally, we find that the brick of largest volume with surface area $2c$ is the cube of side $(c/3)^{1/2}$ and volume $(c/3)^{3/2}$.

PROBLEMS

1. By investigating all critical points, find the local extrema, if any, of
   (a) $z = 3x + 6y - x^2 - xy + y^2$;   (b) $z = x^2 - xy + y^2 - 2x + y$;
   (c) $z = 2x^3 - xy^2 + 5x^2 + y^2$.
2. Do the same for
   (a) $z = 2x^3 + xy^2 - 216x$;   (b) $z = 3x^2 - 2x\sqrt{y} + y - 8x + 8$;
   (d) $z = (x - y + 1)^2$.
3. Find the global extrema of the function $z = x^2 - y^2$ in the closed disk $x^2 + y^2 \leqslant 4$.
4. Find the global extrema of the function $z = (2x^2 + 3y^2)e^{-x^2 - y^2}$ in the closed disk $x^2 + y^2 \leqslant 1$.
5. Let V be the same as in Example 6.42e. Show that $V \to 0$ as $z \to \infty$.
6. Suppose that in Example 6.42e we *subtract* λ times the constraint (8) from the function (7), obtaining the new function

   $$V^* = xyz - \lambda(xy + xz + yz - c)$$

   instead of (11). Does this have any effect on the final answer?
7. Use a Lagrange multiplier to show that c^n is the largest value of the product of n positive numbers $x_1, x_2, \ldots, x_n$ with a given sum nc.
*8. Use a Lagrange multiplier to show that

   $$\sqrt[n]{x_1 x_2 \cdots x_n} \leqslant \frac{x_1 + x_2 + \cdots + x_n}{n}$$

   for arbitrary positive numbers $x_1, x_2, \ldots, x_n$, thereby generalizing Sec. 1.4, Problem 14.
*9. Solve Example 6.42e by maximizing (10).
*10. Use Theorem 6.41c to confirm that the function $V = xyz$ has a strict local maximum at the point (16), subject to the constraint (8).
*11. Use a Lagrange multiplier to derive formula (11), p. 34, for the distance d between a point $P_1 = (x_1, y_1)$ and the line $Ax + By + C = 0$.
*12. Suppose a firm produces two commodities. Then the *total cost* to the firm of producing a quantity Q_1 of the first commodity and a quantity Q_2 of the second commodity is some function of Q_1 and Q_2, called the *cost function* and denoted by

$C(Q_1, Q_2)$. There are now two *marginal costs*, $MC_1(Q_1, Q_2) = \partial C(Q_1, Q_2)/\partial Q_1$, the marginal cost of the first commodity, and $MC_2(Q_1, Q_2) = \partial C(Q_1, Q_2)/\partial Q_2$, the marginal cost of the second commodity. This is, of course, just the natural extension of the considerations of Sec. 3.22a to the case of a two-commodity firm.

Suppose the firm's cost function is

$$C(Q_1, Q_2) = 3Q_1^2 + 2Q_1 Q_2 + 3Q_2^2.$$

Find the marginal costs $MC_1(Q_1, Q_2)$ and $MC_2(Q_1, Q_2)$. Suppose further that the two commodities are sold at predetermined prices P_1 and P_2, chosen to make them sell in a competitive market. Write an expression for the firm's profit $\Pi(Q_1, Q_2)$, which is now a function of two variables. What output levels of the two commodities maximize this profit? Verify that your answer actually leads to a maximum. What condition must the prices satisfy?

*13. In the preceding problem, suppose profit is to be maximized subject to a constraint of the form $Q_1 + Q_2 = q > 0$. For example, an automobile manufacturer may want to make a given total number of sedans and station wagons. What output levels maximize the profit in this case? What condition must now be satisfied by the prices and the number q?

TABLES

Table 1. GREEK ALPHABET

Letter	Name	Letter	Name
A α	Alpha	N ν	Nu
B β	Beta	Ξ ξ	Xi
Γ γ	Gamma	O o	Omicron
Δ δ	Delta	Π π	Pi
E ε	Epsilon	P ρ	Rho
Z ζ	Zeta	Σ σ	Sigma
H η	Eta	T τ	Tau
Θ θ (ϑ)	Theta	Υ υ	Upsilon
I ι	Iota	Φ φ (ϕ)	Phi
K κ	Kappa	X χ	Chi
Λ λ	Lambda	Ψ ψ	Psi
M μ	Mu	Ω ω	Omega

Table 2.   EXPONENTIAL FUNCTIONS

x	e^x	e^{-x}	x	e^x	e^{-x}
0.00	1.0000	1.0000	2.5	12.182	0.0821
0.05	1.0513	0.9512	2.6	13.464	0.0743
0.10	1.1052	0.9048	2.7	14.880	0.0672
0.15	1.1618	0.8607	2.8	16.445	0.0608
0.20	1.2214	0.8187	2.9	18.174	0.0550
0.25	1.2840	0.7788	3.0	20.086	0.0498
0.30	1.3499	0.7408	3.1	22.198	0.0450
0.35	1.4191	0.7047	3.2	24.533	0.0408
0.40	1.4918	0.6703	3.3	27.113	0.0369
0.45	1.5683	0.6376	3.4	29.964	0.0334
0.50	1.6487	0.6065	3.5	33.115	0.0302
0.55	1.7333	0.5769	3.6	36.598	0.0273
0.60	1.8221	0.5488	3.7	40.447	0.0247
0.65	1.9155	0.5220	3.8	44.701	0.0224
0.70	2.0138	0.4966	3.9	49.402	0.0202
0.75	2.1170	0.4724	4.0	54.598	0.0183
0.80	2.2255	0.4493	4.1	60.340	0.0166
0.85	2.3396	0.4274	4.2	66.686	0.0150
0.90	2.4596	0.4066	4.3	73.700	0.0136
0.95	2.5857	0.3867	4.4	81.451	0.0123
1.0	2.7183	0.3679	4.5	90.017	0.0111
1.1	3.0042	0.3329	4.6	99.484	0.0101
1.2	3.3201	0.3012	4.7	109.95	0.0091
1.3	3.6693	0.2725	4.8	121.51	0.0082
1.4	4.0552	0.2466	4.9	134.29	0.0074
1.5	4.4817	0.2231	5	148.41	0.0067
1.6	4.9530	0.2019	6	403.43	0.0025
1.7	5.4739	0.1827	7	1096.6	0.0009
1.8	6.0496	0.1653	8	2981.0	0.0003
1.9	6.6859	0.1496	9	8103.1	0.0001
2.0	7.3891	0.1353	10	22026	0.00005
2.1	8.1662	0.1225			
2.2	9.0250	0.1108			
2.3	9.9742	0.1003			
2.4	11.023	0.0907			

Table 3. NATURAL LOGARITHMS

n	$\ln n$	n	$\ln n$	n	$\ln n$
0.0		4.5	1.5041	9.0	2.1972
0.1	*7.6974	4.6	1.5261	9.1	2.2083
0.2	*8.3906	4.7	1.5476	9.2	2.2192
0.3	*8.7960	4.8	1.5686	9.3	2.2300
0.4	*9.0837	4.9	1.5892	9.4	2.2407
0.5	*9.3069	5.0	1.6094	9.5	2.2513
0.6	*9.4892	5.1	1.6292	9.6	2.2618
0.7	*9.6433	5.2	1.6487	9.7	2.2721
0.8	*9.7769	5.3	1.6677	9.8	2.2824
0.9	*9.8946	5.4	1.6864	9.9	2.2925
1.0	0.0000	5.5	1.7047	10	2.3026
1.1	0.0953	5.6	1.7228	11	2.3979
1.2	0.1823	5.7	1.7405	12	2.4849
1.3	0.2624	5.8	1.7579	13	2.5649
1.4	0.3365	5.9	1.7750	14	2.6391
1.5	0.4055	6.0	1.7918	15	2.7081
1.6	0.4700	6.1	1.8083	16	2.7726
1.7	0.5306	6.2	1.8245	17	2.8332
1.8	0.5878	6.3	1.8405	18	2.8904
1.9	0.6419	6.4	1.8563	19	2.9444
2.0	0.6931	6.5	1.8718	20	2.9957
2.1	0.7419	6.6	1.8871	25	3.2189
2.2	0.7885	6.7	1.9021	30	3.4012
2.3	0.8329	6.8	1.9169	35	3.5553
2.4	0.8755	6.9	1.9315	40	3.6889
2.5	0.9163	7.0	1.9459	45	3.8067
2.6	0.9555	7.1	1.9601	50	3.9120
2.7	0.9933	7.2	1.9741	55	4.0073
2.8	1.0296	7.3	1.9879	60	4.0943
2.9	1.0647	7.4	2.0015	65	4.1744
3.0	1.0986	7.5	2.0149	70	4.2485
3.1	1.1314	7.6	2.0281	75	4.3175
3.2	1.1632	7.7	2.0412	80	4.3820
3.3	1.1939	7.8	2.0541	85	4.4427
3.4	1.2238	7.9	2.0669	90	4.4998
3.5	1.2528	8.0	2.0794	95	4.5539
3.6	1.2809	8.1	2.0919	100	4.6052
3.7	1.3083	8.2	2.1041		
3.8	1.3350	8.3	2.1163		
3.9	1.3610	8.4	2.1282		
4.0	1.3863	8.5	2.1401		
4.1	1.4110	8.6	2.1518		
4.2	1.4351	8.7	2.1633		
4.3	1.4586	8.8	2.1748		
4.4	1.4816	8.9	2.1861		

* Take tabular value − 10.

Table 4.   ELEMENTARY DIFFERENTIATION RULES*

Function	Derivative	Proved in
c	0	Sec. 2.43b
x	1	Sec. 2.43b
x^r	rx^{r-1}	Sec. 4.45c
$f + g$	$f' + g'$	Sec. 2.71a
$f - g$	$f' - g'$	Sec. 2.71a
fg	$f'g + fg'$	Sec. 2.72a
$\dfrac{f}{g}$	$\dfrac{f'g - fg'}{g^2}$	Sec. 2.73
f^{-1}	$\dfrac{1}{f'}$	Sec. 2.81a
$F(y)$	$\dfrac{dF}{dy} y'$	Sec. 2.82a
$\ln x$	$\dfrac{1}{x}$	Sec. 4.32
$\log_a x$	$\dfrac{1}{x} \log_a e$	Sec. 4.36c
e^x	e^x	Sec. 4.43b
a^x	$a^x \ln a$	Sec. 4.44c
$\ln y$	$\dfrac{y'}{y}$	Sec. 4.53a
$\cosh x$	$\sinh x$	Sec. 4.5, Prob. 21
$\sinh x$	$\cosh x$	Sec. 4.5, Prob. 21

* Here c, r and $a > 0$ are arbitrary constants, f and g are functions of the independent variable x, f^{-1} is the inverse of f, and F is a function of the dependent variable y. The prime denotes differentiation with respect to x.

SELECTED HINTS AND ANSWERS

Chapter 1

Sec. 1.2

1. $\{a\}, \{b\}, \{c\}, \{a, b\}, \{a, c\}, \{b, c\}, \emptyset$.
2. (a) $\{0\}$; (c) $\{3, -3\}$; (e) $\{a, c, l, s, u\}$.
3. Only (a) is true. A has 4 elements.
4. All but (d) are true.
5. (a) $\{a, b, c, d\}$.
6. (a) $\{3, 4\}$.
7. Trivial, but give details anyway.
10. (a) $\{3\}$; (c) $\{1, 2, 3\}$.
12. Only (c) and (e) are empty.
13. 16.

Sec. 1.3

1. Yes. $-1 - (-2) = 1, (-1)(-1) = 1, -1 \div -1 = 1$.
2. (a) $\frac{1}{2} + \frac{1}{3} = \frac{5}{6}$; (c) $\frac{1}{2} \cdot \frac{1}{3} = \frac{1}{6}$.
3. Let m/n and m'/n' be two rational numbers, where m, n ($\neq 0$), m', n' ($\neq 0$) are integers. Then $\dfrac{m}{n} + \dfrac{m'}{n'} = \dfrac{mn' + m'n}{nn'}$, where $mn' + m'n$ and nn' ($\neq 0$) are again integers, and similarly for subtraction and multiplication. If $m' \neq 0$ as well, so that $m'/n' \neq 0$, then $\dfrac{m}{n} \div \dfrac{m'}{n'} = \dfrac{mn'}{m'n}$, where mn' and $m'n$ ($\neq 0$) are again integers.
4. All but (a) exist.
5. Irrational.
6. $(1 - \sqrt{2}) + \sqrt{2} = (1 - \sqrt{2}) - (-\sqrt{2}) = 1$, where both terms are irrational.
7. $\sqrt{2} \cdot \sqrt{2} = 2$ is rational, and so is $\sqrt{2} \div \sqrt{2} = 1$.
8. The set of irrational numbers is not closed under any of the operations.
9. On the one hand, $0 \cdot c + 1 \cdot c = 0 \cdot c + c$, while on the other hand, $0 \cdot c + 1 \cdot c = (0 + 1) \cdot c = 1 \cdot c = c$, so that $0 \cdot c + c = c$. Now subtract c from both sides.
12. The formula holds for $n = 1$, since $1 = \dfrac{1(1 + 1)}{2}$. Suppose the formula is true for $n = k$, so that $1 + 2 + \cdots + k = \dfrac{k(k + 1)}{2}$. Then $1 + 2 + \cdots + k + (k + 1) = \dfrac{k(k + 1)}{2} + (k + 1) = \dfrac{k(k + 1) + 2k + 2}{2} = \dfrac{(k + 1)(k + 2)}{2}$, so that the formula also holds for $n = k + 1$.

13. By definition, $a/0$ is the number c such that $0 \cdot c = a$. But $0 \cdot c = 0$ for all c (Prob. 9), and hence there is no such c unless $a = 0$. If $a = 0$, we get $0/0$ which is meaningless, not because there is *no* number c such that $0 \cdot c = 0$, but rather because *every* number c has this property!

14. $\frac{311}{1000}$.

15. $\frac{1}{3} + \frac{1}{6} = 0.3333\ldots + 0.1666\ldots = 0.4999\ldots$ Let $x = 0.4999\ldots$ Then $10x = 4.9999\ldots$, and hence $9x = 10x - x = 4.9999\ldots - 0.4999\ldots = 4.5000\ldots = 4.5$, so that $x = 4.5/9 = 0.5$. Thus a decimal with an endless run of nines after a certain place represents the same rational number as the "next highest decimal," which always terminates.

16. (a) $\frac{23}{25}$;   (c) $\frac{1219}{990}$. Use the same reasoning as in Prob. 15.

18. $1.414214673\ldots$

20. Suppose k is a sum of threes and fives exclusively. Then this sum either contains a five or it does not. In the first case, replace a five by 2 threes. In the second case, there are at least 3 threes, since k exceeds 7, by hypothesis, and we can replace 3 threes by 2 fives. To start the induction (Sec. 1.37c), note that $8 = 5 + 3$.

Sec. 1.4

1. (a) $a - b > 0$, and hence $-a - (-b) < 0$.

2. $p > p'$ means $p - p' > 0$, or equivalently $\dfrac{mn' - m'n}{nn'} > 0$. But $nn' > 0$, and hence $mn' - m'n > 0$, or equivalently $mn' > m'n$.

3. (b) $-\frac{33}{10}$.

4. By Theorem 1.43 twice, $ac > bc$ and $bc > bd$. Hence $ac > bd$, by Theorem 1.45.

5. Clearly $a \neq b$. If $a > b$, then, by Prob. 4, with $c = a, d = b$, we have $a^2 > b^2$, contrary to $b^2 > a^2$.

6. An immediate consequence of Theorem 1.43. If $a^2 = a$, then $a = 0$ or $a = 1$.

10. (a) 0;   (c) 1;   (e) -1.

11. (b) n.

12. If $p < q$, then $\frac{1}{2}p < \frac{1}{2}q$. Adding first $\frac{1}{2}p$ and then $\frac{1}{2}q$ to both sides of the last inequality, we get $p = \frac{1}{2}p + \frac{1}{2}p < \frac{1}{2}p + \frac{1}{2}q$ and $\frac{1}{2}p + \frac{1}{2}q < \frac{1}{2}q + \frac{1}{2}q = q$. Let r and s be rational numbers such that $r < 1, s > 0$. Then $r' = \frac{1}{2}(r + 1)$ and $s' = \frac{1}{2}(0 + s) = \frac{1}{2}s$ are rational numbers such that $r < r' < 1, 0 < s' < s$. The argument works equally well for real p, q, r, s, r', s'.

13. (b) Show that $(a + b)^2 \geq 4ab$ is equivalent to $(a - b)^2 \geq 0$.

Sec. 1.5

1. Examine cases. For example, if $x > 0, y < 0$, then $|x| = x, |y| = -y, |xy| = -xy$.

4. By two applications of the inequality (3), $|x + y + z| = |(x + y) + z| \leq |x + y| + |z| \leq |x| + |y| + |z|$.

7. $0, -\frac{8}{3}$.

8. x^2 lies to the right of x if $x > 1$ or if $x < 0$, and to the left of x if $0 < x < 1$. x^2 and x coincide if $x = 0$ or $x = 1$.

9. $|x_1 - \frac{1}{2}(x_1 + x_2)| = |x_2 - \frac{1}{2}(x_1 + x_2)| = \frac{1}{2}|x_1 - x_2| = \frac{1}{2}|P_1P_2|$.

10. First replace x by $x - y$ in (3), and then replace y by $y - x$. Equality occurs under the same conditions as for (3) itself, namely if x and y have the same sign or if one (or both) of the numbers x and y is zero.

12. The point moves from a to b.

13. (a) If n is any integer greater than $1/|x_1 - x_2|$, and if c is the irrational number $1/\sqrt{2} < 1$, then at least one of the rational numbers..., $-3/2n, -1/2n, 1/2n, 3/2n, ...$ and at least one of the irrational numbers..., $-3c/2n, -c/2n, c/2n, 3c/2n, ...$ falls between x_1 and x_2; (b) Apply (a) repeatedly.

Sec. 1.6

1. $(0, 2)$. $[-3, 3]$.
3. The interval $1 \leqslant x \leqslant 2$.
5. $(-2, 1)$. $[-1, 2]$.
6. (a) $(1, 3]$; (c) $(-\infty, \infty)$.
7. (b) $\{1\}$.
8. $(0, \infty), (-\infty, 1]$.

Sec. 1.7

1. A six-pointed star.
2. $A' = (3, 2), B' = (3, 4), C' = (1, 5), D' = (-1, 4), E' = (-1, 2), F' = (1, 1)$. Each abscissa is increased by 1 and each ordinate by 2.
3. $x < 0, y > 0$ in the second quadrant, $x < 0, y < 0$ in the third, $x > 0, y < 0$ in the fourth.
4. (a) $4\sqrt{2}$; (c) 1.
7. 5.
8. $|AB| = |BC| = |CD| = |DA| = \sqrt{17}$ and $|AC| = |BD| (= \sqrt{34})$.
10. $(3, 3), (15, 15)$.
11. 21.
12. $C = (x + x', y + y')$.

Sec. 1.8

1. (a) 11; (c) Slope undefined.
2. When $m = m'$.
3. (a) $45°$; (c) $135°$.
4. (a) $\tan 20° = 0.36397$; (c) $\tan 165° = -\tan 15° = -0.26795$.
6. -1.
8. The lines are perpendicular.

Sec. 1.9

1. (a) $y = 2x - 2$; (c) $y = 2x - 1$; (e) $y = 2x - 3$.
2. (a) $y = 7x - 19$; (c) $y = x + 4$.
3. (a) $y = -x + 1$; (c) $y = -2$.
4. (a) $m = 3, a = 2, b = -6$; (c) $m = -1, a = 3, b = 3$.
5. (a) $m = 5, a = -\frac{4}{5}, b = 4$; (c) $m = 0, a$ undefined, $b = 3$.
6. $y = 2x - 8$.

7. $y = -x + 3.$ $(\frac{1}{2}, \frac{5}{2})$.
8. 20.
11. $y = -2x.$
12. The line has slope $m = -b/a$. Substitute this value of m in formula (2).
13. (b) $x + 3y + 3 = 0.$
14. $y = -x + 5, y = \frac{3}{2}x.$
15. $\frac{5}{2}$.
17. (a) 2;   (c) 0.

Chapter 2

Sec. 2.1

1. $f(0) = 6, f(1) = 10, f(2) = 16, f(\sqrt{2}) = 8 + 3\sqrt{2}.$
3. $g(-1) = -\frac{1}{2}, g(0) = -1, g(1) = \frac{3}{2}, g(1/\sqrt{2}) = 2\sqrt{2} + 2, g(1/\sqrt{3})$ does not exist.
4. (a) Domain all x such that $|x| \geq 3$, range all $y \geq 0$;   (c) Domain all $x \neq 3$, range all $y \neq 0$.
5. $f(3, 1) = \frac{1}{5}, f(0, 1) = 2, f(1, 0) = \frac{1}{2}, f(a, a) = -1, f(a, -a) = 1.$
6. Domain all points in the xy-plane except the origin, range $-\infty < z < \infty$.
7. Yes.
8. No. Yes.
10. No.
11. Yes.
13. True.
14. False.
16. $f(1, 1, 1) = 3, f(4, 1, 9) = \frac{11}{6}, f(1, 9, 1) = \frac{7}{3}, f(4, 9, 16) = \frac{13}{12}.$
18. Yes.
19. The inverse function has domain Y and range X.
20. All but (e). The inverses are: (a) $x = y$ (b) $x = 1/y$;   (c) $x = 1 + (1/y)$;
    (d) $x = y^2$ $(y \geq 0)$.
21. False.
22. Domain $-\infty < x < \infty$, range the set of all integers.
24. No, only when every $y \in Y$ is the second element of a pair $(x, y) \in f$.
25. 2^n.
26. When no two ordered pairs in f have the same second element. To get f^{-1}, write all the pairs in f in reverse order.
27. Delete either the point x_1 and the arrow joining x_1 to y_1 or the point x_1' and the arrow joining x_1' to y_1.
28. Reverse the directions of all the arrows.
30. (a) Finite;   (c) Infinite.

Sec. 2.2

1. (a) $x \neq 0$;   (b) $x \geq 0$;   (c) $x \geq 0$.
3. $\frac{13}{6}, -\frac{11}{6}, \frac{1}{3}, \frac{1}{216}, \frac{1}{12}.$
5. 2.
6. False.

7. (a) $\frac{1}{2}, \frac{2}{3}, \frac{3}{4}, \frac{4}{5}, \frac{5}{6}$; (c) $1, -\frac{1}{2}, \frac{1}{3}, -\frac{1}{4}, \frac{1}{5}$.

8. $1, 4, 9, \ldots, n^2, \ldots$ (recall Sec. 1.37a).

9. $1, 1, 2, 3, 5, 8, 13, 21$.

11. Examine the three cases $-\infty < x < -1$, $-1 \leqslant x \leqslant 1$ and $1 < x < \infty$ separately.

13. $\frac{1}{18}$.

Sec. 2.3

1. The pair of intersecting lines $y = x$ and $y = -x$.

3. $x^2 + y^2 + 4x - 6y + 9 = 0$.

4. True. If the line $x = c$ intersects the graph in more than one point, then the function takes more than one value at $x = c$, contrary to the definition of a function.

5. False. Consider circles, for example.

6. See Prob. 4.

7. No line parallel to either the x-axis or the y-axis can intersect the graph in more than one point.

9. (a), (d) and (f) are even, (c) and (e) are odd, (b) is neither even nor odd.

10. Note that $(-1)^n = 1$ if n is even, while $(-1)^n = -1$ if n is odd.

12. True.

13. $x^2 + y^2 - x - y = 0$.

14. Reflect G in the line $y = x$. Then interchange the labelling of the coordinate axes.

15. If $x \neq x'$, then either $x < x'$ or $x' < x$. Since f is increasing, $f(x) < f(x')$ in the first case and $f(x') < f(x)$ in the second case, so that in any event $f(x) \neq f(x')$. Therefore f is one-to-one, with an inverse f^{-1}. Let $y = f(x)$, $y' = f(x')$, so that $x = f^{-1}(y)$, $x' = f^{-1}(y')$, and suppose $y < y'$. Then $x \neq x'$, since f^{-1} is itself one-to-one, but $x' < x$ is impossible, since then $y' < y$. Therefore $x < x'$, so that f^{-1} is also increasing.

16. Use Probs. 12 and 15.

17. See Figure 1A.

18. See Figure 1B.

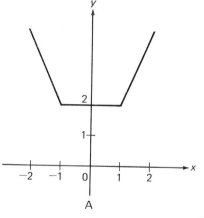

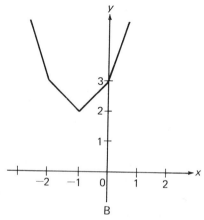

Figure 1.

19. Suppose $0 < x < x'$ and $0 < x^k < x'^k$. Then $0 < x^{k+1} < x'^{k+1}$, by Sec. 1.4, Prob. 4. The result now follows by mathematical induction (Sec. 1.37).

20. Use Prob. 19 and the symmetry of the curve $y = x^n$ (recall Prob. 10).

Sec. 2.4

3. No, unless $m = 0$. Yes.

4. Yes.

5. If $|f(x) - A|$ is "arbitrarily small," then so is $|[f(x) - A] - ^\wedge|$, and conversely.

6. (a) 0;   (c) 1;   (e) No limit.

7. No. Yes.

8. (b) 0.

9. In general, yes. No.

10. If $|f(x) - 0|$ is "arbitrarily small," then so is $||f(x)| - 0| = ||f(x)|| = |f(x)|$, and conversely.

11. No.

12. Suppose both $f(x) \to A_1$ as $x \to x_0$ and $f(x) \to A_2$ as $x \to x_0$, where $A_1 \neq A_2$. Then, choosing $\varepsilon = \frac{1}{2}|A_1 - A_2| > 0$, we can find numbers $\delta_1 > 0$ and $\delta_2 > 0$ such that $|f(x_1) - A_1| < \varepsilon$ whenever $0 < |x - x_0| < \delta_1$ and $|f(x) - A_2| < \varepsilon$ whenever $0 < |x - x_0| < \delta_2$. Let δ be the smaller of the two numbers δ_1 and δ_2. Then $|A_1 - A_2| = |A_1 - f(x) + f(x) - A_2| \leq |A_1 - f(x)| + |f(x) - A_2| < 2\varepsilon = |A_1 - A_2|$ whenever $0 < |x - x_0| < \delta$. But this is impossible!

13. Let $f(x) \to A$ as $x \to x_0$, and let $f_1(x) = f(x)$ everywhere except at $x = x_1$. Then, given any $\varepsilon > 0$, there is a $\delta > 0$ such that $|f(x) - A| < \varepsilon$ whenever $0 < |x - x_0| < \delta$. Let δ_1 be the smaller of the numbers δ and $|x_0 - x_1| \neq 0$. Then, given any $\varepsilon > 0$, we have $|f_1(x) - A| = |f(x) - A| < \varepsilon$ whenever $0 < |x - x_0| < \delta_1$, so that $f_1(x) \to A$ as $x \to x_0$.

14. Choosing $\varepsilon = 1$, we can find a $\delta > 0$ such that $|f(x) - A| < 1$ whenever $0 < |x - x_0| < \delta$. But $|f(x) - A| \geq |f(x)| - |A|$, with the help of Sec. 1.5, Prob. 10. Therefore $0 < |x - x_0| < \delta$ implies $|f(x)| - |A| < 1$, or equivalently $|f(x)| < |A| + 1$.

15. If $A > 0$, choose $\varepsilon = \frac{1}{2}A$. Then there is a $\delta > 0$ such that $0 < |x - x_0| < \delta$ implies $|f(x) - A| < \frac{1}{2}A$, or equivalently $0 < \frac{1}{2}A < f(x) < \frac{3}{2}A$, so that, in particular, $f(x) > 0$ and $\frac{1}{2}|A| < f(x) = |f(x)|$. If $A < 0$, choose $\varepsilon = -\frac{1}{2}A$. Then there is a $\delta > 0$ such that $0 < |x - x_0| < \delta$ implies $|f(x) - A| < -\frac{1}{2}A$, or equivalently $\frac{3}{2}A < f(x) < \frac{1}{2}A < 0$, so that, in particular, $f(x) < 0$ and $-|f(x)| = f(x) < \frac{1}{2}A$, or equivalently $|f(x)| > -\frac{1}{2}A = \frac{1}{2}|A|$.

Sec. 2.5

1. $\Delta x = -0.009$, $\Delta y = 990{,}000$.

4. No. Yes, the tangents at any pair of points (x_0, x_0^3), $(-x_0, -x_0^3)$ are parallel.

6. $y = 0$ or $y = 8x - 16$.

7. $(2, 4)$.

8. -2.

9. (a) $\Delta y = 7$, $dy = 3$, $E = 4$, about 57% of Δy;   (c) $\Delta y = 0.030301$, $dy = 0.03$, $E = 0.000301$, about 1% of Δy. The approximation of Δy by dy improves as Δx gets smaller.

10. When $|\Delta x|$ is too large or when $f'(x) = 0$.

11. $\Delta(uv) = \Delta u(x)v(x + \Delta x) + u(x)\,\Delta v(x) = u(x + \Delta x)\,\Delta v(x) + \Delta u(x)v(x)$.

12. $b = -3$, $c = 4$.

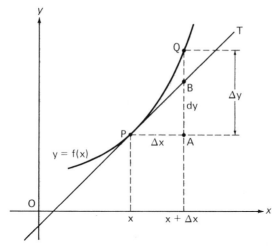

Figure 2.

13. Recall Sec. 2.3, Prob. 17.
14. About 19 square miles, almost as large as Manhattan Island.
15. In Figure 2, drawn for the case $\Delta x > 0$, $0 < dy < \Delta y$, we have $dy = |AB|$, $\Delta y = |AQ|$. Thus dy is the increment of the ordinate of the tangent T to the curve $y = f(x)$, while Δy is the increment of the ordinate of the curve itself.

Sec. 2.6

2. Apply formulas (2) and (4) repeatedly.
4. (a) 0;   (c) $\frac{2}{3}$.
5. At $x = 1, 2$.
7. $\lim\limits_{x \to 2-} f(x) = 3$, $\lim\limits_{x \to 2+} f(x) = 5$.
8. See Figure 3. At $x = 0, \pm 1, \pm 2, \ldots$

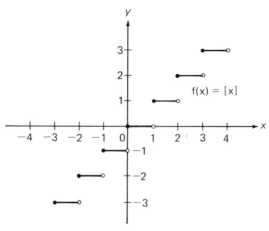

Figure 3.

9. Examine the figure, noting that the solid dots belong to the graph.

10. No. Yes. No.

13. No, since such intervals do not contain their end points.

15. We have

$$\begin{aligned}
|f(x)g(x) - AB| &= |f(x)g(x) - Bf(x) + Bf(x) - AB| \\
&= |f(x)[g(x) - B] + B[f(x) - A]| \\
&\leqslant |f(x)[g(x) - B]| + |B[f(x) - A]| \\
&< |f(x)| \, |g(x) - B| + (|B| + 1)|f(x) - A|.
\end{aligned}$$

Given any $\varepsilon > 0$, there are positive numbers $\delta_1, \delta_2, \delta_3$ such that $|f(x)| < |A| + 1$ whenever $0 < |x - x_0| < \delta_1, |f(x) - A| < \dfrac{\varepsilon}{2(|B| + 1)}$ whenever $0 < |x - x_0| < \delta_2$, and $|g(x) - B| < \dfrac{\varepsilon}{2(|A| + 1)}$ whenever $0 < |x - x_0| < \delta_3$. Let δ be the smallest of the numbers $\delta_1, \delta_2, \delta_3$. Then

$$|f(x)g(x) - AB| < (|A| + 1)\dfrac{\varepsilon}{2(|A| + 1)} + (|B| + 1)\dfrac{\varepsilon}{2(|B| + 1)} = \varepsilon$$

whenever $0 < |x - x_0| < \delta$.

Similarly,

$$\left| \dfrac{f(x)}{g(x)} - \dfrac{A}{B} \right| = \left| \dfrac{Bf(x) - Ag(x)}{Bg(x)} \right| = \dfrac{|Bf(x) - AB + AB - Ag(x)|}{|Bg(x)|}$$

$$< \dfrac{|B| \, |f(x) - A| + (|A| + 1)|g(x) - B|}{|B| \, |g(x)|}.$$

Given any $\varepsilon > 0$, there are positive numbers $\delta_1, \delta_2, \delta_3$ such that $|g(x)| > \frac{1}{2}|B|$ whenever $0 < |x - x_0| < \delta_1, |f(x) - A| < \dfrac{\varepsilon|B|}{4}$ whenever $0 < |x - x_0| < \delta_2$, and $|g(x) - B| < \dfrac{\varepsilon|B|^2}{4(|A| + 1)}$ whenever $0 < |x - x_0| < \delta_3$. Let δ_3 be the smallest of the numbers $\delta_1, \delta_2, \delta_3$. Then

$$\left| \dfrac{f(x)}{g(x)} - \dfrac{A}{B} \right| < \dfrac{2}{|B|} \dfrac{\varepsilon|B|}{4} + \dfrac{2(|A| + 1)}{|B|^2} \dfrac{\varepsilon|B|^2}{4(|A| + 1)} = \varepsilon$$

whenever $0 < |x - x_0| < \delta$.

Sec. 2.7

1. (a) $4x^3 + 6x$;   (c) $1 + \dfrac{1}{x^2}$.

2. (a) $2x - (a + b)$;   (c) $32x^3 + 12x$.

3. (a) $\dfrac{2a}{(x + a)^2}$;   (c) $\dfrac{-x^4 + 10x^3 + 6x - 15}{(x^3 + 3)^2}$.

4. True.

8. (a) $\dfrac{1}{3\sqrt[3]{x^2}}$; (c) $\dfrac{3\sqrt{x}}{2}$.

9. True.

10. $y' = -\dfrac{1}{x^2},\;\; y'' = \dfrac{2}{x^3},\;\; y''' = -\dfrac{3\cdot 2}{x^4},\ldots, y^{(n)} = (-1)^n \dfrac{n!}{x^{n+1}}$.

12. $y^{(6)} = 4\cdot 6!,\; y^{(7)} = 0$.

Sec. 2.8

1. If $y = \sqrt[3]{x} = x^{1/3}$, then $x = y^3$ and $\dfrac{dy}{dx} = \dfrac{1}{dx/dy} = \dfrac{1}{3y^2} = \dfrac{1}{3(\sqrt[3]{x})^2} = \dfrac{1}{3}x^{-2/3}$

2. $\sqrt{3} + \sqrt{5}$.

3. (b) $\dfrac{(x+2)(x+4)}{(x+3)^2}$

4. $200\cdot 3^{99}$.

7. (a) $\dfrac{x}{\sqrt{x^2+a^2}}$; (c) $\dfrac{a^2}{(\sqrt{a^2-x^2})^3}$.

8. $-\dfrac{(1-x)^{r-1}[(r+s)+(r-s)x]}{(1+x)^{s+1}}$.

9. Start from $\dfrac{d}{dx}\dfrac{1}{f(x)} = -\dfrac{f'(x)}{f^2(x)}$.

11. In the interval $0 \leqslant x < \infty$. In the whole interval $-\infty < x < \infty$.

13. $y' = \dfrac{2x-y}{x-2y},\;\; y'|_{x=1,y=0} = 2,\;\; y'|_{x=1,y=1} = -1$.

14. $y'' = \dfrac{m(m+n)}{n^2x^2}x^{-m/n}$.

15. $\dfrac{d}{dx}x^{m/n} = \dfrac{dy}{dx} = \dfrac{dy}{dt}\dfrac{dt}{dx} = mt^{m-1}\dfrac{dt}{dx} = \dfrac{mt^{m-1}}{dx/dt} = \dfrac{mt^{m-1}}{nt^{n-1}} = \dfrac{m}{n}(x^{1/n})^{m-n} = \dfrac{m}{n}x^{(m/n)-1}$,

where we make free use of formulas (15) and (16), p. 78.

17. $y' = -\tfrac{3}{4},\; y'' = -\tfrac{25}{64},\; y''' = -\tfrac{225}{1024}$.

19. Solving (16) for y^2, we get $y^2 = \tfrac{1}{2}(x^2 \pm \sqrt{-3x^4})$. Hence the only real solution of (15) is $x = y = 0$, and it is meaningless to talk about y'.

Sec. 2.9

1. (a) 1; (c) $\tfrac{2}{5}$.
2. (a) $-\infty$; (c) ∞.
3. $f(x) \to \infty$ as $x \to 0+,\, 1-,\, 2+$, while $f(x) \to -\infty$ as $x \to 0-,\, 1+,\, 2-$.

5. $(\frac{3}{2})^{30}$.

6. $x > 2998$.

8. $x < \frac{3000}{999}, x > \frac{3000}{1001}$.

9. (a) $x = -2, y = \frac{1}{2}$;   (c) $x = \pm 2, y = 1$.

11. See Figure 16.

13. (a) 1;   (c) $\frac{2}{9}$.

14. None.

16. The formula obviously holds for $n = 1$. Suppose it holds for $n = k$, so that $(1 + x)^k \geqslant 1 + kx$. Then it holds for $n = k + 1$, since $(1 + x)^{k+1} = (1 + x)^k(1 + x) \geqslant (1 + kx)(1 + x) = 1 + (k + 1)x + kx^2 \geqslant 1 + (k + 1)x$.

17. 1 if $|a| > 1, 0$ if $|a| < 1, \frac{1}{2}$ if $a = 1$.

18. (a) $1 + \frac{1}{2} + \frac{1}{3} + \frac{1}{4} + \frac{1}{5} = \frac{137}{60}$;   (c) $2^1 + 3^2 + 4^3 + 5^4 = 700$.

19. (a) 2;   (c) 1. Note that $\dfrac{1}{n(n + 1)} = \dfrac{1}{n} - \dfrac{1}{n + 1}$.

20. There are no points at which $f(x) \to \pm\infty$, and $f(x)$ does not approach a finite limit as $x \to \pm\infty$. However,

$$\lim_{x \to \pm\infty} \left| \frac{x^3}{2x^2 + 1} - \frac{x}{2} \right| = \lim_{x \to \pm\infty} \left| \frac{-x}{2(2x^2 + 1)} \right| = 0.$$

21. Let x_n equal $\sqrt{2}$ to n decimal places.

23. Note that $x_n = s_{n+1} - s_n \to s - s = 0$ as $n \to \infty$.

Chapter 3

Sec. 3.1

1. The average velocities are 215, 210.5 and 210.05 ft/sec. The instantaneous velocity is 210 ft/sec.

3. The acceleration is variable.

4. The stone hits the ground 3 seconds later, travelling at a speed of 64 ft/sec.

6. The car is accelerating with a constant acceleration of $k = 8.8$ ft/sec². Equation (7) can only be valid when the car's speed is well below its top speed.

7. The car is decelerating with a constant deceleration of $k = 4$ ft/sec².

9. The stone's motion during the last 3 seconds is the "reverse" of its motion during the first 3 seconds (make this precise).

10. The flywheel has angular velocity $\theta'(t) = b - 2ct$ and angular acceleration $\theta''(t) = -2c$.

Sec. 3.2

1. $5/8\pi \approx 0.2$ ft/min $= 2.4$ in/min.

3. No. Yes.

5. Increasing, at 40 in²/sec.

7. $MR(Q) = \dfrac{d}{dQ} QAR(Q).$

 9. $Q(P)$ is a decreasing function of P.
10. Since $Q = Q(P)$ is decreasing, it has a decreasing inverse $P = P(Q)$, by Sec. 2.3, Prob. 16, so that $PQ = PQ(P) = QP(Q)$.
12. $-\frac{25}{3}$ ft/min^2.

Sec. 3.3

 1. (a) No extrema;   (c) A maximum equal to 1 at $x = 1$, a minimum equal to $\frac{1}{2}$ at $x = 2$;   (e) No maximum, a minimum equal to -1 at $x = -1$.
 2. (a) A maximum and a minimum, both equal to 0, at every point of $(0, 1)$;   (c) A maximum equal to 0 at $x = 0$, a minimum equal to -1 at every point of $(-1, 0)$;   (e) No extrema.
 4. True, by the intermediate value theorem.
 5. No. Yes.
 7. No maximum, a minimum equal to -1 at $x = 0$.
 8. The function $f(x) = 1/x$ maps $(0, 1)$ into $(1, \infty)$.
 9. The function graphed in Figure 14, p. 91, maps $(-\infty, \infty)$ into $(1, 2)$.
10. Let

$$f(x) = x, \qquad g(x) = \begin{cases} 0 & \text{if } x < 0, \\ x & \text{if } x \geqslant 0, \end{cases} \qquad h(x) = \begin{cases} -1 & \text{if } x < -1, \\ x & \text{if } -1 \leqslant x \leqslant 1, \\ 1 & \text{if } x > 1. \end{cases}$$

These functions are all continuous in $(-2, 2)$, say; f maps $(-2, 2)$ into $(-2, 2)$, g maps $(-2, 2)$ into $[0, 2)$, and h maps $(-2, 2)$ into $[-1, 1]$.

Sec. 3.4

 1. $f'(x) = 3x^2 - 12x + 11 = 0$ if $x = 2 \pm 1/\sqrt{3} \approx 1.42, 2.58$.
 3. The formula $f(a) - f(b) = f'(c)(a - b)$ is equivalent to (10), but now $c \in (b, a)$.
 5. $c = \sqrt{2}$.
 7. Apply formula (9) to the train's distance function $s = s(t)$.
 9. Apply the mean value theorem to the function $\sqrt{x}$.

Sec. 3.5

 1. The domain of f is not an interval.
 2. (b) $-(1/x) + C$.
 3. Trivial, but worthy of note.
 4. Use the chain rule.
 5. (a) $\frac{1}{5}x^5 - x^3 + \frac{1}{2}x^2 - 4x + C$;   (c) $\frac{2}{3}x\sqrt{x} + 2\sqrt{x} + C$.
 6. True.
 8. (b) Increasing in $[-1, 1]$, decreasing in $(-\infty, -1]$ and $[1, \infty)$.
 9. f is a polynomial of degree less than n.

Sec. 3.6

 1. Reread Sec. 3.62a.
 2. (a) Minimum $y = 0$ at $x = 0$;   (c) No extrema.

3. (b) Maximum $y = -2$ at $x = -1$, minimum $y = 2$ at $x = 1$.
4. (a) Maximum $y = 66$ at $x = 10$, minimum $y = 2$ at $x = 2$;
    (c) Maximum $y = 3$ at $x = -1$, minimum $y = 1$ at $x = 1$.
6. $y' \neq 0$ if $ad - bc \neq 0$, $y \equiv$ constant if $ad - bc = 0$.
8. $c = -\frac{1}{2}$.
10. $y_{10,000} = \frac{1}{200}$.

Sec. 3.7

2. Nothing. Explain.
4. Inflection points at $x = \pm\frac{1}{2}$, concave upward in $(-\infty, -\frac{1}{2})$, concave downward in $(-\frac{1}{2}, \frac{1}{2})$, concave upward in $(\frac{1}{2}, \infty)$.
6. $c = -3$.
8. See Figure 4.
9. See Figure 5.
11. Let $y = T(x)$ be the tangent to the curve $y = f(x)$ at $x = p$. Then, by the mean value theorem in increment form (Sec. 3.43b),

$$f(x) - T(x) = f(x) - f(p) - f'(p)\,\Delta x$$
$$= f(p + \Delta x) - f(p) - f'(p)\,\Delta x = [f'(p + \alpha\Delta x) - f'(p)]\,\Delta x, \quad (i)$$

where $\Delta x = x - p$ and $0 < \alpha < 1$. If f' is increasing in a δ-neighborhood of p, then $f'(p + \alpha\Delta x) - f'(p) < 0$ if $-\delta < \Delta x < 0$, while $f'(p + \alpha\Delta x) - f'(p) > 0$ if $0 < \Delta x < \delta$, so that, in either case, the right side of (i) is positive. Therefore $f(x) > T(x)$ in the δ-neighborhood, so that f is concave upward at p. The proof for decreasing f' is virtually the same.
12. Again we start from (i). Suppose the extremum is a maximum. Then $f'(p + \alpha\Delta x) - f'(p) < 0$ if $-\delta < \Delta x < 0$, while $f'(p + \alpha\Delta x) - f'(p) < 0$ if $0 < \Delta x < \delta$, so that the right side of (i) is positive to the left of p and negative

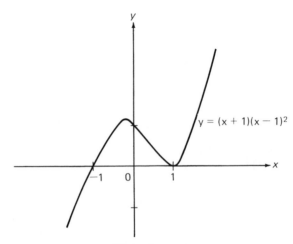

$$y = (x + 1)(x - 1)^2$$

Figure 4.

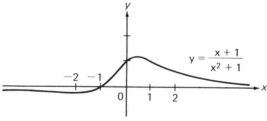

Figure 5.

to the right of p. Therefore $f(x) > T(x)$ to the left of p, while $f(x) < T(x)$ to the right of p, so that p is an inflection point of f. The proof for the case of a minimum is virtually the same.

Sec. 3.8

1. The square of side $\sqrt{A}$.
2. The triangle with legs $c/3$ and $c/\sqrt{3}$.
3. $2\pi l^3/9\sqrt{3}$.
5. $P = (\frac{3}{2}, 0)$.
6. $3\sqrt[3]{\pi V^2}$.
9. $t = \dfrac{\alpha a + \beta b}{\alpha^2 + \beta^2}, d = \dfrac{|\alpha b - \beta a|}{\sqrt{\alpha^2 + \beta^2}}$.
10. The line $(x/2a) + (y/2b) = 1$, with x-intercept $2a$ and y-intercept $2b$.
12. $Q = 30$.
14. The chord whose distance from the point A equals $\frac{3}{4}$ of the diameter of the circle.
15. $\pi R^2(1 + \sqrt{5})$.

Chapter 4

Sec. 4.1

1. (a) 2;   (c) max A does not exist.
4. $(b - a)/n$. No, although $\lambda < b - a$.
7. Yes. Define the area A between the curve $y = f(x)$ and the x-axis from $x = a$ to $x = b$ by the integral $A = \int_a^b f(x)\, dx$, whether or not $f(x)$ is nonnegative. Then $A < 0$ if more area lies "above" the curve than "below" it.
9. max $A = a$ if $0 \leqslant a \leqslant 1$, max A does not exist if $a > 1$, max $A = a^2$ if $-1 \leqslant a < 0$, max A does not exist if $a < -1$.
10. Show that $0 \leqslant \sigma \leqslant \lambda$, where σ is the sum (3), regardless of the choice of the points $\xi_1, \xi_2, \ldots, \xi_n$. Therefore $\sigma \to 0$ as $\lambda \to 0$.
11. (b) Choosing all the points $\xi_1, \xi_2, \ldots, \xi_n$ in the sum (3) to be rational, we have $\sigma = 1$, and choosing them all to be irrational, we have $\sigma = -1$, regardless of the size of λ. Therefore σ cannot approach a limit as $\lambda \to 0$.

Sec. 4.2

2. (a) $\frac{14}{3}$;   (c) $\frac{17}{6}$.

4. $\frac{3}{4}$.

5. $A = \int_0^1 (\sqrt{x} - x^2)\,dx = \frac{1}{3}$. See Sec. 2.3, Prob. 14.

7. $c = 4$.

9. Use the fundamental theorem of calculus, noting that $v = ds/dt$.

11. An immediate consequence of formula (11).

12. Given any $\varepsilon > 0$, there is a $\delta > 0$ such that $|\sigma(\lambda) - \sigma_0| < \varepsilon$, or equivalently $\sigma_0 - \varepsilon < \sigma(\lambda) < \sigma_0 + \varepsilon$, whenever $0 < |\lambda| < \delta$. If $\sigma_0 < 0$, choose $\varepsilon = -\sigma_0 > 0$. Then there is a $\delta > 0$ such that $2\sigma_0 < \sigma(\lambda) < 0$ whenever $0 < |\lambda| < \delta$, which contradicts $\sigma(\lambda) \geqslant 0$. Therefore $\sigma_0 \geqslant 0$.

13. Apply Prob. 12 to the functions $\sigma(\lambda) - A$ and $B - \sigma(\lambda)$.

14. (b) $-f(a)$.

15. Follow the argument used to prove formula (7).

17. Apply Prob. 16 to the function $f = f_2 - f_1$.

21. Note that $\frac{1}{12} \leqslant 1/(10 + x) \leqslant \frac{1}{10}$ if $0 \leqslant x \leqslant 2$, where equality occurs only if $x = 0$ or $x = 2$.

Sec. 4.3

2. No, since $\ln(x^2)$ is defined for all $x \neq 0$, while $2 \ln x$ is defined only for $x > 0$.

3. (a) $4 \leqslant x \leqslant 6$;   (c) $x > e$.

4. (a) $\dfrac{3x^2 - 2}{x^3 - 2x + 5}$;   (c) $\dfrac{3}{x}(\ln x)^2$.

5. (a) $\dfrac{2}{1 - x^2}$;   (c) $\dfrac{1}{1 - x^2}$.

6. $c = \dfrac{1}{\ln 2}$.

8. $y^{(4)} = -2/x^2$.

10. 1.

12. At $x = 1$. No.

14. (a) $x \geqslant 1$ if $a > 1$, $0 < x \leqslant 1$ if $0 < a < 1$.

16. Use (7).

18. Note that $\log_x a = \dfrac{\ln a}{\ln x}$.

Sec. 4.4

2. (a) $4e^{4x+5}$;   (c) $e^x(1 + x)$.

3. (a) $2xe^{x^2}$;   (c) $\dfrac{e^{\sqrt{x+1}}}{2\sqrt{x+1}}$.

4. $\alpha = \ln(e - 1) \approx 0.54$.

6. $\frac{5}{6}$.

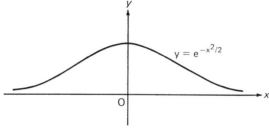

Figure 6.

8. Inflection points at $x = \pm 1$, concave upward in $(-\infty, -1)$, concave downward in $(-1, 1)$, concave upward in $(1, \infty)$. The graph is the 'bell-shaped" curve shown in Figure 6.

9. They are inverses of each other.

11. The advisor had the effrontery to ask for more than 18 billion billion billion grains of rice! Show this by using formula (13), p. 97, to evaluate the sum $1 + 2 + 2^2 + \cdots + 2^{63}$.

12. (a) $10^x(1 + x \ln 10)$;   (c) $a^x x^a \left(\dfrac{a}{x} + \ln a \right)$.

13. Use (14), (11) and (12), noting that $\ln a > 0$ if $a > 1$, $\ln a < 0$ if $0 < a < 1$.

14. Use (17), (11) and (12), together with formulas (8) and (9), p. 155.

16. $c \leqslant -e/6, c > 0$.

Sec. 4.5

1. (a) $1/e$;   (c) $1/e$.
2. (a) e;   (c) e^2.

4. $\displaystyle \lim_{x \to 0} \frac{a^x - 1}{x} = \lim_{t \to 0} \frac{t}{\log_a (1 + t)} = \frac{1}{\log_a e} = \ln a$.

6. (a) $\frac{3}{5}$;   (c) $(\log_{10} e)^2$.

7. $\$1,485.95$.

9. About 13 years and a month.

11. $P(1 + r_E)^t = P \left(1 + \dfrac{r}{N} \right)^{Nt}$.

12. One dollar grows to e dollars in one year if compounded continuously at an annual interest rate of 100%.

13. (a) $\dfrac{(x + 1)^2}{(x + 2)^3(x + 3)^4} \left(\dfrac{2}{x + 1} - \dfrac{3}{x + 2} - \dfrac{4}{x + 3} \right)$.

14. (a) $x^{x^2}(2x \ln x + x)$;   (c) $(\ln x)^x \left(\ln \ln x + \dfrac{1}{\ln x} \right)$.

16. By the ordinary chain rule, $\varepsilon_{zx} = \dfrac{x}{z} \dfrac{dz}{dx} = \dfrac{x}{z} \dfrac{dz}{dy} \dfrac{dy}{dx} = \left(\dfrac{y}{z} \dfrac{dz}{dy} \right) \left(\dfrac{x}{y} \dfrac{dy}{dx} \right) = \varepsilon_{zy} \varepsilon_{yx}$.

18. x^r.

20. Note that $\varepsilon_C = \dfrac{MC(Q)}{AC(Q)}$, while $\dfrac{d}{dQ} AC(Q) = \dfrac{C}{Q^2}(\varepsilon_C - 1)$.

21. See Figure 7. The various properties of cosh x and sinh x are easy consequences of those of e^x and e^{-x}.

Sec. 4.6

1. If $x = -t$, then $\displaystyle\int_{-a}^{0} f(x)\,dx = -\int_{a}^{0} f(-t)\,dt = \int_{0}^{a} f(t)\,dt = \int_{0}^{a} f(x)\,dx$.

4. (a) $\frac{2}{3}\sqrt{x^3 + 1} + C$;   (c) $-\dfrac{1}{\ln x} + C$.

5. Let $t = f(x)$, noting that if $t < 0$, then $\dfrac{d}{dt}\ln|t| = \dfrac{d}{dt}\ln(-t) = \dfrac{(-t)'}{-t} = \dfrac{-1}{-t} = \dfrac{1}{t}$.

6. (a) $\ln(1 + x^2) + C$;   (c) $\ln|\ln x| + C$.

8. (a) $\dfrac{xa^x}{\ln a} - \dfrac{a^x}{(\ln a)^2} + C$;   (c) $\frac{1}{4}x^4 \ln x - \frac{1}{16}x^4 + C$.

9. Let $u = \ln(x + \sqrt{1 + x^2})$, $dv = dx$.

10. (a) $1 - \dfrac{2}{e}$;   (c) $2 - \dfrac{3}{4\ln 2}$.

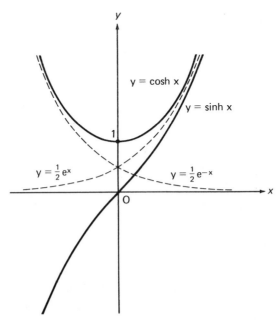

Figure 7.

11. $\dfrac{4^x}{\ln 4} + 2\dfrac{6^x}{\ln 6} + \dfrac{9^x}{\ln 9} + C$. Note that $a^x b^x = e^{x \ln a} e^{x \ln b} = e^{x(\ln a + \ln b)} = e^{x \ln (ab)}$

$= (ab)^x$.

12. $\displaystyle\int \frac{dx}{(x + a)(x + b)} = \int \frac{1}{a - b}\left(\frac{1}{x + b} - \frac{1}{x + a}\right) dx$

$\qquad\qquad = \dfrac{1}{a - b}\,(\ln |x + b| - \ln |x + a|) + C$

$\qquad\qquad = \dfrac{1}{a - b}\,\ln\left|\dfrac{x + b}{x + a}\right| + C.$

13. (a) $\ln\left|\dfrac{x - 3}{x - 2}\right| + C.$

14. $3 - e$.

15. See Sec. 4.5, Prob. 21d.

17. (b) $\displaystyle\int \frac{1 + x}{1 - x}\,dx = \int\left(-1 + \frac{2}{1 - x}\right) dx = -x - 2\ln|1 - x| + C.$

18. The formulas (23) are equivalent to

$$F(x) \equiv G(t(x)) + C, \qquad F(x(t)) \equiv G(t) + C. \qquad\qquad \text{(i)}$$

The substitution $x = x(t)$ transforms the first of these formulas into the second, while the substitution $t = t(x)$ transforms the second into the first, since $t(x(t)) \equiv t$, $x(t(x)) \equiv x$ (Sec. 2.22b).

19. By the fundamental theorem of calculus,

$$\int_a^b f(x)\,dx = F(b) - F(a) = G(t(b)) - G(t(a)) = \int_{t(a)}^{t(b)} g(t)\,dt,$$

$$\int_{x(\alpha)}^{x(\beta)} f(x)\,dx = F(x(\beta)) - F(x(\alpha)) = G(\beta) - G(\alpha) = \int_\alpha^\beta g(t)\,dt,$$

with the help of (i).

20. Repeated integration by parts gives $\displaystyle\int_0^1 x^m(1 - x)^n\,dx = \dfrac{m!n!}{(m + n + 1)!}.$

21. Integrate by parts repeatedly, noting that $P^{(n+1)}(x) \equiv 0$, since $P(x)$ is of degree n.

Sec. 4.7

1. The integral is divergent if $r \leqslant 1$ and equals $a^{1-r}/(r - 1)$ if $r > 1$.

2. (b) Divergent.

3. The integral is divergent if $r \geqslant 1$ and equals $a^{1-r}/(1 - r)$ if $r < 1$.

4. (b) $\frac{9}{2}$.

5. Note that

$$\lim_{X \to \infty} \int_c^X f(x)\,dx = \int_c^{c'} f(x)\,dx + \lim_{X \to \infty} \int_{c'}^X f(x)\,dx,$$

so that one limit is finite when the other is finite, and only then. Similarly

$$\lim_{X \to -\infty} \int_X^{c'} f(x)\,dx = \lim_{X \to -\infty} \int_X^c f(x)\,dx + \int_c^{c'} f(x)\,dx.$$

Moreover,

$$\int_{-\infty}^{c} + \int_{c}^{\infty} = \int_{-\infty}^{c} + \left(\int_{c}^{c'} + \int_{c'}^{\infty} \right) = \left(\int_{-\infty}^{c} + \int_{c}^{c'} \right) + \int_{c'}^{\infty} = \int_{-\infty}^{c'} + \int_{c'}^{\infty},$$

where, for brevity, we omit the expression $f(x)\, dx$ behind the integral signs.

6. For example, $s(X) = \int_0^X f(x)\, dx$ stands in the same relation to the improper integral $\int_0^{\infty} f(x)\, dx$ as the partial sum $s_N = \sum_{n=1}^{N} f_n$ to the infinite series $\sum_{n=1}^{\infty} f_n$. Develop the analogy further.

7. $A = \int_0^1 (x^{-1/2} - x^{-1/3})\, dx$.

Chapter 5

Sec. 5.1

1. Let $\Phi(x, y, z) = z - F(x, y)$.

3. $y = e^{-x^2}$.

5. Take the square root and then separate variables. The extra solution is $y \equiv 0$.

7. We have $p(dp/dy) = 2y^3$, and hence $\int p\, dp = \int 2y^3\, dy + C_1$, or $p^2 = y^4 + C_1$. Application of the initial conditions gives $C_1 = 0$. Therefore $p^2 = y^4$, or $y' = y^2$, so that $y = 1/(1 - x)$, after solving this first-order equation and applying the initial conditions again.

8. The general solution is $\ln |x| = \Phi(y/x) + C$, where $\Phi(u)$ is any antiderivative of the function $1/[f(u) - u]$.

10. Solving the differential equation $x - (y/y') = x^2$, subject to the initial condition $y|_{x=-1} = -1$, we get $y = 2x/(1 - x)$.

Sec. 5.2

1. $\dfrac{3 \ln 10}{\ln 2} \approx 10$ hrs. $N = 1000 \cdot 2^t$.

3. $\dfrac{5 \ln 2}{\ln (1.2)} \approx 19$ yrs.

5. No.

8. 1 hr.

10. To get the solution of (21), change e^{-rt} to e^{rt} in formula (13).

11. $N = N_0 e^{rt} + \dfrac{s}{r}(e^{rt} - 1)$.

12. If the fresh specimen has N_0 radioactive atoms, the old specimen has $N_0 e^{-kt}$ atoms, where $k = (\ln 2)/5570$ is the decay constant of radiocarbon. Therefore $n = \alpha N_0$, $m = \alpha N_0 e^{-k\tau}$, where α is some constant of proportionality. But then $n/m = e^{k\tau}$.

Sec. 5.3

1. $s = Ft^2/2m$.

3. $s = (kt^3/6m) + v_0 t$.

5. 32 ft.

7. $V = \int_0^s kx \, dx = \frac{1}{2}ks^2$.

9. As in Example 5.33f, the work done on the rocket by the earth's gravitational pull is

$$W = -\int_R^{R+h} \frac{kMm}{s^2} \, ds = \frac{kMm}{R+h} - \frac{kMm}{R} = -\frac{kMmh}{(R+h)R} = -\frac{mgRh}{R+h},$$

with the help of (24). Therefore $\frac{1}{2}mv_0^2 = mgRh/(R+h)$.

11. About 1.5 mi/sec.

Chapter 6

Sec. 6.1

1. The point P in R_3 with x, y and z-coordinates a, b and c is the unique point of intersection of the planes $x = a$, $y = b$ and $z = c$.

3. (a) 6; (c) 25.

5. All but $(\sqrt{2}, \sqrt{2}, 1)$ and $(\sqrt{2}, \sqrt{2}, \sqrt{2})$.

7. Yes. Consider a sphere.

8. (a) The plane $y = a$.

9. (a) $z = 1 - |x|$ $(-1 \leqslant x \leqslant 1)$.

10. Intervals are connected.

12. $(5, 0, 0), (-11, 0, 0)$.

14. A pair of right circular cones with their common vertex at the origin (make a sketch).

15. The domain of f is the set of all points (x, y) such that $|x| > |y|$. This set is not connected (why not?).

Sec. 6.2

1. Formula (2) means that, given any $\varepsilon > 0$, there is a $\delta > 0$ such that $|f(x, y) - A| < \varepsilon$ whenever $0 < \sqrt{(x-a)^2 + (y-b)^2} < \delta$. Therefore $|f(x, b) - A| < \varepsilon$ whenever $0 < \sqrt{(x-a)^2} = |x - a| < \delta$, and $|f(a, y) - A| < \varepsilon$ whenever $0 < \sqrt{(y-b)^2} = |y - b| < \delta$.

2. Use Prob. 1, first setting $y = 0$ and then $x = 0$.

3. Yes. Use the inequalities $|x - a| \leqslant \sqrt{(x-a)^2 + (y-b)^2}$,
$|y - b| \leqslant \sqrt{(x-a)^2 + (y-b)^2}$, $\sqrt{(x-a)^2 + (y-b)^2} \leqslant |x - a| + |y - b|$.

4. By Prob. 3, $\lim\limits_{\substack{x \to a \\ y \to b}} f(x, y) = \lim\limits_{x \to a} g(x) = g(a) = f(a, b)$.

5. See Prob. 4.

6. Use the analogue of Theorem 2.63c for functions of two variables.

7. (b) 1.

8. The triple limit is another way of writing $(x, y, z) \to (0, 1, e)$. By the three-dimensional analogue of Prob. 6, the function $\ln \dfrac{z}{\sqrt{x^2 + y^2}}$ is continuous at $(0, 1, e)$. Therefore $A = \ln e = 1$.

10. On the cylinder $x^2 + y^2 = 1$.

11. (b) $\dfrac{\partial z}{\partial x} = -\dfrac{2y}{(x-y)^2}$, $\dfrac{\partial z}{\partial y} = \dfrac{2x}{(x-y)^2}$.

12. (b) $\dfrac{\partial u}{\partial x} = \dfrac{1}{x}, \ \dfrac{\partial u}{\partial y} = \dfrac{1}{y}, \ \dfrac{\partial u}{\partial z} = -\dfrac{1}{z}.$

13. (a) $-1/x^2.$

14. (a) $\dfrac{\partial^2 z}{\partial x \, \partial y} = \dfrac{\partial}{\partial x}\dfrac{1}{x+y} = -\dfrac{1}{(x+y)^2} = \dfrac{\partial}{\partial y}\dfrac{1}{x+y} = \dfrac{\partial^2 z}{\partial y \, \partial x}.$

15. (a) $\Delta f(x, y) = (x+\Delta x)^2 + (y+\Delta y)^2 - x^2 - y^2 = 2x\,\Delta x + 2y\,\Delta y + (\Delta x)^2 + (\Delta y)^2$
    is of the form (8) with $A = 2x$, $B = 2y$, $\alpha(\Delta x, \Delta y) = \Delta x$, $\beta(\Delta x, \Delta y) = \Delta y.$

16. (a) $dz = (y - 2xy^3 + 3x^2 y)\,dx + (x - 3x^2 y^2 + x^3)\,dy.$

17. (a) 5.022.

20. First solve for each of the three variables as a function of the other two.

Sec. 6.3

1. (a) $2e^{2t} + \left(\dfrac{1}{t^2} - \dfrac{2}{t^3}\right) e^t.$

2. (a) $\dfrac{\partial z}{\partial x} = 2x + \dfrac{2\ln(x+y)}{x+y}, \ \dfrac{\partial z}{\partial y} = \dfrac{2\ln(x+y)}{x+y}.$

3. (a) $\dfrac{\partial z}{\partial x} = p\dfrac{\partial z}{\partial u} + r\dfrac{\partial z}{\partial v}, \ \dfrac{\partial z}{\partial y} = q\dfrac{\partial z}{\partial u} + s\dfrac{\partial z}{\partial v}.$

4. $z_x = \dfrac{yz}{z - xy}, \ z_y = \dfrac{xz}{z - xy}. \ \ z_x|_{x=1,y=2,z=1} = -2, \ \ z_y|_{x=1,y=2,z=1} = -1,$
   $z_x|_{x=1,y=2,z=3} = 6, \ \ z_y|_{x=1,y=2,z=3} = 3.$

5. (a) 2; (c) $\frac{1}{2}$; (e) $-1.$

6. Differentiate (19) with respect to t, and then set $t = 1.$

7. $\dfrac{\partial^2 z}{\partial x^2} = -\dfrac{2}{5}, \ \dfrac{\partial^2 z}{\partial y \, \partial x} = \dfrac{\partial^2 z}{\partial y \, \partial x} = -\dfrac{1}{5}, \ \dfrac{\partial^2 z}{\partial y^2} = -\dfrac{394}{125}$ at the indicated point.

Sec. 6.4

1. (b) Minimum $z = -1$ at $(x, y) = (1, 0).$

2. (b) Minimum $z = 0$ at $(x, y) = (2, 4).$

4. Maximum $z = 3/e$ at $(x, y) = (0, \pm 1)$, minimum $z = 0$ at $(x, y) = (0, 0).$

5. If $z \to \infty$, then $x \to 0$, $y \to 0$, since otherwise $xz + yz \to \infty$. Therefore

$$\lim_{z \to \infty} V = \lim_{\substack{x \to 0 \\ y \to 0}} (c - xy)\dfrac{xy}{x+y} = c\lim_{\substack{x \to 0 \\ y \to 0}} \dfrac{xy}{x+y} = 0,$$

with the help of (10) and the inequality $\left|\dfrac{xy}{x+y}\right| = \left|\dfrac{x}{x+y}\right| |y| \le |y|$, valid for positive x and $y.$

6. No.

9. Solve the equations

$$\dfrac{\partial V}{\partial x} = \dfrac{cy^2 - x^2 y^2 - 2xy^3}{(x+y)^2} = 0, \qquad \dfrac{\partial V}{\partial y} = \dfrac{cx^2 - x^2 y^2 - 2x^3 y}{(x+y)^2} = 0 \quad \text{(i)}$$

for positive x and $y.$

10. It follows from (i) that

$$A = \frac{\partial^2 V}{\partial x^2} = -2y^2 \frac{y^2 + c}{(x + y)^3},$$

$$B = \frac{\partial^2 V}{\partial x \, \partial y} = 2xy \frac{c - x^2 - 3xy - y^2}{(x + y)^3},$$

$$C = \frac{\partial^2 V}{\partial y^2} = -2x^2 \frac{x^2 + c}{(x + y)^3}.$$

Let

$$f(x, y) = \frac{(x + y)^6}{4x^2 y^2} (AC - B)^2 = (x^2 + c)(y^2 + c) - (c - x^2 - 3xy - y^2)^2.$$

Then

$$f\left(\sqrt{\frac{c}{3}}, \sqrt{\frac{c}{3}}\right) = \left(\frac{4c}{3}\right)^2 - \left(\frac{2c}{3}\right)^2 > 0,$$

so that $D = AC - B^2$ is positive at the point $(\sqrt{c/3}, \sqrt{c/3})$, while A is negative.

11. The (perpendicular) distance d between P_1 and the line L with equation $Ax + By + C = 0$ is, of course, also the minimum distance between P_1 and a variable point $P = (x, y)$ of L. Minimizing $\delta = \sqrt{(x - x_1)^2 + (y - y_1)^2}$ subject to the condition $Ax + By + C = 0$ is equivalent to minimizing δ^2 subject to the same condition. Let $u = \delta^2 - \lambda(Ax + By + C)$, where λ is a Lagrange multiplier. Setting the partial derivatives of u with respect to x, y and λ equal to zero, we get

$$\frac{\partial u}{\partial x} = 2(x - x_1) - \lambda A = 0,$$

$$\frac{\partial u}{\partial y} = 2(y - y_1) - \lambda B = 0,$$

$$\frac{\partial u}{\partial \lambda} = Ax + By + C = 0.$$

The last equation is just the equation of L. It follows from the first two equations that $(x_2 - x_1)/A = (y_2 - y_1)/B$, where $P_2 = (x_2, y_2)$ is the point of L minimizing u and hence δ. Let this last ratio be denoted by q. Then $x_2 - x_1 = Aq$, $y_2 - y_1 = Bq$, and hence d, the minimum value of δ, equals $\sqrt{A^2 q^2 + B^2 q^2} = \sqrt{A^2 + B^2}|q|$. But P_2 lies on L, and hence $Ax_2 + By_2 + C = A(Aq + x_1) + B(Bq + y_1) + C = 0$, so that $q = -(Ax_1 + By_1 + C)/(A^2 + B^2)$. Substituting this value of q into the formula for d, we get the required answer.

12. $MC_1(Q_1, Q_2) = 6Q_1 + 2Q_2$, $MC_2(Q_1, Q_2) = 2Q_1 + 6Q_2$. The profit $\Pi(Q_1, Q_2) = P_1 Q_1 + P_2 Q_2 - C(Q_1, Q_2)$ is maximized when $Q_1 = \frac{1}{16}(3P_1 - P_2)$, $Q_2 = \frac{1}{16}(3P_2 - P_1)$. Note that $\partial^2 \Pi/\partial Q_1^2 = -6 < 0$, while $(\partial^2 \Pi/\partial Q_1^2)(\partial^2 \Pi/\partial Q_2^2) - (\partial^2 \Pi/\partial Q_1 \, \partial Q_2)^2 = 32 > 0$. The larger price must not exceed three times the smaller price.

SUPPLEMENTARY
HINTS AND ANSWERS

Chapter 1

2. (b) {5}; (d) {2,3}.

5. (b) {-1,0,1,2,3,4}.

6. (b) $\emptyset$.

8. If x belongs to A, then x certainly belongs to A or B. If x belongs to both A and B, then x certainly belongs to A.

9. If x belongs to A and B, then x certainly belongs to A or B, in fact to both. Yes, if A = B.

10. (b) {1,2,3}; (d) $\emptyset$.

11. (b) and (d).

12. (a) The triangle with sides 3, 4, 5 is a right triangle, and so is the triangle with sides 3n, 4n, 5n, where n is any positive integer; (b) Note that $5^2 + 12^2 = 13^2$; (c) The interior angles of a regular polygon all equal $\frac{n-2}{n} \cdot 180°$, and hence cannot be smaller than 60°; (d) The square is a regular polygon; (e) There is no positive integer n such that $\frac{n-2}{n} \cdot 180° = 100°$.

Sec. 1.3

2. (b) $\frac{1}{2} - \frac{1}{3} = \frac{1}{6}$; (d) $\frac{1}{2} \div \frac{1}{3} = \frac{3}{2}$.

5. If $1 - \sqrt{2}$ were rational, then $1 - \sqrt{2} = \frac{m}{n}$, where m and n are

    integers. But then $\sqrt{2} = 1 - \frac{m}{n} = \frac{n - m}{n}$, where $n - m$ and n are

    integers. This is impossible, since $\sqrt{2}$ is irrational.

10. $\frac{1}{4} = 0.25$, $\frac{1}{20} = 0.05$.

11. $\frac{1}{9} = 0.\overline{1}$, $\frac{1}{11} = 0.\overline{09}$.

16. (b) $\frac{31}{99}$;   (d) $- \frac{139}{333}$.

17. Let the rational number be $\frac{m}{n}$, and carry out the long division.

    Each step of the division gives a remainder less than n. If 0

    is obtained at any step, the decimal representing $\frac{m}{n}$ terminates.

    Otherwise, since there are at most $n - 1$ nonzero remainders,

    one of the remainders must eventually repeat. But then the

    same group of digits must repeat in the quotient, provided we

    are in the part of the quotient past the decimal point.

19. The formula holds for $n = 1$, since $1^2 = \frac{2 \cdot 1^3 + 3 \cdot 1^2 + 1}{6}$

    $= \frac{2 + 3 + 1}{6}$. Suppose the formula is true for $n = k$, so that

    $1^2 + 2^2 + \cdots + k^2 = \frac{2k^3 + 3k^2 + k}{6}$. Then

    $1^2 + 2^2 + \cdots + k^2 + (k + 1)^2 = \frac{2k^3 + 3k^2 + k}{6} + (k + 1)^2$

    $= \frac{2k^3 + 3k^2 + k + 6(k^2 + 2k + 1)}{6} = \frac{2k^3 + 9k^2 + 13k + 6}{6}$

    $= \frac{2(k + 1)^3 + 3(k + 1)^2 + (k + 1)}{6}$, so that the formula also

    holds for $n = k + 1$.

Sec. 1.4

1. (b) $a - b > 0$, $c - d > 0$, and hence $(a - b) + (c - d)$

    $= (a + c) - (b + d) > 0$.

<u>3</u>. (a) $\frac{10}{3}$; (c) $\frac{167}{50}$.

<u>7</u>. Note that $1^2 = 1$, $\frac{1-1}{2} = 0$, $\sqrt{2} > 1 > 0 > -3$. We can also write $-3 \le 0 \le \frac{1-1}{2} \le 1 \le 1^2 \le \sqrt{2}$.

<u>8</u>. (a) Use Prob. 1a and the fact that $a = b$ implies $-a = -b$; (b) Examine cases, using Theorem 1.45; (c) Same hint.

<u>9</u>. Use Theorem 1.43 and the fact that $a = b$ implies $ac = bc$.

<u>10</u>. (b) 1; (d) 1; (f) -2.

<u>11</u>. (a) n; (c) n $-$ 1.

<u>13</u>. (a) Start from $(a - b)^2 \ge 0$; (c) Start from $(a - 1)^2 \ge 0$, and use Theorem 1.43 to divide by a.

<u>14</u>. Use Prob. 13b, noting that equality occurs when $a = b$ and only then. Also use Prob. 5.

<u>15</u>. A rectangle of length x and width y has perimeter $p = 2(x + y)$ and area $A = xy$. In terms of the arithmetic mean a and geometric mean g, we have $a = \frac{1}{4} p$, $g = \sqrt{A}$. Holding a (or p) fixed, we get the greatest value of g (or $\sqrt{A}$), and hence of $A = (\sqrt{A})^2$, when $x = y$.

<u>Sec. 1.5</u>

<u>2</u>. An immediate consequence of formula (1).

<u>3</u>. If $x \ge 0$, then $|x| = x$ and $|x|^2 = x^2$, while if $x < 0$, then $|x| = -x$ and $|x|^2 = (-x)^2 = x^2$.

<u>5</u>. Use mathematical induction (Sec. 1.37) and an argument like that in Prob. 4.

<u>6</u>. -3, 1.

<u>11</u>. (b) $x = 2$; (d) $x = -2$, $\frac{2}{3}$.

Sec. 1.6

2. $[2,8)$. $(\sqrt{2} - 2, \sqrt{3} - 2]$.

4. $(3 - \sqrt{2}, 3 + \sqrt{2})$. $(3 - \sqrt{2}, 3) \cup (3, 3 + \sqrt{2})$.

6. (b) $[1,\infty)$.

7. (a) $[-1,1]$;   (c) $(-1,1]$.

Sec. 1.7

4. (b) 5;   (d) 3.

5. $(1,1)$, $(-1,1)$, $(-1,-1)$, $(1,-1)$.

6. $\sqrt{[x_1 - \frac{1}{2}(x_1 + x_2)]^2 + [y_1 - \frac{1}{2}'y_1 + y_2)]^2}$

   $= \sqrt{[x_2 - \frac{1}{2}(x_1 + x_2)]^2 + [y_2 - \frac{1}{2}(y_1 + y_2)]^2}$

   $= \frac{1}{2}\sqrt{(x_1 - x_2)^2 + (y_1 - y_2)^2} = \frac{1}{2}|P_1P_2|$.

9. $(\frac{7}{2},3)$ is the midpoint of AB, $(1,\frac{9}{2})$ is the midpoint of BC, and so on.

Sec. 1.8

1. (b) $-\frac{1}{3}$;   (d) $\sqrt{3} - \sqrt{2}$.

3. (b) $90°$;   (d) $0°$.

4. (b) $\tan 100° = -\tan 80° = -5.67128$.

5. -3.

7. Let L have slope m and L' slope m'. Then $m = 2$, $m' = -\frac{1}{2}$, so that $m' = -\frac{1}{m}$.

Sec. 1.9

1.   (b) $y = 2x + 1$;   (d) $y = 2x$.

2.   (b) $y = \frac{1}{2} x + \frac{1}{4}$;   (d) $y = -\frac{1}{2} x + \frac{11}{2}$.

3.   (b) $y = 3x$;   (d) $y = -\frac{1}{2} x + \frac{3}{2}$.

4.   (b) $m = 2$, $a = -2$, $b = 4$;   (d) $m = 0$, a undefined, $b = 2$.

5.   (b) $m = -\frac{3}{2}$, $a = 0$, $b = 0$;   (d) $m = -1$, $a = -1$, $b = -1$.

9.   Below it.

10. The first two lines are parallel, the second two are perpendicular.

13. (a) $2x + y - 2 = 0$;   (c) $4x + 8y - 1 = 0$.

16. Let L be the line $Ax + by + C = 0$, and let $P_2 = (x_2, y_2)$ be the foot of the perpendicular dropped from P_1 to L. Then $d = |P_1 P_2|$. Since the slope of L equals $-\frac{A}{B}$, the slope of the line L' through P_1 and P_2 equals $\frac{B}{A}$. Hence the equation of L' is $y - y_1 = \frac{B}{A}(x - x_1)$. Since P_2 lies on L', we have $y_2 - y_1 = \frac{B}{A}(x_2 - x_1)$, or $\frac{x_2 - x_1}{A} = \frac{y_2 - y_1}{B}$. Let this last ratio be denoted by q. Then $x_2 - x_1 = Aq$, $y_2 - y_1 = Bq$, and

$$d = |P_1 P_2| = \sqrt{(x_2 - x_1)^2 + (y_2 - y_1)^2} = \sqrt{A^2 q^2 + B^2 q^2}$$
$$= \sqrt{A^2 + B^2}\,|q| .$$

But P_2 also lies on L, and hence $Ax_2 + By_2 + C = A(Aq + x_1) + B(Bq + y_1) + C = 0$, so that $q = -(Ax_1 + By_1 + C)/(A^2 + B^2)$. Substituting this value of q into the formula for d, we get the required answer.

17. (b) 5.

Chapter 2

Sec. 2.1

2. $\varphi(-2) = 14$, $\varphi(-1) = 4$, $\varphi(0) = 0$, $\varphi(\sqrt{3}) = 9 + \sqrt{3}$.

4. (b) Domain $-3 \le x \le 3$, range $0 \le y \le 3$; (d) Domain all
   $x \ne -5$, range all $y \ne 0$.

9. Take an evening paper dated d, and look up P in the financial
   section. The function is undefined on days when the exchange
   is closed.

12. No.

15. $V = \ell wh$.

17. Let x be the temperature in degrees Centigrade and y the
    temperature in degrees Fahrenheit. Then $y = \frac{9}{5} x + 32$,
    $x = \frac{5}{9} (y - 32)$. The missing entries are $x = 40$ and $y = 176$.

23. Convince yourself that every "rule" or "procedure" associating
    a unique value of y with each given value of x is in effect a
    set of ordered pairs of the type described.

29. The one-to-one function $f(n) = n + 1$ maps the even numbers into
    the odd numbers.

30. (b) Finite; (d) Infinite.

Sec. 2.2

2. $a = 4$, $b = -1$.

4. $f(f(x)) = x$, $f(g(x)) = 1/x^2$, $g(f(x)) = 1/x^2$, $g(g(x)) = x^4$.

7. (b) $\frac{1}{2}, \frac{1}{6}, \frac{1}{12}, \frac{1}{20}, \frac{1}{30}$; (d) $1, 1, \frac{1}{3}, 1, \frac{1}{5}$.

10. $a_1 = 4$, $a_3 = 4$, $a_4 = 2$, $a_7 = 5$.

12. No. For example, let $f(x) = \begin{cases} 0 \text{ if } x \geq 0, \\ 1 \text{ if } x < 0, \end{cases}$ $g(x) = \begin{cases} 1 \text{ if } x \geq 0, \\ 0 \text{ if } x < 0. \end{cases}$

Sec. 2.3

2. The circle of radius 1 with its center at the point $(-1,1)$.

8. The graph of $f(x) + c$ is obtained by shifting G a distance c upward if $c > 0$ and a distance $|c|$ downward if $c < 0$. The graph of $f(x + c)$ is obtained by shifting G a distance c to the left if $c > 0$ and a distance $|c|$ to the right if $c < 0$.

11. For example, if $f(x)$ and $g(x)$ are odd, then $f(-x) \equiv -f(x)$, $g(-x) \equiv -g(x)$, and hence $f(-x)g(-x) \equiv f(x)g(x)$, so that $f(x)g(x)$ is even.

17. Yes, The function is increasing in the interval $1 \leq x < \infty$, decreasing in the interval $-\infty < x \leq -1$, and constant in the interval $-1 \leq x \leq 1$.

18. Yes. The graph has corners at the points $(-2,3)$, $(-1,2)$ and $(0,3)$. The function is increasing in the interval $-1 \leq x < \infty$ and decreasing in the interval $-\infty < x \leq 1$.

Sec. 2.4

1. $f'(x_0) = \lim\limits_{h \to 0} \dfrac{a(x_0 + h)^2 + b(x_0 + h) + c - ax_0^2 - bx_0 - c}{h}$

$= \lim\limits_{h \to 0} \dfrac{2ax_0h + ah^2 + bh}{h} = \lim\limits_{h \to 0} (2ax_0 + b + ah) = 2ax_0 + b.$

2. $f'(x_0) = \lim_{h \to 0} \dfrac{(x_0 + h)^3 - x_0^3}{h} = \lim_{h \to 0} \dfrac{3x_0^2 h + 3x_0 h^2 + h^3}{h}$

$\quad = \lim_{h \to 0} (3x_0^2 + 3x_0 h + h^2) = 3x_0^2.$

4. For example, if $f(x) = x^2$, then $f'(x_0) = f(x_0)$ for $x_0 = 0$ or 2.

6. (b) 1;   (d) 0.

8. (a) 1;   (c) 2.

11. If $|f(x) - A|$ is "arbitrarily small," then so $\big||f(x)| - |A|\big|$,

    since $\big||f(x)| - |A|\big| \le |f(x) - A|$, by Sec. 1.5, Prob. 10.   The

    converse is false; for example, $\lim_{x \to 0} \dfrac{|x|}{x}$ does not exist

    (Example 2.45e), but $\lim_{x \to 0} \left|\dfrac{|x|}{x}\right| = 1$.

Sec. 2.5

2. $\Delta(u + v) = [u(x + \Delta x) + v(x + \Delta x)] - [u(x) + v(x)]$

    $\quad = [u(x + \Delta x) - u(x)] + [v(x + \Delta x) - v(x)] = \Delta u + \Delta v.$

3. Here $f(x) = mx + b$, $f(x_0) = mx_0 + b$, $f'(x_0) = m$, so that (3)

    becomes $y = m(x - x_0) + mx_0 + b = mx + b$.

5. Yes, the tangents at any pair of points (x_1, x_1^2), (x_2, x_2^2) such

    that $x_1 x_2 = -\dfrac{1}{4}$ are perpendicular.   No.

9. (b) $\Delta y = 0.331$, $dy = 0.3$, $E = 0.031$, about 9% of Δy;

    (d) $\Delta y = 0.003003001$, $dy = 0.003$, $E = 0.000003001$, about 0.1%

    of Δy.

13. In the intervals $-\infty < x < -1$, $-1 < x < 1$, $1 < x < \infty$.   At the

    points $(-1, 2)$, $(1, 2)$.

14. If S is the surface area of the earth and R its radius (≈ 4000

    miles), then $S = 4\pi R^2$.   Therefore $\Delta S \approx dS = 8\pi R \Delta R = \dfrac{4000}{5280}$ square

    miles.

Sec. 2.6

1. Let $g(x) \equiv c$.

3. We get the indeterminate form $\frac{0}{0}$.

4. (b) -3;   (d) 23.

6. $-\frac{3}{2}$.

11. If $|f(x) - A|$ is "arbitrarily small" both for all "sufficiently
    small" $x_0 - x > 0$ and all "sufficiently small" $x - x_0 > 0$, then
    $|f(x) - A|$ is "arbitrarily small" for all "sufficiently small"
    $|x - x_0| > 0$, and conversely.

12. Use Prob. 11.

14. Use Sec. 2.4, Prob. 11.

Sec. 2.7

1. (b) $6ax^2 - 2bx$;   (d) $-\dfrac{1}{x^2} - \dfrac{4}{x^3} - \dfrac{9}{x^4}$

2. (b) $3x^2 - 2(a + b)x + ab$;   (d) $6x^2 - 26x + 12$.

3. (b) $\dfrac{2(1 + x^2)}{(1 - x^2)^2}$;   (d) $\dfrac{x^2 - 4x + 1}{(x - 2)^2}$.

5. If n is odd and $t^n = x$, then $(-t)^n = -x$, so that $\sqrt[n]{-x} = -\sqrt[n]{x}$.

6. If n is odd, then $(\sqrt[n]{-x})^m = (-\sqrt[n]{x})^m = (-1)^m (\sqrt[n]{x})^m$.

7. Use Prob. 6 and the fact that division by zero is impossible.

8. (b) $-\dfrac{1}{3\sqrt[3]{x^4}}$;   (d) $\dfrac{2}{3\sqrt[3]{x}}$.

11. $f'''g + 3f''g' + 3f'g'' + g'''$.

13. The tangent T to the curve $y = 1/x$ at the point $P_0 = (x_0, 1/x_0)$
    has equation $y = -\dfrac{1}{x_0^2} (x - x_0) + \dfrac{1}{x_0} = -\dfrac{x}{x_0^2} + \dfrac{2}{x_0}$.   Therefore T
    has x-intercept $2x_0$ and y-intercept $2/x_0$.   Now use Sec. 1.7,
    Prob. 6.

14. By Sec. 2.66, g is continuous at x. But $g(x) \neq 0$, by hypothesis, and hence $g(x + \Delta x) \neq 0$ for all "sufficiently small" $|\Delta x|$, by Sec. 2.4, Prob. 15.

Sec. 2.8

2. Use the fact that $\sqrt{x}$ is an increasing function.

3. (a) $2(x + 2)(x + 3)^2(3x^2 + 11x + 9)$;   (c) $\dfrac{7 - 3x}{(1 - x)^3}$.

5. For example, if f is even, then $f(-x) \equiv f(x)$, so that $\dfrac{d}{dx} f(-x) \equiv \dfrac{d}{dx} f(x)$, and hence $-f'(-x) \equiv f'(x)$ or $f'(-x) \equiv -f'(x)$.

6. $(\frac{f}{g})' = (f \frac{1}{g})' = f' \frac{1}{g} + f(-\frac{1}{g^2})g' = \dfrac{f'g - fg'}{g^2}$.

7. (b) $\dfrac{1}{(1 - x)\sqrt{1 - x^2}}$.

10. Note that $\dfrac{1}{x(1 - x)} = \dfrac{1}{x} + \dfrac{1}{1 - x}$.

12. $y' = -\dfrac{2x + y}{x + 3y^2}$, $y'|_{x=1,y=0} = -2$.

16. $ny^{n-1}y' = mx^{m-1}$, and therefore $y' = \dfrac{mx^{m-1}}{ny^{n-1}} = \dfrac{m}{n} \dfrac{x^{m-1}}{(x^{m/n})^{n-1}}$

$= \dfrac{m}{n} x^{(m/n)-1}$.

18. Solving the quadratic equation (14) for y, we get

$y = \frac{1}{2}(x \pm \sqrt{4 - 3x^2})$, and hence $y' = \frac{1}{2}\left(1 \mp \dfrac{3x}{\sqrt{4 - 3x^2}}\right)$, so that

$y'|_{x=1} = \frac{1}{2}(1 \mp 3) = -1, 2$ when $y|_{x=1} = \frac{1}{2}(1 \pm 1) = 1, 0$.

20. In the theory of equations, it is shown that the cubic equation $y^3 + ay + b = 0$ has three distinct real roots if $\dfrac{b^2}{4} + \dfrac{a^3}{27} < 0$. There is only one real root if $\dfrac{b^2}{4} + \dfrac{a^3}{27} > 0$. Use this to investigate equations (11) and (13).

Sec. 2.9

<u>1</u>.   (b) -1.

<u>2</u>.   (b) ∞.

<u>4</u>.   The limit is a product of five limits, all equal to $\frac{1}{5}$.

<u>7</u>. $x < -1000$.

<u>9</u>.   (b) $x = -d/c$, $y = a/c$.

<u>10</u>. $f(x) = \dfrac{1}{(x - a_1)(x - a_2)\cdots(x - a_n)}$, where the constants

$a_1, a_2, \cdots, a_n$ are all different.

<u>12</u>.   Consider the separate branches of the function graphed in

Figure 15.

<u>13</u>.   (b) 0;   (d) No limit.

<u>15</u>.   20.

<u>18</u>.   (b) $1! + 2! + 3! + 4! + 5! + 6! = 1 + 2 + 6 + 24 + 120 + 720$

$= 873$.

<u>19</u>.   (b) 1.

<u>22</u>.   Let $x_n = \frac{1}{n}$.   Then $x_1 = 1$, $x_2 = \frac{1}{2}$, $x_3 + x_4 = \frac{1}{3} + \frac{1}{4} > \frac{2}{4} = \frac{1}{2}$,

$x_5 + x_6 + x_7 + x_8 = \frac{1}{5} + \frac{1}{6} + \frac{1}{7} + \frac{1}{8} > \frac{1}{8} + \frac{1}{8} + \frac{1}{8} + \frac{1}{8} = \frac{4}{8} = \frac{1}{2}$,

and so on.   Therefore $s_1 = x_1 > \frac{1}{2}$, $s_2 = x_1 + x_2 > 2 \cdot \frac{1}{2}$,

$s_4 = x_1 + x_2 + x_3 + x_4 > 3 \cdot \frac{1}{2}$, $s_8 = x_1 + x_2 + \cdots + x_8 > 4 \cdot \frac{1}{2}$,

and so on.   Thus the sequence s_n is unbounded and hence

divergent.

<u>23</u>.   No, as shown by the example of the harmonic series.

Chapter 3

Sec. 3.1

2. $v = t^2 - 4t + 3$, $a = 2t - 4$. The direction of motion changes
   at $t = 1$ and $t = 3$. The particle returns to its initial
   position at $t = 3$.

5. 128 ft. 240 ft.

8. Differentiating (8), we find that the velocity after braking is
   $v_0 - kt$. Hence it takes a time equal to v_0/k to bring the car
   to a stop, during which it travels a distance equal to
   $$v_0(v_0/k) - \frac{1}{2} k(v_0/k)^2 = \frac{1}{2} (v_0^2/k).$$

10. The flywheel stops rotating when $t = b/2c$.

Sec. 3.2

2. $\dfrac{90}{\sqrt{13}} \approx 25$ mi/hr.

4. x increases faster if $x < 4$, y increases faster if $x > 4$,
   x and y increase at the same rate if $x = 4$.

6. 2 ft/sec. No.

8. The curve of marginal cost is the straight line
   $MC(Q) = a - 2mQ$.

11. 5 ft^2/min.

Sec. 3.3

1. (b) No maximum, a minimum equal to 1 at $x = 1$; (d) No extrema.

2. (b) A maximum equal to 1 at $x = 1$, a minimum equal to 0 at every point of $(0,1)$; (d) No maximum, a minimum equal to 0 at every point of $(0,1)$.

3. If f is increasing in $[a,b]$, f has its minimum at a and its maximum at b, while if f is decreasing in $[a,b]$, f has its maximum at a and its minimum at b.

6. No.

Sec. 3.4

2. f is not differentiable at $x = 0$.

4. $(1,1)$, $(-1,-1)$.

6. $\alpha = 1 - \dfrac{1}{\sqrt{3}} \approx 0.42$.

8. $c = \dfrac{1}{2}$, $\sqrt{2}$.

Sec. 3.5

2. (a) $\dfrac{1}{4} x^4 + C$; (c) $\dfrac{2}{3} \sqrt{x^3} + C$.

5. (b) $x - 3x^2 + \dfrac{11}{3} x^3 - \dfrac{3}{2} x^4 + C$;

   (d) $\dfrac{1}{3} x^3 - \dfrac{6}{7} x^{7/2} + \dfrac{3}{4} x^4 - \dfrac{2}{9} x^{9/2} + C$.

7. No.

8. (a) Increasing in $(-\infty, \dfrac{1}{2}]$, decreasing in $[\dfrac{1}{2}, \infty)$; (c) Increasing in $[-1,1]$, decreasing in $(-\infty,-1]$ and $[1,\infty)$.

Sec. 3.6

2. (b) Maximum $y = \dfrac{9}{4}$ at $x = \dfrac{1}{2}$.

3. (a) Maximum $y = 1$ at $x = \pm 1$, minimum $y = 0$ at $x = 0$; (c) No

extremum at $x = 0$, maximum $y = \frac{1}{3}\sqrt[3]{4}$ at $x = \frac{1}{3}$, minimum $y = 0$ at

$x = 1$.

4. (b) Maximum $y = 100.01$ at $x = 0.01$, 100, minimum $y = 2$ at $x = 1$;

(d) Maximum $y = 132$ at $x = -10$, minimum $y = 0$ at $x = 1$, 2.

5. $|3x - x^3|$ has its maximum in $[-2,2]$ at the points $x = \pm 1$, ± 2.

7. Minimum $y = 0$ at $x = 0$ if m is even and no extremum at $x = 0$

if m is odd, maximum $y = m^n n^n / (m + n)^{m+n}$ at $x = m/(m + n)$,

minimum $y = 0$ at $x = 1$ if n is even and no extremum at $x = 1$

if n is odd.

9. Solve the equations $y|_{x=2} = -1$, $y'|_{x=2} = 0$ to get $a = 1$, $b = 0$.

Then show that $y''|_{x=2} < 0$.

Sec. 3.7

1. True.

2. For example, if $f(x) = x^3$, $g(x) = x^4$, $h(x) = -x^4$, then $f''(0)$

$= g''(0) = h''(0) = 0$, but f has an inflection point at $x = 0$,

g is concave upward at $x = 0$, h is concave downward at $x = 0$.

3. No.

5. Inflection points at $x = 0$, $\pm 3a$, concave upward in $(-\infty,-3a)$,

concave downward in $(-3a,0)$, concave upward in $(0,3a)$, concave

downward in $(3a,\infty)$.

7. $a = -\frac{3}{2}$, $b = \frac{9}{2}$.

10. The points have abscissas 1, $-2\pm\sqrt{3}$, the solutions of the

equation $x^3 + 3x^2 - 3x - 1 = 0$.

13. Reread Secs. 3.33b, 3.64a and 3.72.

Sec. 3.8

4. $4\pi R^3/3\sqrt{3}$

7. No, even if the buggy is much faster than the boat.

8. $x = \frac{1}{n} (x_1 + x_2 + \cdots + x_n)$.

11. If $\Pi(Q)$ has a local extremum at $Q = Q_0$, then $\Pi'(Q_0)$
$= R'(Q_0) - C'(Q_0) = MR(Q_0) - MC(Q_0) = 0$, so that $MR(Q_0)$
$= MC(Q_0)$. This extremum will be a maximum if $\Pi''(Q_0)$
$= R''(Q_0) - C''(Q_0) < 0$, that is, if $MR'(Q_0) < MC'(Q_0)$.

13. Overhead is positive, and hence $d > 0$. The marginal cost is
$MC(Q) = 3aQ^2 + 2bQ + c$, with first derivative $MC'(Q)$
$= 6aQ + 2b$ and second derivative $MC''(Q) = 6a$. Therefore $MC(Q)$
has a local minimum at $Q_0 = -b/3a$ if $a > 0$. But $Q_0 > 0$, and
hence $b < 0$. Moreover, $MC(Q_0) = \frac{3ac - b^2}{3a} > 0$, and hence
$b^2 < 3ac$, which, in particular, implies $c > 0$.

16. Let the sides of the angle be the x-axis and the line $y = mx$.
The line through $P = (a,b)$ with slope λ intersects the x-axis
in the point $\left(\frac{a\lambda - b}{\lambda}, 0\right)$ and the line $y = mx$ in the point
$\left(\frac{a\lambda - b}{\lambda - m}, m\frac{a\lambda - b}{\lambda - m}\right)$, forming a triangle of area
$m(a\lambda - b)^2/2\lambda(\lambda - m)$.

17. Choosing $A = (a,b)$, $P = (x,0)$ and $B = (c,d)$, minimize
$|AP| + |PB| = \sqrt{(x - a)^2 + b^2} + \sqrt{(x - c)^2 + d^2}$ (the speed of
light can be cancelled out). The minimum is achieved when x
satisfies the condition $\frac{x - a}{|AP|} = \frac{c - x}{|PB|}$. Now use similar
triangles.

Chapter 4

Sec. 4.1

1. (b) 1.

2. The global maximum of f in [a,b], whose existence is guaranteed by Theorem 3.32c.

3. No. Yes, since the number of subintervals cannot be less than the integral part of $(b - a)/\lambda$.

5. If $f(x) = c$, the region bounded by the curve $y = f(x)$, the x-axis, and the lines $x = a$ and $x = b$ is a rectangle of length $b - a$ and width c; then $A = \int_a^b f(x)\,dx = \int_a^b c\,dx = c(b - a)$, by (6) and (11). If $a = 0$, $f(x) = cx$, the region bounded by the curve $y = f(x)$, the x-axis, and the lines $x = 0$ and $x = b$ is a right triangle with legs b and cb; then $A = \int_a^b f(x)\,dx = \int_0^b cx\,dx = \frac{1}{2}\,cb^2 = \frac{1}{2}\,b \cdot cb$, by (6) and (12). Both results are in keeping with elementary geometry.

6. $A = \int_2^4 [f(x) - g(x)]\,dx = \int_2^4 (x - 1)\,dx = \int_2^4 x\,dx - \int_2^4 dx = \frac{1}{2}(4^2 - 2^2) - (4 - 2) = 6 - 2 = 4$. But $A = \frac{1}{2} \cdot 2(1 + 3)$, by elementary geometry.

8. $f(x)$ is continuous in every such interval.

11. (a) Use the same argument as in Example 2.45e, noting that $f(x)$ takes both values 1 and -1 in every deleted neighborhood of c, since every such neighborhood contains both rational and irrational points (this is a consequence of Sec. 1.5, Prob. 13).

Sec. 4.2

1. $0 = -\frac{1}{4} + \frac{1}{4}$.

2. (b) 6;   (d) 1.

3. $\frac{n+1}{2} \int_{-1}^{1} x^n dx = \frac{1}{2} x^{n+1} \Big|_{-1}^{1} = \frac{1}{2} [1 - (-1)^{n+1}]$.

6. $\frac{9}{2}$.

8. $\frac{1}{b-a} \int_{a}^{b} x dx = \frac{b^2 - a^2}{2(b-a)} = \frac{1}{2} (a + b)$.

10. Yes, provided that $x > 0$.

14. (a) 0;   (c) $f(b)$.

16. Choose $A = 0$ in Prob. 15.

18. Clearly $f(c) > 0$. Suppose $a < c < b$. Then there is an interval $[c - \delta, c + \delta]$ such that $f(x) > 0$ for every $x \in [c - \delta, c + \delta]$ (why?). By Theorem 4.21a,

$$\int_{a}^{b} f(x)dx = \int_{a}^{c-\delta} f(x)dx + \int_{c-\delta}^{c+\delta} f(x)dx + \int_{c+\delta}^{b} f(x)dx .$$

The first and third integrals on the right are nonnegative, by Prob. 16, while the second integral is positive, by the mean value theorem for integrals. Therefore the integral on the left is also positive. The proof is even simpler if $c = a$ or $c = b$.

19. Use Prob. 18.

20. Apply Prob. 18 to the function $f = f_1 - f_2$.

22. Use Prob. 17, noting that $-|f(x)| \le f(x) \le |f(x)|$, where, by Sec. 2.6, Prob. 14, $|f(x)|$ is continuous and hence integrable in $[a,b]$.

23. The assertion is obviously true if f is a constant function.
Otherwise f takes values between its maximum M and its
minimum m in [a,b]. But then M - f(x) > 0 at some point in
[a,b], and hence $\int_a^b [M - f(x)]dx > 0$, by Prob. 18, or

equivalently $\frac{1}{b - a} \int_a^b f(x)dx < M$. In the same way, we find that

$m < \frac{1}{b - a} \int_a^b f(x)dx$. Continuing as in the proof of Theorem

4.22a, we observe that the point c is now known to lie between
the points p and q at which f takes its maximum and minimum,
so that $c \in (a,b)$.

Sec. 4.3

1. Because r = -1.

3. (b) x > 1.

4. (b) ln x + 1; (d) $\frac{1}{x \ln x}$.

5. (b) $\frac{4x}{1 - x^4}$; (d) $\frac{1}{\sqrt{1 + x^2}}$.

7. The tangent has equation y = x/e.

9. x ln x - x is an antiderivative of ln x.

11. Increasing in [-1,0) and [1,∞), decreasing in (0,1] and
(-∞,-1].

13. Inflection points at x = ±1, concave downward in (-∞,-1),
concave upward in (-1,1), concave downward in (1,∞).

14. (b) 2 < x < 3.

15. Use (12), (8) and (9), noting that $\ln a > 0$ if $a > 1$, while $\ln a < 0$ if $0 < a < 1$.

17. Let $f(x) = \ln x$. Then $\ln b - \ln a = (b - a)f'(c) = \dfrac{b - a}{c}$, where $a < c < b$.

Sec. 4.4

1. If $c = \ln k$, then $ke^x = e^c e^x = e^{c+x}$.

2. (b) $-3e^{-3x}$;   (d) $e^x(1 - 2x - x^2)$.

3. (b) $\dfrac{2e^x}{(e^x + 1)^2}$;   (d) $\dfrac{e^x}{2\sqrt{1 + e^x}}$.

5. e^{ax}/a is an antiderivative of e^{ax}.

7. Maximum $y = 10^{10}e^{-9}$ at $x = 9$, minimum $y = 0$ at $x = 1$.

10. $x = 1, 2$.

12. (b) $2xe^{x^2}\ln a$;   (d) $-\dfrac{10^x \ln 100}{(1 + 10^x)^2}$.

15. $x = \frac{1}{2}(2^y - 2^{-y})$.

17. The function $\dfrac{x^{10}}{2^x}$ has its maximum at $x = \dfrac{10}{\ln 2} \approx 14.4$.   Now compare y_{14} with y_{15}.

Sec. 4.5

1. (b) e.

2. (b) e^2.

3. $\displaystyle\lim_{x \to 0} \dfrac{\log_a(1 + x)}{x} = \dfrac{1}{\ln a}\lim_{x \to 0}\dfrac{\ln(1 + x)}{x} = \dfrac{1}{\ln a} = \log_a e$ (Sec. 4.36b).

5. $\lim\limits_{x\to 1} \dfrac{x^r - 1}{x - 1} = \lim\limits_{x\to 1} \dfrac{e^{r\ln x} - 1}{r \ln x} \dfrac{r \ln x}{x - 1} = r \ln e \cdot \lim\limits_{x\to 1} \dfrac{\ln x}{x - 1}$

   $= r \lim\limits_{t\to 0} \dfrac{\ln (1 + t)}{t} = r.$

6. (b) $\ln 4$.

8. $\dfrac{1}{10} \ln 2 \approx 6.93\%$.

10. $7,408.18

13. (b) $\dfrac{e^{x^2+2x}}{x^{4/3} \ln x} \left(2x + 2 - \dfrac{4}{3x} - \dfrac{1}{x \ln x} \right).$

14. (b) $x^{1/x} \dfrac{1 - \ln x}{x^2}$; (d) $e^{x^x} x^x (\ln x + 1).$

15. y'' is nonvanishing.

17. $ax.$

19. If $y = f(x)$, then $\varepsilon_{yx} = \dfrac{x}{y} \dfrac{dy}{dx}$. The function $xf(x) = xy$ has

   elasticity $\dfrac{x}{xy} (xy)' = \dfrac{1}{y} (y + xy') = 1 + \dfrac{x}{y} \dfrac{dy}{dx} = 1 + \varepsilon_{yx}.$

22. The sum of 8 terms of the series is $2.71825...$

Sec. 4.6

2. Let $x = -t.$

3. Let $x = 1 - t.$

4. (b) $\dfrac{1}{2} e^{x^2} + C$; (d) $\dfrac{2}{3} (1 + \ln x)^{3/2} + C.$

5. (b) $\dfrac{1}{2} \ln |x^2 + 2x - 3| + C$; (d) $\dfrac{1}{2} \ln (e^{2x} + 1) + C.$

7. $\dfrac{1}{e - 1}.$

8. (b) $x^3 e^x - 3x^2 e^x + 6xe^x - 6e^x + C$;

   (d) $\dfrac{2}{3} x^{3/2} \ln x - \dfrac{4}{9} x^{3/2} + C.$

10. (b) 1; (d) $e - 2.$

13. (b) $\dfrac{1}{8} \ln \left| \dfrac{2x - 1}{2x + 3} \right| + C.$

16. Note that $x = \dfrac{t^2 - 1}{2t}$, $\sqrt{1 + x^2} = \dfrac{t^2 + 1}{2t}$, $dx = \dfrac{t^2 + 1}{2t^2} dt$.

17. (a) $\displaystyle\int \dfrac{e^{3x} + 1}{e^x + 1} dx = \int (e^{2x} - e^x + 1) dx = \dfrac{1}{2} e^{2x} - e^x + x + C$;

   (c) $\displaystyle\int \dfrac{x^2}{1 - x^2} dx = \int \left(-1 + \dfrac{1}{2} \dfrac{1}{1 + x} + \dfrac{1}{2} \dfrac{1}{1 - x} \right) dx$

   $= -x + \dfrac{1}{2} \ln \left| \dfrac{1 + x}{1 - x} \right| + C.$

22. Divide the price range $[P_1, P_0]$ into n equal small units
   $\Delta P = \dfrac{1}{n} (P_0 - P_1)$, where ΔP is just large enough so that each
   successive price drop causes more of the commodity to be sold.
   Let ΔQ_i be the extra quantity sold when the price is lowered
   from $P_0 - (i - 1) \Delta P$ to $P_0 - i \Delta P$. Then the total revenue
   received in the course of the staged price drop is just
   $\displaystyle\sum_{i=1}^{n} (P_0 - i\Delta P) \Delta Q_i$, which approximates the integral in (25). To
   convert (25) into (26), integrate by parts.

23. $100 P_0 (1 - \dfrac{2}{e}) \approx 26.4 P_0.$

Sec. 4.7

2. (a) $\dfrac{1}{2}$; (c) $\dfrac{1}{\ln 2}$.

4. (a) Divergent; (c) 6.

7. By Prob. 3, $A = \dfrac{1}{1 - \dfrac{1}{2}} - \dfrac{1}{1 - \dfrac{1}{3}} = 2 - \dfrac{3}{2} = \dfrac{1}{2}$.

8. By Sec. 4.6, Prob. 15, $A = \displaystyle\int_0^{\infty} (\cosh x - \sinh x) dx$

   $= \displaystyle\lim_{X \to \infty} \int_0^X (\cosh x - \sinh x) dx = \lim_{X \to \infty} \left[\sinh X - \cosh X \right]_0^X$

   $= 1 + \displaystyle\lim_{X \to \infty} (\sinh X - \cosh X) = 1 + \lim_{X \to \infty} (-e^{-X}) = 1.$

Chapter 5

Sec. 5.1

2. $y = 2/x$.

4. $y = (x \ln x - x + 1)^2$.

6. $y = \frac{1}{2} x^2 (\ln x - \frac{3}{2}) + C_1 x + C_2$.

9. $y = \frac{1 - x^2}{2x}$.

Sec. 5.2

2. 4.8 billion. 6.9 billion.

4. $\dfrac{20 \ln 2}{\ln \frac{4}{3}} \approx 48$ yrs.

6. (a), (b) and (c) follow at once from formula (13). To prove
    (d), differentiate (11), obtaining $N'' = rN' - 2sNN'$
    $= (r - 2sN)N' = (r - 2sN)(r - sN)N$, where the expression on
    the right is positive for $N < r/2s$, zero for $N = r/2s$, and
    negative for $r/2s < N < r/s$. Now use Sec. 3.72, Proposition (4).

7. Per capita consumption is constant if $r = s$, grows
    exponentially at the rate of $r - s$ percent per year if $r > s$,
    and decays exponentially at the rate of $s - r$ percent per year
    if $r < s$.

9. $\mu = \dfrac{3 \ln 10}{5} \approx 1.4$ per meter.

11. Let $N^* = N + \frac{s}{r}$. Then $\frac{dN^*}{dt} = rN^*$, and hence $N^* = Ce^{rt}$, or
    $N = Ce^{rt} - \frac{s}{r}$. Applying the initial condition $N|_{t=0} = N_0$, we
    get $C = N_0 + \frac{s}{r}$.

13. About 2310 yrs.

Sec. 5.3

2. $v_0 = \dfrac{s_1 - s_0}{t_1 - t_0} - \dfrac{F}{2m}(t_1 - t_0)$.

4. 64 ft/sec. 4 sec.

6. The truck has more kinetic energy in the first case, the bullet in the second.

8. Let the fixed points be s = ±a. Then the force is F(s)
   = -k(s + a) - k(s - a) = -2ks. Hence the work done in going
   from -a to a is $-2k\displaystyle\int_{-a}^{a} s\,ds = 0$.

10. About 84 mi.

12. $v_1 = \sqrt{v_0^2 + 2gh}$, $t_1 = (-v_0 + v_1)/g$, $t_2 = (v_0 + v_1)g$, $\Delta t = 2v_0/g$.

13. Since the spider's weight mg stretches the strand by an amount
    s, the tension ks in the strand satisfies the condition
    ks = mg, so that k = mg/s. Therefore the potential energy of
    the stretched strand is $\frac{1}{2}ks^2 = \frac{1}{2}mgs$ (Prob. 7). As a result
    of the spider's climb, the potential energy of the system
    consisting of the spider and the strand changes by
    $2mgs - \frac{1}{2}mgs = \frac{3}{2}mgs$, since the strand is no longer stretched
    after the climb. This is the work W_1 done by the spider in
    climbing up the strand, to be compared with the work $W_2 = 2mgs$
    done by the spider in climbing up an inelastic strand of
    length 2s.

Chapter 6

Sec. 6.1

2. $(1,1,-1)$, $(1,-1,1)$, $(-1,1,1)$, $(-1,-1,1)$.

3. (b) 14.

4. Show that the side lengths satisfy the Pythagorean theorem.

6. Complete the squares, as in Sec. 2.32b.

8. (b) The cylinder $x^2 + y^2 = a^2$.

9. (b) $z = 1 - |y|$ $(-1 \le y \le 1)$.

11. $-1 < x < 1$, $-1 < y < 1$. The region is an "open square."

13. $|P_1P_2| = 2$. P_1 is closer to the origin.

Sec. 6.2

2. If the limit in question exists, then $\lim_{x \to 0} \frac{x}{x} = \lim_{y \to 0} \frac{-y}{y}$, which is impossible.

7. (a) $\frac{1}{2}$;   (c) $-\ln 2$.

9. In the first and third quadrants of the xy-plane.

11. (a) $\frac{\partial z}{\partial x} = 2xy^3 + 3x^2y^2$, $\frac{\partial z}{\partial y} = 3x^2y^2 + 2x^3y$;   (c) $\frac{\partial z}{\partial x} = -\frac{1}{y} e^{-x/y}$,

   $\frac{\partial z}{\partial y} = \frac{x}{y^2} e^{-x/y}$.

12. (a) $\frac{\partial u}{\partial x} = yze^{xyz}$, $\frac{\partial u}{\partial y} = xze^{xyz}$, $\frac{\partial u}{\partial z} = xye^{xyz}$;   (c) $\frac{\partial u}{\partial x} = \frac{z}{x} (xy)^z$,

   $\frac{\partial u}{\partial y} = \frac{z}{y} (xy)^z$, $\frac{\partial u}{\partial z} = (xy)^z \ln (xy)$.

13. (b) $(1 + 3xyz + x^2y^2z^2)e^{xyz}$.

14. (b) $\frac{\partial^2 z}{\partial x \partial y} = \frac{\partial}{\partial x} \frac{1}{y} = 0 = \frac{\partial}{\partial y} \frac{1}{x} = \frac{\partial^2 z}{\partial y \partial x}$.

15. (b) $\Delta f(x,y) = (x + \Delta x)(y + \Delta y) - xy = y\Delta x + x\Delta y$ is of the form

 (8) with $A = y$, $B = x$, $\alpha(\Delta x, \Delta y) = \beta(\Delta x, \Delta y) = 0$.

16. (b) $dz = y^x \ln y \, dx + xy^{x-1} dy$.

17. (b) 108.432.

18. Let $r = \sqrt{x^2 + y^2}$. Then $\dfrac{\partial u}{\partial x} = -\dfrac{x}{r^2}$, $\dfrac{\partial^2 u}{\partial x^2} = -\dfrac{1}{r^2} + \dfrac{2x^2}{r^4}$

 $= \dfrac{x^2 - y^2}{r^4}$. Interchanging the roles of x and y, we get

 $\dfrac{\partial^2 u}{\partial y^2} = \dfrac{y^2 - x^2}{r^4}$.

19. No, since the partial derivatives $f_x(0,0)$ and $f_y(0,0)$ do not

 exist (why not?).

Sec. 6.3

1. (b) 0.

2. (b) $\dfrac{\partial z}{\partial x} = \dfrac{2}{x}$, $\dfrac{\partial z}{\partial y} = -\dfrac{2}{y}$.

3. (b) $\dfrac{\partial z}{\partial x} = y \dfrac{\partial z}{\partial u} - \dfrac{y}{x^2} \dfrac{\partial z}{\partial v}$, $\dfrac{\partial z}{\partial y} = x \dfrac{\partial z}{\partial u} + \dfrac{1}{x} \dfrac{\partial z}{\partial v}$.

5. (b) 1; (d) 0.

8. If $F(u,v) = \displaystyle\int_u^v f(t) dt$, then $\dfrac{\partial F}{\partial v} = f(v)$, by Theorem 4.23a, while

 $\dfrac{\partial F}{\partial u} = -\dfrac{\partial}{\partial u} \displaystyle\int_v^u f(t) dt = -f(u)$. But $\dfrac{dF}{dx} = \dfrac{\partial F}{\partial u} \dfrac{du}{dx} + \dfrac{\partial F}{\partial v} \dfrac{dv}{dx}$, by the

 chain rule and hence $\dfrac{dF}{dx} = \dfrac{d}{dx} \displaystyle\int_{u(x)}^{v(x)} f(t) dt$

 $= f(v(x)) \dfrac{dv}{dx} - f(u(x)) \dfrac{du}{dx}$.

Sec. 6.4

1. (a) No extrema;   (c) Minimum $z = 0$ at $(x,y) = (0,0)$, no

    extrema at $(x,y) = (1,\pm 4)$, $(-\frac{5}{3},0)$.

2. (a) Maximum $z = 864$ at $(x,y) = (-6,0)$, minimum $z = -864$ at

    $(x,y) = (6,0)$, no extrema at $(x,y) = (0,\pm 6\sqrt{6})$;   (c) Minimum

    $z = 0$ at every point of the line $y = x + 1$.

3. Maximum $z = 4$ at $(x,y) = (\pm 2,0)$, minimum $z = -4$ at

    $(x,y) = (0,\pm 2)$.

7. Let $u = x_1 x_2 \cdots x_n$, $u^* = u - \lambda(x_1 + x_2 + \cdots + x_n - nc)$.
    Then, at any critical point,

    $$\frac{\partial u^*}{\partial x_1} = x_2 \cdots x_n - \lambda = \frac{u}{x_1} - \lambda = 0,$$

    $$\bullet \quad \bullet \quad \bullet \quad \bullet \quad \bullet \quad \bullet \quad \bullet \quad \bullet \quad \bullet \quad \bullet \quad \bullet \quad \bullet$$

    $$\frac{\partial u^*}{\partial x_n} = x_1 \cdots x_{n-1} - \lambda = \frac{u}{x_n} - \lambda = 0,$$

    so that

    $$\frac{u}{x_1} = \cdots = \frac{u}{x_n} = \lambda,$$

    which implies $x_1 = \cdots = x_n = c$.

8. Maximize $u = \sqrt[n]{x_1 x_2 \cdots x_n}$ subject to the condition

    $x_1 + x_2 + \cdots + x_n = nc$.

13. The profit is maximized when $Q_1 = \frac{1}{8}(P_1 - P_2 + 4q)$,

    $Q_2 = \frac{1}{8}(P_2 - P_1 + 4q)$. The absolute value of the price

    difference must not exceed $4q$.

INDEX

A CATALOG OF SELECTED
DOVER BOOKS
IN SCIENCE AND MATHEMATICS

A CATALOG OF SELECTED
DOVER BOOKS
IN SCIENCE AND MATHEMATICS

QUALITATIVE THEORY OF DIFFERENTIAL EQUATIONS, V.V. Nemytskii and V.V. Stepanov. Classic graduate-level text by two prominent Soviet mathematicians covers classical differential equations as well as topological dynamics and ergodic theory. Bibliographies. 523pp. 5⅜ × 8½.   65954-2 Pa. $10.95

MATRICES AND LINEAR ALGEBRA, Hans Schneider and George Phillip Barker. Basic textbook covers theory of matrices and its applications to systems of linear equations and related topics such as determinants, eigenvalues and differential equations. Numerous exercises. 432pp. 5⅜ × 8½.   66014-1 Pa. $9.95

QUANTUM THEORY, David Bohm. This advanced undergraduate-level text presents the quantum theory in terms of qualitative and imaginative concepts, followed by specific applications worked out in mathematical detail. Preface. Index. 655pp. 5⅜ × 8½.   65969-0 Pa. $13.95

ATOMIC PHYSICS (8th edition), Max Born. Nobel laureate's lucid treatment of kinetic theory of gases, elementary particles, nuclear atom, wave-corpuscles, atomic structure and spectral lines, much more. Over 40 appendices, bibliography. 495pp. 5⅜ × 8½.   65984-4 Pa. $11.95

ELECTRONIC STRUCTURE AND THE PROPERTIES OF SOLIDS: The Physics of the Chemical Bond, Walter A. Harrison. Innovative text offers basic understanding of the electronic structure of covalent and ionic solids, simple metals, transition metals and their compounds. Problems. 1980 edition. 582pp. 6⅛ × 9¼.   66021-4 Pa. $14.95

BOUNDARY VALUE PROBLEMS OF HEAT CONDUCTION, M. Necati Özisik. Systematic, comprehensive treatment of modern mathematical methods of solving problems in heat conduction and diffusion. Numerous examples and problems. Selected references. Appendices. 505pp. 5⅜ × 8½.   65990-9 Pa. $11.95

A SHORT HISTORY OF CHEMISTRY (3rd edition), J.R. Partington. Classic exposition explores origins of chemistry, alchemy, early medical chemistry, nature of atmosphere, theory of valency, laws and structure of atomic theory, much more. 428pp. 5⅜ × 8½. (Available in U.S. only)   65977-1 Pa. $10.95

A HISTORY OF ASTRONOMY, A. Pannekoek. Well-balanced, carefully reasoned study covers such topics as Ptolemaic theory, work of Copernicus, Kepler, Newton, Eddington's work on stars, much more. Illustrated. References. 521pp. 5⅜ × 8½.   65994-1 Pa. $11.95

PRINCIPLES OF METEOROLOGICAL ANALYSIS, Walter J. Saucier. Highly respected, abundantly illustrated classic reviews atmospheric variables, hydrostatics, static stability, various analyses (scalar, cross-section, isobaric, isentropic, more). For intermediate meteorology students. 454pp. 6⅛ × 9¼. 65979-8 Pa. $12.95

RELATIVITY, THERMODYNAMICS AND COSMOLOGY, Richard C. Tolman. Landmark study extends thermodynamics to special, general relativity; also applications of relativistic mechanics, thermodynamics to cosmological models. 501pp. 5⅜ × 8½. 65383-8 Pa. $12.95

APPLIED ANALYSIS, Cornelius Lanczos. Classic work on analysis and design of finite processes for approximating solution of analytical problems. Algebraic equations, matrices, harmonic analysis, quadrature methods, much more. 559pp. 5⅜ × 8½. 65656-X Pa. $12.95

SPECIAL RELATIVITY FOR PHYSICISTS, G. Stephenson and C.W. Kilmister. Concise elegant account for nonspecialists. Lorentz transformation, optical and dynamical applications, more. Bibliography. 108pp. 5⅜ × 8½. 65519-9 Pa. $4.95

INTRODUCTION TO ANALYSIS, Maxwell Rosenlicht. Unusually clear, accessible coverage of set theory, real number system, metric spaces, continuous functions, Riemann integration, multiple integrals, more. Wide range of problems. Undergraduate level. Bibliography. 254pp. 5⅜ × 8½. 65038-3 Pa. $7.95

INTRODUCTION TO QUANTUM MECHANICS With Applications to Chemistry, Linus Pauling & E. Bright Wilson, Jr. Classic undergraduate text by Nobel Prize winner applies quantum mechanics to chemical and physical problems. Numerous tables and figures enhance the text. Chapter bibliographies. Appendices. Index. 468pp. 5⅜ × 8½. 64871-0 Pa. $11.95

ASYMPTOTIC EXPANSIONS OF INTEGRALS, Norman Bleistein & Richard A. Handelsman. Best introduction to important field with applications in a variety of scientific disciplines. New preface. Problems. Diagrams. Tables. Bibliography. Index. 448pp. 5⅜ × 8½. 65082-0 Pa. $11.95

MATHEMATICS APPLIED TO CONTINUUM MECHANICS, Lee A. Segel. Analyzes models of fluid flow and solid deformation. For upper-level math, science and engineering students. 608pp. 5⅜ × 8½. 65369-2 Pa. $13.95

ELEMENTS OF REAL ANALYSIS, David A. Sprecher. Classic text covers fundamental concepts, real number system, point sets, functions of a real variable, Fourier series, much more. Over 500 exercises. 352pp. 5⅜ × 8½. 65385-4 Pa. $9.95

PHYSICAL PRINCIPLES OF THE QUANTUM THEORY, Werner Heisenberg. Nobel Laureate discusses quantum theory, uncertainty, wave mechanics, work of Dirac, Schroedinger, Compton, Wilson, Einstein, etc. 184pp. 5⅜ × 8½. 60113-7 Pa. $4.95

INTRODUCTORY REAL ANALYSIS, A.N. Kolmogorov, S.V. Fomin. Translated by Richard A. Silverman. Self-contained, evenly paced introduction to real and functional analysis. Some 350 problems. 403pp. 5⅜ × 8½. 61226-0 Pa. $9.95

PROBLEMS AND SOLUTIONS IN QUANTUM CHEMISTRY AND PHYSICS, Charles S. Johnson, Jr. and Lee G. Pedersen. Unusually varied problems, detailed solutions in coverage of quantum mechanics, wave mechanics, angular momentum, molecular spectroscopy, scattering theory, more. 280 problems plus 139 supplementary exercises. 430pp. 6½ × 9¼. 65236-X Pa. $11.95

ASYMPTOTIC METHODS IN ANALYSIS, N.G. de Bruijn. An inexpensive, comprehensive guide to asymptotic methods—the pioneering work that teaches by explaining worked examples in detail. Index. 224pp. 5⅜ × 8½.   64221-6 Pa. $6.95

OPTICAL RESONANCE AND TWO-LEVEL ATOMS, L. Allen and J.H. Eberly. Clear, comprehensive introduction to basic principles behind all quantum optical resonance phenomena. 53 illustrations. Preface. Index. 256pp. 5⅜ × 8½.
65533-4 Pa. $7.95

COMPLEX VARIABLES, Francis J. Flanigan. Unusual approach, delaying complex algebra till harmonic functions have been analyzed from real variable viewpoint. Includes problems with answers. 364pp. 5⅜ × 8½.   61388-7 Pa. $7.95

ATOMIC SPECTRA AND ATOMIC STRUCTURE, Gerhard Herzberg. One of best introductions; especially for specialist in other fields. Treatment is physical rather than mathematical. 80 illustrations. 257pp. 5⅜ × 8½.   60115-3 Pa. $5.95

APPLIED COMPLEX VARIABLES, John W. Dettman. Step-by-step coverage of fundamentals of analytic function theory—plus lucid exposition of five important applications: Potential Theory; Ordinary Differential Equations; Fourier Transforms; Laplace Transforms; Asymptotic Expansions. 66 figures. Exercises at chapter ends. 512pp. 5⅜ × 8½.   64670-X Pa. $10.95

ULTRASONIC ABSORPTION: An Introduction to the Theory of Sound Absorption and Dispersion in Gases, Liquids and Solids, A.B. Bhatia. Standard reference in the field provides a clear, systematically organized introductory review of fundamental concepts for advanced graduate students, research workers. Numerous diagrams. Bibliography. 440pp. 5⅜ × 8½.   64917-2 Pa. $11.95

UNBOUNDED LINEAR OPERATORS: Theory and Applications, Seymour Goldberg. Classic presents systematic treatment of the theory of unbounded linear operators in normed linear spaces with applications to differential equations. Bibliography. 199pp. 5⅜ × 8½.   64830-3 Pa. $7.95

LIGHT SCATTERING BY SMALL PARTICLES, H.C. van de Hulst. Comprehensive treatment including full range of useful approximation methods for researchers in chemistry, meteorology and astronomy. 44 illustrations. 470pp. 5⅜ × 8½.   64228-3 Pa. $10.95

CONFORMAL MAPPING ON RIEMANN SURFACES, Harvey Cohn. Lucid, insightful book presents ideal coverage of subject. 334 exercises make book perfect for self-study. 55 figures. 352pp. 5⅜ × 8¼.   64025-6 Pa. $8.95

OPTICKS, Sir Isaac Newton. Newton's own experiments with spectroscopy, colors, lenses, reflection, refraction, etc., in language the layman can follow. Foreword by Albert Einstein. 532pp. 5⅜ × 8½.   60205-2 Pa. $9.95

GENERALIZED INTEGRAL TRANSFORMATIONS, A.H. Zemanian. Graduate-level study of recent generalizations of the Laplace, Mellin, Hankel, K. Weierstrass, convolution and other simple transformations. Bibliography. 320pp. 5⅜ × 8½.   65375-7 Pa. $7.95

CATALOG OF DOVER BOOKS

NUMERICAL METHODS FOR SCIENTISTS AND ENGINEERS, Richard Hamming. Classic text stresses frequency approach in coverage of algorithms, polynomial approximation, Fourier approximation, exponential approximation, other topics. Revised and enlarged 2nd edition. 721pp. 5⅜ × 8½.
65241-6 Pa. $14.95

THEORETICAL SOLID STATE PHYSICS, Vol. I: Perfect Lattices in Equilibrium; Vol. II: Non-Equilibrium and Disorder, William Jones and Norman H. March. Monumental reference work covers fundamental theory of equilibrium properties of perfect crystalline solids, non-equilibrium properties, defects and disordered systems. Appendices. Problems. Preface. Diagrams. Index. Bibliography. Total of 1,301pp. 5⅜ × 8½. Two volumes. Vol. I 65015-4 Pa. $12.95
Vol. II 65016-2 Pa. $12.95

OPTIMIZATION THEORY WITH APPLICATIONS, Donald A. Pierre. Broad-spectrum approach to important topic. Classical theory of minima and maxima, calculus of variations, simplex technique and linear programming, more. Many problems, examples. 640pp. 5⅜ × 8½. 65205-X Pa. $13.95

THE MODERN THEORY OF SOLIDS, Frederick Seitz. First inexpensive edition of classic work on theory of ionic crystals, free-electron theory of metals and semiconductors, molecular binding, much more. 736pp. 5⅜ × 8½.
65482-6 Pa. $15.95

ESSAYS ON THE THEORY OF NUMBERS, Richard Dedekind. Two classic essays by great German mathematician: on the theory of irrational numbers; and on transfinite numbers and properties of natural numbers. 115pp. 5⅜ × 8½.
21010-3 Pa. $4.95

THE FUNCTIONS OF MATHEMATICAL PHYSICS, Harry Hochstadt. Comprehensive treatment of orthogonal polynomials, hypergeometric functions, Hill's equation, much more. Bibliography. Index. 322pp. 5⅜ × 8½. 65214-9 Pa. $9.95

NUMBER THEORY AND ITS HISTORY, Oystein Ore. Unusually clear, accessible introduction covers counting, properties of numbers, prime numbers, much more. Bibliography. 380pp. 5⅜ × 8½. 65620-9 Pa. $8.95

THE VARIATIONAL PRINCIPLES OF MECHANICS, Cornelius Lanczos. Graduate level coverage of calculus of variations, equations of motion, relativistic mechanics, more. First inexpensive paperbound edition of classic treatise. Index. Bibliography. 418pp. 5⅜ × 8½. 65067-7 Pa. $10.95

MATHEMATICAL TABLES AND FORMULAS, Robert D. Carmichael and Edwin R. Smith. Logarithms, sines, tangents, trig functions, powers, roots, reciprocals, exponential and hyperbolic functions, formulas and theorems. 269pp. 5⅜ × 8½. 60111-0 Pa. $5.95

THEORETICAL PHYSICS, Georg Joos, with Ira M. Freeman. Classic overview covers essential math, mechanics, electromagnetic theory, thermodynamics, quantum mechanics, nuclear physics, other topics. First paperback edition. xxiii + 885pp. 5⅜ × 8½. 65227-0 Pa. $18.95

HANDBOOK OF MATHEMATICAL FUNCTIONS WITH FORMULAS, GRAPHS, AND MATHEMATICAL TABLES, edited by Milton Abramowitz and Irene A. Stegun. Vast compendium: 29 sets of tables, some to as high as 20 places. 1,046pp. 8 × 10½. 61272-4 Pa. $22.95

MATHEMATICAL METHODS IN PHYSICS AND ENGINEERING, John W. Dettman. Algebraically based approach to vectors, mapping, diffraction, other topics in applied math. Also generalized functions, analytic function theory, more. Exercises. 448pp. 5⅜ × 8¼. 65649-7 Pa. $8.95

A SURVEY OF NUMERICAL MATHEMATICS, David M. Young and Robert Todd Gregory. Broad self-contained coverage of computer-oriented numerical algorithms for solving various types of mathematical problems in linear algebra, ordinary and partial, differential equations, much more. Exercises. Total of 1,248pp. 5⅜ × 8½. Two volumes. Vol. I 65691-8 Pa. $14.95
Vol. II 65692-6 Pa. $14.95

TENSOR ANALYSIS FOR PHYSICISTS, J.A. Schouten. Concise exposition of the mathematical basis of tensor analysis, integrated with well-chosen physical examples of the theory. Exercises. Index. Bibliography. 289pp. 5⅜ × 8½.
65582-2 Pa. $7.95

INTRODUCTION TO NUMERICAL ANALYSIS (2nd Edition), F.B. Hildebrand. Classic, fundamental treatment covers computation, approximation, interpolation, numerical differentiation and integration, other topics. 150 new problems. 669pp. 5⅜ × 8½. 65363-3 Pa. $14.95

INVESTIGATIONS ON THE THEORY OF THE BROWNIAN MOVEMENT, Albert Einstein. Five papers (1905–8) investigating dynamics of Brownian motion and evolving elementary theory. Notes by R. Fürth. 122pp. 5⅜ × 8½.
60304-0 Pa. $4.95

NUMERICAL METHODS FOR SCIENTISTS AND ENGINEERS, Richard Hamming. Classic text stresses frequency approach in coverage of algorithms, polynomial approximation, Fourier approximation, exponential approximation, other topics. Revised and enlarged 2nd edition. 721pp. 5⅜ × 8½. 65241-6 Pa. $14.95

AN INTRODUCTION TO STATISTICAL THERMODYNAMICS, Terrell L. Hill. Excellent basic text offers wide-ranging coverage of quantum statistical mechanics, systems of interacting molecules, quantum statistics, more. 523pp. 5⅜ × 8½. 65242-4 Pa. $11.95

ELEMENTARY DIFFERENTIAL EQUATIONS, William Ted Martin and Eric Reissner. Exceptionally clear, comprehensive introduction at undergraduate level. Nature and origin of differential equations, differential equations of first, second and higher orders. Picard's Theorem, much more. Problems with solutions. 331pp. 5⅜ × 8½. 65024-3 Pa. $8.95

STATISTICAL PHYSICS, Gregory H. Wannier. Classic text combines thermodynamics, statistical mechanics and kinetic theory in one unified presentation of thermal physics. Problems with solutions. Bibliography. 532pp. 5⅜ × 8½.
65401-X Pa. $11.95

ORDINARY DIFFERENTIAL EQUATIONS, Morris Tenenbaum and Harry Pollard. Exhaustive survey of ordinary differential equations for undergraduates in mathematics, engineering, science. Thorough analysis of theorems. Diagrams. Bibliography. Index. 818pp. 5⅜ × 8½. 64940-7 Pa. $16.95

STATISTICAL MECHANICS: Principles and Applications, Terrell L. Hill. Standard text covers fundamentals of statistical mechanics, applications to fluctuation theory, imperfect gases, distribution functions, more. 448pp. 5⅜ × 8½. 65390-0 Pa. $9.95

ORDINARY DIFFERENTIAL EQUATIONS AND STABILITY THEORY: An Introduction, David A. Sánchez. Brief, modern treatment. Linear equation, stability theory for autonomous and nonautonomous systems, etc. 164pp. 5⅜ × 8¼. 63828-6 Pa. $5.95

THIRTY YEARS THAT SHOOK PHYSICS: The Story of Quantum Theory, George Gamow. Lucid, accessible introduction to influential theory of energy and matter. Careful explanations of Dirac's anti-particles, Bohr's model of the atom, much more. 12 plates. Numerous drawings. 240pp. 5⅜ × 8½. 24895-X Pa. $5.95

THEORY OF MATRICES, Sam Perlis. Outstanding text covering rank, non-singularity and inverses in connection with the development of canonical matrices under the relation of equivalence, and without the intervention of determinants. Includes exercises. 237pp. 5⅜ × 8½. 66810-X Pa. $7.95

GREAT EXPERIMENTS IN PHYSICS: Firsthand Accounts from Galileo to Einstein, edited by Morris H. Shamos. 25 crucial discoveries: Newton's laws of motion, Chadwick's study of the neutron, Hertz on electromagnetic waves, more. Original accounts clearly annotated. 370pp. 5⅜ × 8½. 25346-5 Pa. $9.95

INTRODUCTION TO PARTIAL DIFFERENTIAL EQUATIONS WITH AP-PLICATIONS, E.C. Zachmanoglou and Dale W. Thoe. Essentials of partial differential equations applied to common problems in engineering and the physical sciences. Problems and answers. 416pp. 5⅜ × 8½. 65251-3 Pa. $10.95

BURNHAM'S CELESTIAL HANDBOOK, Robert Burnham, Jr. Thorough guide to the stars beyond our solar system. Exhaustive treatment. Alphabetical by constellation: Andromeda to Cetus in Vol. 1; Chamaeleon to Orion in Vol. 2; and Pavo to Vulpecula in Vol. 3. Hundreds of illustrations. Index in Vol. 3. 2,000pp. 6⅛ × 9¼. 23567-X, 23568-8, 23673-0 Pa., Three-vol. set $41.85

ASYMPTOTIC EXPANSIONS FOR ORDINARY DIFFERENTIAL EQUA-TIONS, Wolfgang Wasow. Outstanding text covers asymptotic power series, Jordan's canonical form, turning point problems, singular perturbations, much more. Problems. 384pp. 5⅜ × 8½. 65456-7 Pa. $9.95

AMATEUR ASTRONOMER'S HANDBOOK, J.B. Sidgwick. Timeless, compre-hensive coverage of telescopes, mirrors, lenses, mountings, telescope drives, micrometers, spectroscopes, more. 189 illustrations. 576pp. 5⅜ × 8¼. (USO) 24034-7 Pa. $9.95

ROTARY-WING AERODYNAMICS, W.Z. Stepniewski. Clear, concise text covers aerodynamic phenomena of the rotor and offers guidelines for helicopter performance evaluation. Originally prepared for NASA. 537 figures. 640pp. 6⅛ × 9¼.
64647-5 Pa. $14.95

DIFFERENTIAL GEOMETRY, Heinrich W. Guggenheimer. Local differential geometry as an application of advanced calculus and linear algebra. Curvature, transformation groups, surfaces, more. Exercises. 62 figures. 378pp. 5⅜ × 8½.
63433-7 Pa. $7.95

INTRODUCTION TO SPACE DYNAMICS, William Tyrrell Thomson. Comprehensive, classic introduction to space-flight engineering for advanced undergraduate and graduate students. Includes vector algebra, kinematics, transformation of coordinates. Bibliography. Index. 352pp. 5⅜ × 8½. 65113-4 Pa. $8.95

A SURVEY OF MINIMAL SURFACES, Robert Osserman. Up-to-date, in-depth discussion of the field for advanced students. Corrected and enlarged edition covers new developments. Includes numerous problems. 192pp. 5⅜ × 8½.
64998-9 Pa. $8.95

ANALYTICAL MECHANICS OF GEARS, Earle Buckingham. Indispensable reference for modern gear manufacture covers conjugate gear-tooth action, gear-tooth profiles of various gears, many other topics. 263 figures. 102 tables. 546pp. 5⅜ × 8½. 65712-4 Pa. $11.95

SET THEORY AND LOGIC, Robert R. Stoll. Lucid introduction to unified theory of mathematical concepts. Set theory and logic seen as tools for conceptual understanding of real number system. 496pp. 5⅜ × 8¼. 63829-4 Pa. $10.95

A HISTORY OF MECHANICS, René Dugas. Monumental study of mechanical principles from antiquity to quantum mechanics. Contributions of ancient Greeks, Galileo, Leonardo, Kepler, Lagrange, many others. 671pp. 5⅜ × 8½.
65632-2 Pa. $14.95

FAMOUS PROBLEMS OF GEOMETRY AND HOW TO SOLVE THEM, Benjamin Bold. Squaring the circle, trisecting the angle, duplicating the cube: learn their history, why they are impossible to solve, then solve them yourself. 128pp. 5⅜ × 8½. 24297-8 Pa. $3.95

MECHANICAL VIBRATIONS, J.P. Den Hartog. Classic textbook offers lucid explanations and illustrative models, applying theories of vibrations to a variety of practical industrial engineering problems. Numerous figures. 233 problems, solutions. Appendix. Index. Preface. 436pp. 5⅜ × 8½. 64785-4 Pa. $9.95

CURVATURE AND HOMOLOGY, Samuel I. Goldberg. Thorough treatment of specialized branch of differential geometry. Covers Riemannian manifolds, topology of differentiable manifolds, compact Lie groups, other topics. Exercises. 315pp. 5⅜ × 8½. 64314-X Pa. $8.95

HISTORY OF STRENGTH OF MATERIALS, Stephen P. Timoshenko. Excellent historical survey of the strength of materials with many references to the theories of elasticity and structure. 245 figures. 452pp. 5⅜ × 8½. 61187-6 Pa. $10.95

GEOMETRY OF COMPLEX NUMBERS, Hans Schwerdtfeger. Illuminating, widely praised book on analytic geometry of circles, the Moebius transformation, and two-dimensional non-Euclidean geometries. 200pp. 5⅜ × 8¼.
63830-8 Pa. $6.95

MECHANICS, J.P. Den Hartog. A classic introductory text or refresher. Hundreds of applications and design problems illuminate fundamentals of trusses, loaded beams and cables, etc. 334 answered problems. 462pp. 5⅜ × 8½. 60754-2 Pa. $8.95

TOPOLOGY, John G. Hocking and Gail S. Young. Superb one-year course in classical topology. Topological spaces and functions, point-set topology, much more. Examples and problems. Bibliography. Index. 384pp. 5⅜ × 8¼.
65676-4 Pa. $8.95

STRENGTH OF MATERIALS, J.P. Den Hartog. Full, clear treatment of basic material (tension, torsion, bending, etc.) plus advanced material on engineering methods, applications. 350 answered problems. 323pp. 5⅜ × 8½. 60755-0 Pa. $7.50

ELEMENTARY CONCEPTS OF TOPOLOGY, Paul Alexandroff. Elegant, intuitive approach to topology from set-theoretic topology to Betti groups; how concepts of topology are useful in math and physics. 25 figures. 57pp. 5⅜ × 8½.
60747-X Pa. $2.95

ADVANCED STRENGTH OF MATERIALS, J.P. Den Hartog. Superbly written advanced text covers torsion, rotating disks, membrane stresses in shells, much more. Many problems and answers. 388pp. 5⅜ × 8½. 65407-9 Pa. $9.95

COMPUTABILITY AND UNSOLVABILITY, Martin Davis. Classic graduate-level introduction to theory of computability, usually referred to as theory of recurrent functions. New preface and appendix. 288pp. 5⅜ × 8½. 61471-9 Pa. $6.95

GENERAL CHEMISTRY, Linus Pauling. Revised 3rd edition of classic first-year text by Nobel laureate. Atomic and molecular structure, quantum mechanics, statistical mechanics, thermodynamics correlated with descriptive chemistry. Problems. 992pp. 5⅜ × 8½. 65622-5 Pa. $19.95

AN INTRODUCTION TO MATRICES, SETS AND GROUPS FOR SCIENCE STUDENTS, G. Stephenson. Concise, readable text introduces sets, groups, and most importantly, matrices to undergraduate students of physics, chemistry, and engineering. Problems. 164pp. 5⅜ × 8½. 65077-4 Pa. $6.95

THE HISTORICAL BACKGROUND OF CHEMISTRY, Henry M. Leicester. Evolution of ideas, not individual biography. Concentrates on formulation of a coherent set of chemical laws. 260pp. 5⅜ × 8½. 61053-5 Pa. $6.95

THE PHILOSOPHY OF MATHEMATICS: An Introductory Essay, Stephan Körner. Surveys the views of Plato, Aristotle, Leibniz & Kant concerning propositions and theories of applied and pure mathematics. Introduction. Two appendices. Index. 198pp. 5⅜ × 8½. 25048-2 Pa. $6.95

THE DEVELOPMENT OF MODERN CHEMISTRY, Aaron J. Ihde. Authoritative history of chemistry from ancient Greek theory to 20th-century innovation. Covers major chemists and their discoveries. 209 illustrations. 14 tables. Bibliographies. Indices. Appendices. 851pp. 5⅜ × 8½. 64235-6 Pa. $17.95

DE RE METALLICA, Georgius Agricola. The famous Hoover translation of greatest treatise on technological chemistry, engineering, geology, mining of early modern times (1556). All 289 original woodcuts. 638pp. 6¾ × 11.
60006-8 Pa. $17.95

SOME THEORY OF SAMPLING, William Edwards Deming. Analysis of the problems, theory and design of sampling techniques for social scientists, industrial managers and others who find statistics increasingly important in their work. 61 tables. 90 figures. xvii + 602pp. 5⅜ × 8½.
64684-X Pa. $15.95

THE VARIOUS AND INGENIOUS MACHINES OF AGOSTINO RAMELLI: A Classic Sixteenth-Century Illustrated Treatise on Technology, Agostino Ramelli. One of the most widely known and copied works on machinery in the 16th century. 194 detailed plates of water pumps, grain mills, cranes, more. 608pp. 9 × 12. (EBE)
25497-6 Clothbd. $34.95

LINEAR PROGRAMMING AND ECONOMIC ANALYSIS, Robert Dorfman, Paul A. Samuelson and Robert M. Solow. First comprehensive treatment of linear programming in standard economic analysis. Game theory, modern welfare economics, Leontief input-output, more. 525pp. 5⅜ × 8½.
65491-5 Pa. $13.95

ELEMENTARY DECISION THEORY, Herman Chernoff and Lincoln E. Moses. Clear introduction to statistics and statistical theory covers data processing, probability and random variables, testing hypotheses, much more. Exercises. 364pp. 5⅜ × 8½.
65218-1 Pa. $9.95

THE COMPLEAT STRATEGYST: Being a Primer on the Theory of Games of Strategy, J.D. Williams. Highly entertaining classic describes, with many illustrated examples, how to select best strategies in conflict situations. Prefaces. Appendices. 268pp. 5⅜ × 8½.
25101-2 Pa. $6.95

MATHEMATICAL METHODS OF OPERATIONS RESEARCH, Thomas L. Saaty. Classic graduate-level text covers historical background, classical methods of forming models, optimization, game theory, probability, queueing theory, much more. Exercises. Bibliography. 448pp. 5⅜ × 8¼.
65703-5 Pa. $12.95

CONSTRUCTIONS AND COMBINATORIAL PROBLEMS IN DESIGN OF EXPERIMENTS, Damaraju Raghavarao. In-depth reference work examines orthogonal Latin squares, incomplete block designs, tactical configuration, partial geometry, much more. Abundant explanations, examples. 416pp. 5⅜ × 8¼.
65685-3 Pa. $10.95

THE ABSOLUTE DIFFERENTIAL CALCULUS (CALCULUS OF TENSORS), Tullio Levi-Civita. Great 20th-century mathematician's classic work on material necessary for mathematical grasp of theory of relativity. 452pp. 5⅜ × 8½.
63401-9 Pa. $9.95

VECTOR AND TENSOR ANALYSIS WITH APPLICATIONS, A.I. Borisenko and I.E. Tarapov. Concise introduction. Worked-out problems, solutions, exercises. 257pp. 5⅜ × 8¼.
63833-2 Pa. $6.95

THE FOUR-COLOR PROBLEM: Assaults and Conquest, Thomas L. Saaty and Paul G. Kainen. Engrossing, comprehensive account of the century-old combinatorial topological problem, its history and solution. Bibliographies. Index. 110 figures. 228pp. 5⅜ × 8½. 65092-8 Pa. $6.95

CATALYSIS IN CHEMISTRY AND ENZYMOLOGY, William P. Jencks. Exceptionally clear coverage of mechanisms for catalysis, forces in aqueous solution, carbonyl- and acyl-group reactions, practical kinetics, more. 864pp. 5⅜ × 8½. 65460-5 Pa. $19.95

PROBABILITY: An Introduction, Samuel Goldberg. Excellent basic text covers set theory, probability theory for finite sample spaces, binomial theorem, much more. 360 problems. Bibliographies. 322pp. 5⅜ × 8½. 65252-1 Pa. $8.95

LIGHTNING, Martin A. Uman. Revised, updated edition of classic work on the physics of lightning. Phenomena, terminology, measurement, photography, spectroscopy, thunder, more. Reviews recent research. Bibliography. Indices. 320pp. 5⅜ × 8¼. 64575-4 Pa. $8.95

PROBABILITY THEORY: A Concise Course, Y.A. Rozanov. Highly readable, self-contained introduction covers combination of events, dependent events, Bernoulli trials, etc. Translation by Richard Silverman. 148pp. 5⅜ × 8¼. 63544-9 Pa. $5.95

THE CEASELESS WIND: An Introduction to the Theory of Atmospheric Motion, John A. Dutton. Acclaimed text integrates disciplines of mathematics and physics for full understanding of dynamics of atmospheric motion. Over 400 problems. Index. 97 illustrations. 640pp. 6 × 9. 65096-0 Pa. $17.95

STATISTICS MANUAL, Edwin L. Crow, et al. Comprehensive, practical collection of classical and modern methods prepared by U.S. Naval Ordnance Test Station. Stress on use. Basics of statistics assumed. 288pp. 5⅜ × 8½. 60599-X Pa. $6.95

DICTIONARY/OUTLINE OF BASIC STATISTICS, John E. Freund and Frank J. Williams. A clear concise dictionary of over 1,000 statistical terms and an outline of statistical formulas covering probability, nonparametric tests, much more. 208pp. 5⅜ × 8½. 66796-0 Pa. $6.95

STATISTICAL METHOD FROM THE VIEWPOINT OF QUALITY CONTROL, Walter A. Shewhart. Important text explains regulation of variables, uses of statistical control to achieve quality control in industry, agriculture, other areas. 192pp. 5⅜ × 8½. 65232-7 Pa. $6.95

THE INTERPRETATION OF GEOLOGICAL PHASE DIAGRAMS, Ernest G. Ehlers. Clear, concise text emphasizes diagrams of systems under fluid or containing pressure; also coverage of complex binary systems, hydrothermal melting, more. 288pp. 6½ × 9¼. 65389-7 Pa. $10.95

STATISTICAL ADJUSTMENT OF DATA, W. Edwards Deming. Introduction to basic concepts of statistics, curve fitting, least squares solution, conditions without parameter, conditions containing parameters. 26 exercises worked out. 271pp. 5⅜ × 8½. 64685-8 Pa. $7.95

TENSOR CALCULUS, J.L. Synge and A. Schild. Widely used introductory text covers spaces and tensors, basic operations in Riemannian space, non-Riemannian spaces, etc. 324pp. 5⅜ × 8¼. 63612-7 Pa. $7.95

A CONCISE HISTORY OF MATHEMATICS, Dirk J. Struik. The best brief history of mathematics. Stresses origins and covers every major figure from ancient Near East to 19th century. 41 illustrations. 195pp. 5⅜ × 8½. 60255-9 Pa. $7.95

A SHORT ACCOUNT OF THE HISTORY OF MATHEMATICS, W.W. Rouse Ball. One of clearest, most authoritative surveys from the Egyptians and Phoenicians through 19th-century figures such as Grassman, Galois, Riemann. Fourth edition. 522pp. 5⅜ × 8½. 20630-0 Pa. $10.95

HISTORY OF MATHEMATICS, David E. Smith. Nontechnical survey from ancient Greece and Orient to late 19th century; evolution of arithmetic, geometry, trigonometry, calculating devices, algebra, the calculus. 362 illustrations. 1,355pp. 5⅜ × 8½. 20429-4, 20430-8 Pa., Two-vol. set $23.90

THE GEOMETRY OF RENÉ DESCARTES, René Descartes. The great work founded analytical geometry. Original French text, Descartes' own diagrams, together with definitive Smith-Latham translation. 244pp. 5⅜ × 8½. 60068-8 Pa. $6.95

THE ORIGINS OF THE INFINITESIMAL CALCULUS, Margaret E. Baron. Only fully detailed and documented account of crucial discipline: origins; development by Galileo, Kepler, Cavalieri; contributions of Newton, Leibniz, more. 304pp. 5⅜ × 8½. (Available in U.S. and Canada only) 65371-4 Pa. $9.95

THE HISTORY OF THE CALCULUS AND ITS CONCEPTUAL DEVELOPMENT, Carl B. Boyer. Origins in antiquity, medieval contributions, work of Newton, Leibniz, rigorous formulation. Treatment is verbal. 346pp. 5⅜ × 8½. 60509-4 Pa. $7.95

THE THIRTEEN BOOKS OF EUCLID'S ELEMENTS, translated with introduction and commentary by Sir Thomas L. Heath. Definitive edition. Textual and linguistic notes, mathematical analysis. 2,500 years of critical commentary. Not abridged. 1,414pp. 5⅜ × 8½. 60088-2, 60089-0, 60090-4 Pa., Three-vol. set $29.85

GAMES AND DECISIONS: Introduction and Critical Survey, R. Duncan Luce and Howard Raiffa. Superb nontechnical introduction to game theory, primarily applied to social sciences. Utility theory, zero-sum games, n-person games, decision-making, much more. Bibliography. 509pp. 5⅜ × 8½. 65943-7 Pa. $11.95

THE HISTORICAL ROOTS OF ELEMENTARY MATHEMATICS, Lucas N.H. Bunt, Phillip S. Jones, and Jack D. Bedient. Fundamental underpinnings of modern arithmetic, algebra, geometry and number systems derived from ancient civilizations. 320pp. 5⅜ × 8½. 25563-8 Pa. $8.95

CALCULUS REFRESHER FOR TECHNICAL PEOPLE, A. Albert Klaf. Covers important aspects of integral and differential calculus via 756 questions. 566 problems, most answered. 431pp. 5⅜ × 8½. 20370-0 Pa. $8.95

CHALLENGING MATHEMATICAL PROBLEMS WITH ELEMENTARY SOLUTIONS, A.M. Yaglom and I.M. Yaglom. Over 170 challenging problems on probability theory, combinatorial analysis, points and lines, topology, convex polygons, many other topics. Solutions. Total of 445pp. 5⅜ × 8½. Two-vol. set.

Vol. I 65536-9 Pa. $6.95
Vol. II 65537-7 Pa. $6.95

FIFTY CHALLENGING PROBLEMS IN PROBABILITY WITH SOLUTIONS, Frederick Mosteller. Remarkable puzzlers, graded in difficulty, illustrate elementary and advanced aspects of probability. Detailed solutions. 88pp. 5⅜ × 8½.
65355-2 Pa. $3.95

EXPERIMENTS IN TOPOLOGY, Stephen Barr. Classic, lively explanation of one of the byways of mathematics. Klein bottles, Moebius strips, projective planes, map coloring, problem of the Koenigsberg bridges, much more, described with clarity and wit. 43 figures. 210pp. 5⅜ × 8½.
25933-1 Pa. $5.95

RELATIVITY IN ILLUSTRATIONS, Jacob T. Schwartz. Clear nontechnical treatment makes relativity more accessible than ever before. Over 60 drawings illustrate concepts more clearly than text alone. Only high school geometry needed. Bibliography. 128pp. 6⅛ × 9¼.
25965-X Pa. $5.95

AN INTRODUCTION TO ORDINARY DIFFERENTIAL EQUATIONS, Earl A. Coddington. A thorough and systematic first course in elementary differential equations for undergraduates in mathematics and science, with many exercises and problems (with answers). Index. 304pp. 5⅜ × 8½.
65942-9 Pa. $7.95

FOURIER SERIES AND ORTHOGONAL FUNCTIONS, Harry F. Davis. An incisive text combining theory and practical example to introduce Fourier series, orthogonal functions and applications of the Fourier method to boundary-value problems. 570 exercises. Answers and notes. 416pp. 5⅜ × 8½.
65973-9 Pa. $9.95

THE THEORY OF BRANCHING PROCESSES, Theodore E. Harris. First systematic, comprehensive treatment of branching (i.e. multiplicative) processes and their applications. Galton-Watson model, Markov branching processes, electron-photon cascade, many other topics. Rigorous proofs. Bibliography. 240pp. 5⅜ × 8½.
65952-6 Pa. $6.95

AN INTRODUCTION TO ALGEBRAIC STRUCTURES, Joseph Landin. Superb self-contained text covers "abstract algebra": sets and numbers, theory of groups, theory of rings, much more. Numerous well-chosen examples, exercises. 247pp. 5⅜ × 8½.
65940-2 Pa. $6.95

Prices subject to change without notice.
Available at your book dealer or write for free Mathematics and Science Catalog to Dept. GI, Dover Publications, Inc., 31 East 2nd St., Mineola, N.Y. 11501. Dover publishes more than 175 books each year on science, elementary and advanced mathematics, biology, music, art, literature, history, social sciences and other areas.